SPACE FLIGHT AND RE-ENTRY TRAJECTORIES

INTERNATIONAL SYMPOSIUM

ORGANIZED BY THE
INTERNATIONAL ACADEMY OF ASTRONAUTICS
OF THE IAF

LOUVECIENNES, 19 — 21 JUNE 1961

PROCEEDINGS

WITH 145 FIGURES

WIEN · SPRINGER-VERLAG · 1962

Special edition of the Proceedings
published in "Astronautica Acta"

Vol. VII, Fasc. 5—6, 1961
Vol. VIII, Fasc. 2—3, 1962
Vol. VIII, Fasc. 5, 1962

ISBN-13: 978-3-7091-5472-4 e-ISBN-13: 978-3-7091-5470-0
DOI: 10.1007/978-3-7091-5470-0

Softcover reprint of the hardcover 1st edition 1961

Contents

Astronautica Acta, Vol. VII, Fasc. 5-6, 1961

Astronautica Acta, Vol. VIII, Fasc. 2-3, 1962

From Astronautica Acta, Vol. VIII, Fasc. 5, 1962

Foreword

In this and a following issue (Vol. VIII, 1962, Fasc. 2—3) of "Astronautica Acta" there will appear the papers presented at the first international symposium sponsored by the International Academy of Astronautics of the International Astronautical Federation. The theme of the meeting was "Space Flight and Re-Entry Trajectories." It was held at *Louveciennes* outside of Paris on June 19—21, 1961. Sixteen papers by authors from nine countries were presented; attendees numbered from 80 to 100.

The organizing committee for the symposium was as follows:

Prof. PAUL A. LIBBY, Polytechnic Institute of Brooklyn, U.S.A., Chairman;

Prof. LUIGI BROGLIO, University of Rome, Italy;

Prof. B. FRAEIJS DE VEUBEKE, University of Liége, Belgium;

Dr. D. G. KING-HELE, Royal Aircraft Establishment, Farnborough, Hants, United Kingdom;

Prof. J. M. J. KOOY, Royal Military School, Breda, Netherlands;

Prof. JEAN KOVALEVSKY, Bureau des Longitudes, Paris, France;

Prof. RUDOLF PEŠEK, Academy of Sciences, Prague, Czechoslovakia.

The detailed arrangements for the meeting were made in a most satisfactory manner by Dr. FRANK J. MALINA, Deputy Director of IAA and Mr. A. R. WEILLER, Acting Secretary of IAA.

Prof. THEODORE VON KÁRMÁN, Director of IAA, in his remarks closing the symposium indicated his satisfaction at the interest being shown in "the science of the future." The papers which follow will make a permanent contribution to the literature of this science.

Paul A. Libby

Professor of Aeronautical Engineering
Assistant Director, Aerodynamics Laboratory
Polytechnic Institute of Brooklyn

Possible Paths for Re-Entry Into the Earth's Gravitational Field Avoiding the Radiation Belts

By

W. F. Hilton[1]

(With 2 Figures)

Abstract — Zusammenfassung — Résumé

Possible Paths for Re-Entry Into the Earth's Gravitational Field Avoiding the Radiation Belts. Although it is quite possible to traverse the radiation belts sufficiently rapidly to avoid a lethal radiation dose to a human pilot, the vehicle itself will become radioactive, and act as a secondary source, particularly if the vehicle has thick shielding material.

Most solar system trajectories will lie close to the plane of the earth's orbit (ecliptic). However, it is suggested in this paper that re-entry into the atmosphere should take place in the polar plane, by a very slight inclination of the transfer orbit. This will enable braking ellipses to miss the outer (electron) and the inner (proton) belts altogether, by having successive apogees outside, between and inside these radiation belts. An alternative solution is to have a single intermediate braking ellipse having apogee between the belts at about 5000 miles altitude (8000 km).

Some useful re-entry trajectories have been calculated on a digital computer, both for non lifting ballistic vehicles, and for vehicles having a lift/drag ratio of unity. In the latter case the pilot is able to vary his trajectory, and incidentally to raise his perigee height by the application of positive lift followed by a 180° roll and a period of negative lift.

Rückkehrbahnen zur Erdoberfläche unter Vermeidung der Strahlungsgürtel. Wenn es auch durchaus möglich ist, die Strahlungsgürtel so rasch zu durchqueren, daß tödliche Strahlungsdosen für die Besatzung vermieden werden können, so wird doch das Fahrzeug selbst radioaktiv und sendet Sekundärstrahlung aus.

Die meisten Bahnen in unserem Sonnensystem sind wenig gegen die Ekliptik geneigt. Es wird nun vermutet, daß Rückkehrbahnen in die Atmosphäre ungefähr in einer Meridianebene liegen sollen. Dies verlangt „Bremsellipsen" außerhalb, zwischen und innerhalb der Strahlungsgürtel, wobei auch die entsprechenden Apogäen außerhalb, zwischen und innerhalb dieser Strahlungsgürtel liegen sollen.

Mit Hilfe eines Digitalrechners wurden derartige Bahnen für ballistische Flugkörper ohne Auftrieb und für Fahrzeuge mit einem Auftriebs-Widerstands-Verhältnis Eins ermittelt. In letzterem Falle kann der Pilot die Bahn ändern, insbesondere die Höhe des Perigäums beeinflussen.

Orbites possibles de rentrée dans le champ de gravitation terrestre évitant les ceintures de radiation. Quoique cela soit tout à fait possible de traverser les ceintures de radiation suffisamment vite pour éviter au pilote une dose de radiation mortelle, l'engin lui-même deviendra radioactif, et agira comme une source secondaire, particulièrement s'il possède un épais blindage.

[1] D.Sc., Ph.D., F.B.I.S., F.R.Ae.S.; Head of Astronautics Division, Hawker Siddeley Aviation Ltd., Astronautics Division, Welkin House, 10/11 Charterhouse Sq., London, E. C. 1, England.

La plupart des trajectoires dans le système solaire seront dans un plan proche de celui de l'orbite terrestre (écliptique). Toutefois, il est conseillé dans cet article que la rentrée dans l'atmosphère se fasse dans un plan polaire, ce qui est possible par une légère inclinaison de l'orbite de transfert. Ceci conduira à freiner les ellipses, pour éviter simultanément la ceinture extérieure d'électrons et celle intérieure de protons, en ayant successivement des apogées à l'extérieur des deux ceintures de radiation, entre elles deux puis à l'intérieur de la plus proche. Une solution alternative serait d'avoir une seule ellipse intermédiaire de freinage ayant un apogée entre les ceintures à environ 5000 milles d'altitude (8000 km).

Quelques trajectoires utiles de rentrée ont été calculées tant pour des engins n'offrant pas de portance aérodynamique, que pour des engins possédant un rapport portance sur traînée égal à l'unité. Dans ce dernier cas, le pilote peut faire varier sa trajectoire, et incidemment faire croître son périgée, en se servant d'une portance positive, suivie d'un tonneau de 180° et d'une période à portance négative.

I. Introduction

With a few minor exceptions, the whole of the solar system lies fairly closely in the plane of the ecliptic. For this reason most manned trips into space beyond a few earth diameters will be in this plane of the ecliptic. Hence the solar system and possible transfer orbits between planets are usually represented on a two-dimensional sheet of paper. Straightforward re-entry into the earth's gravitational field will necessarily involve traversing the Van Allen radiation belts at their densest region. However, if the plane of the transfer orbit is tilted very slightly so as to pass above the North or South Poles of the earth, re-entry will lie approximately in the earth's polar plane and be perpendicular to the plane of the ecliptic. This permits avoidance of the Van Allen radiation belts by a suitable choice of apogee heights. It is thus possible to change an interplanetary transfer orbit in the plane of the ecliptic into a re-entry orbit perpendicular to this plane, without the use of rocket fuel. Fuel will only be needed for the adjustment of errors.

The precise solution of this problem involves a solution of the three-body problem in celestial mechanics, or at least the 2-body problem. As a first approximation for the purposes of this paper, the satellite has been assumed to enter the earth's gravitational field with a finite velocity vector due to travelling in the sun's gravitational field. At a distance of about half a million miles from the earth, the effect of the sun may be neglected and the plane of the re-entry will be the plane containing both the velocity vector and the centre of the earth. We have merely postulated that this plane will be nearly perpendicular to the plane of the ecliptic. Two types of drag retarding re-entry are possible [1] namely: lifting and non-lifting. These are considered in Sections III and IV below. The non-lifting case is found to require an impossibly high degree of accuracy in navigation. It is however of interest in the case of lifting re-entry as it shows the target trajectory which we are trying to follow and leaves aerodynamic lift in reserve to make up for any errors in navigation.

The re-entry trajectories have been calculated on a Pegasus digital computer using the method described in [2]. The mass loading of the vehicle has been taken at 20 lbs. per sq. ft., the angle between the flat underside and the direction of motion as 45° and the angle between the shock wave and the direction of motion as 63.5°. The non-lifting case has been calculated by assuming that the vehicle rolls continuously at this angle, thus producing zero net lift.

The earth's atmosphere is assumed to be a perfect vacuum above 200 miles altitude. In space the satellite is assumed to move according to the ordinary equation for elliptical motion:

$$V^2 = \mu \left[\frac{2}{r} - \frac{1}{a} \right] \tag{1}$$

where V is the velocity at a radius r from the centre of the earth and a is the semi major axis of the ellipse. If we put a equal to ∞ this becomes the equation for a parabola.

If a satellite returns to the earth after a mission to another planet such as Mars, it will acquire a velocity due to the transfer orbit, which will correspond to a negative semi axis for the ellipse It is convenient to express these interplanetary missions in this fashion. The resulting curve is of course a hyperbola and not an ellipse due to this change of sign. Using this method and assuming perigee heights between 50 and 60 miles, solutions have been found giving successive braking ellipses followed by landing on the earth's surface. It is necessary to have at least one braking ellipse if excessive heating of the vehicle during the final entry is to be avoided. Solutions have been sought which give the apogees of each braking ellipse above, between or below the VAN ALLEN radiation belts.

The need for a manned vehicle to avoid these belts is discussed in Section II below, together with their effect in making the vehicle itself radioactive. However, re-entry may be delayed after a long interplanetary trip to permit preliminary read-out of data etc. A method of doing this by means of aerodynamic lift is discussed in Section IV. By using a little rocket fuel the satellite could be parked indefinitely in one of these orbits avoiding the VAN ALLEN belts, where it might prove possible to put fresh fuel and a fresh crew on board and use the vehicle again. It would of course be vital that the second crew should not attempt a voyage of many months' duration in a vehicle already rendered radioactive by the VAN ALLEN belts. Hence the need for orbits having VAN ALLEN belt avoidance.

II. Effect of Van Allen Radiation Belts

It has been known for several years that the VAN ALLEN radiation belts can cause permanent damage to solar cells and to transistors. Rapidly moving protons and electrons can also cause permanent damage to human tissues. For example, one traverse of the ellipse shown in Fig. 1 would give a dose of 177.5 Roentgen of which 7.5 R would be due to protons and 170 due to electrons. The lethal human dose is approximately 450 R.

Naturally, the vehicle itself will afford some protection, particularly against the lighter electrons. However, recent reports on the Discoverer satellites which have been recovered after flight in the VAN ALLEN belts show that the structure of the satellite itself becomes radio-active in the process of stopping these radiations. Tritium and Argon 37 were found in the shell of this satellite after recovery.

It therefore follows that an optimum thickness of protection may exist, if the vehicle is to be inhabited for some time after traversing the radiation belts.

It must also be remembered that the re-entry trajectories described in this paper will work equally well in reverse, for take-off at the start of an interplanetary mission, by replacing the drag of the atmosphere at 50 miles by the thrust of rockets at some 150 miles altitude.

It will be essential to minimize the radio-active contamination of the vehicle before setting off on a voyage of some months duration. It is therefore concluded that the orbits describes in this paper will have more than academic interest.

III. Non-Lifting Drag Retarded Re-Entry

There is nothing novel in using a digital computer to calculate re-entry trajectories. The method used in [2] was employed, using the standard atmosphere described in [3] by KING-HELE. This program will also permit elliptical or hyperbolic orbits before re-entry, as the earth's gravitational field is completely included.

An interesting and useful variation was obtained by starting at sub-orbital speed in the atmosphere, and running the orbit backwards, with negative drag, so as to discover the apogee height from which a particular successful re-entry could be made.

A value of Lift/drag = 1 was assumed in all cases, with the vehicle rolling so as to eliminate the effect of lift. The successive apogee and perigee heights are quoted in the following tables, for orbits A and B starting from 40.000 miles, C, D, E and F having infinite apogee, and orbits G and H having a "negative apogee height" of 63,902 miles corresponding to a return from a Mars mission.

Re-Entry from 40,000 Miles Apogee

Orbit A

Apogee height in miles	40,000	8,643	688	
Perigee height in miles	55	54	46	lands

Excellent VAN ALLEN avoidance.

Orbit B

Apogee	40,000	6,531	
Perigee	54	52·93	lands

As illustrated in Fig. 1.

Parabolic Re-Entry

Orbit C

Velocity at Infinity is Zero

Apogee	∞	17,231	3,738	
Perigee	55	54·5	52·8	lands

Close above proton belt. First perigee should be above 55 miles.

Orbit D

Apogee	∞	12,780	1,007	
Perigee	54	53·27	47·33	lands

Close below proton belt. First perigee should be below 54 miles. As illustrated in Fig. 2.

Orbit *E*

Apogee	∞	8,958	lands
Perigee	53	51·97	

Single pass between VAN ALLEN belts.

Orbit *F*

Apogee	∞	301
Perigee	50	lands

Severe heating, hardly cools between first and final re-entries.

Mars Mission

Velocity at infinity corresponds to "apogee height" of $-$ 63,902 miles, i.e. $V = 1.85$ miles/sec. at infinity.

Orbit *G*
Unsuccessful

Apogee	−63,902	−1,302,120
Perigee	60	Does not return

Fails to be captured by earth.

Orbit *H*
Successful

Apogee	−63,902	36,262	7,148	55
Perigee	55	54·53	53·50	lands

First true apogee outside electron belts.
Second apogee between belts. Lands without crossing proton belt.

It will be seen that the whole re-entry corridor is less than 5 miles high, if it be assumed that at least one braking ellipse is necessary to spread the kinetic heating into two equal parts, separated by a time interval to permit cooling by radiation.

The navigation problem will thus be quite severe for non-lifting re-entry. Variable geometry, leading to a variable drag coefficient would be desirable, but we must always have a bluff body shape, if excessive heating due to viscous boundary layers is to be avoided. The bulk of the drag must always be caused by shock-waves, which will heat the air at a distance from the body.

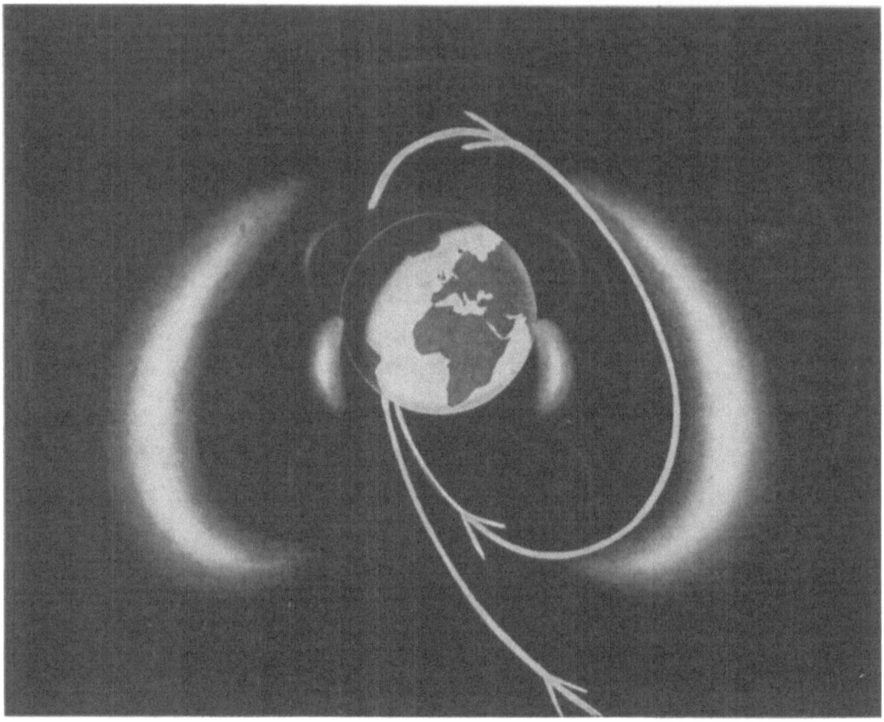

Fig. 1. The VAN ALLEN radiation fields. Orbit *B*, good radiation avoidance

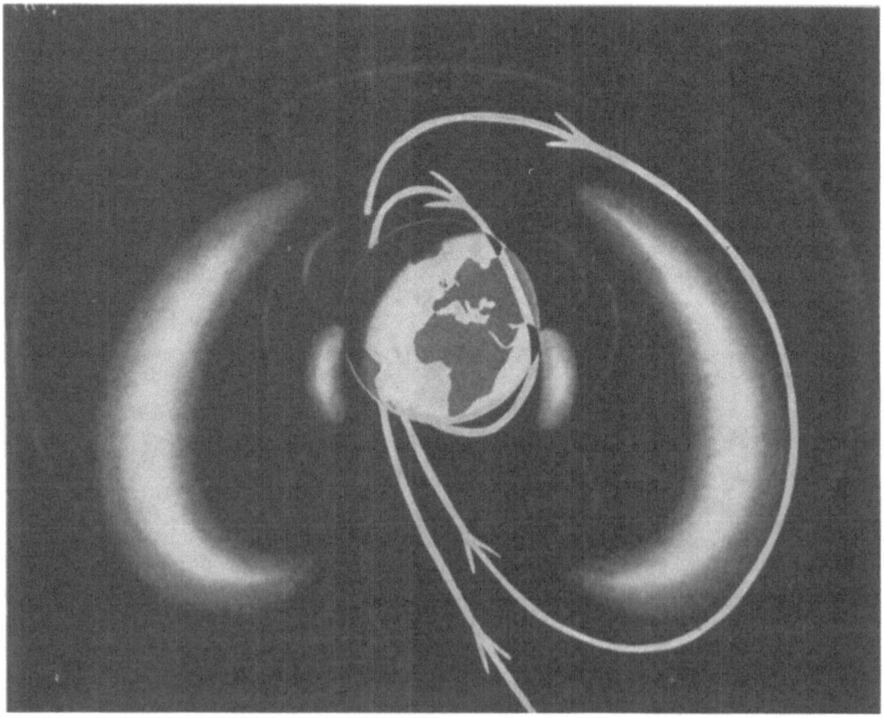

Fig. 2. The VAN ALLEN radiation fields. Orbit *D*, fair radiation avoidance

Orbit A shows an apogee between the radiation belts, followed by one below both belts, while this latter orbit is eliminated if we lower initial perigee height by only one mile, as in orbit B, and Fig. 1.

For parabolic re-entry orbit C makes three circuits, with both apogees too low with respect to the radiation belts. Orbit D is better, and is illustrated in Fig. 2, while orbit E makes a single pass between the belts before landing, and orbit F comes straight in and lands, with severe heating problems.

Returning from a Mars mission, orbit G fails to be captured by the earth, while orbit H, some five miles lower in perigee height, is ideal, making one apogee outside the belts, one between, and then lands.

IV. Drag Retarded Re-Entry of Lifting Vehicles

In the absence of lift, Section III shows the delicate choice of first perigee height. In addition, from interplanetary missions it is obviously desirable to kill the velocity in excess of parabolic velocity, but the figures have been included to cover the case of less successful missions, where fuel reserves have been previously depleted, but return to the earth without rocket braking is still possible.

It has already been shown in [2] how the use of positive aerodynamic lift can assist final re-entry at speeds below 4.8 miles per second, and this method has been used in all cases in this paper. It has also been shown in [1] how negative lift can assist, and indeed in some cases ensure, re-entry from hyperbolic orbits above 7 miles per second.

The question arises, can we employ lift to prolong the final descent? The application of constant lift coefficient when passing through the atmosphere at speeds between 5 and 7 miles/sec. merely results in twisting the axes of the ellipse, and produces a negligible effect on perigee height.

V. Increase of Perigee Height by Use of Positive and Negative Lift

It is of interest to report that perigee height can in fact be raised considerably by entering the atmosphere with positive lift, and rolling to generate negative lift after passing perigee. An example of this is shown in orbit A' below, which has the same initial conditions as orbit A during the first non-lifting re-entry. By the use of this trick it was found possible to make three or four passes between the radiation belts instead of one. It is always necessary to delay the roll manoeuvre until after perigee, since the downward lift must not be so great as to cause an actual descent and loss of altitude.

Orbit A'

Re-Entry from 40,000 Miles Apogee

As orbit A, but employing positive and negative lift to delay landing.

Apogee	40,000	8643	5512	4218	3549	3096
Perigee	55	54	61	66	69	73
Height of roll manoeuvre	—	62	65·2	69·9	71·3	

Since orbit A' enters radiation belt if it has an apogee height of 3549 miles, we reverse the procedure and enter the atmosphere from apogee of 4218 miles with negative lift, and roll to positive lift at 54.7 miles.

Orbit A''

Apogee	4218	728
Perigee	66	lands
Height of reverse roll manoeuvre	54·7	

These tables show how perigee height can be raised very considerably, although necessarily remaining in the sensible atmosphere. Orbit A'' shows that a great deal of decision rests with the pilot as to whether to land a lifting vehicle, or to proceed.

VI. Conclusion

It is hoped that these exact trajectories will have indicated the large degree of control enjoyed by the pilot on re-entry. With the addition of a small amount of rocket fuel, almost any near orbit could be selected for "parking".

Finally the author is indebted to Mr. R. LOTT for the computing effort involved in preparing this paper.

References

1. W. F. HILTON, Recovery after Re-Entry, by the Use of Aerodynamic Lift. High Altitude and Satellite Rocket Symposium, July 1957.
2. W. F. HILTON, Re-Entry Paths for Manned Satellites. Proceedings of the Xth International Astronautical Congress, London 1959, Vol. II, p. 526. Wien: Springer, 1960.
3. D. G. KING-HELE, Determination of Atmospheric Density by Means of Satellites. Nature, May 1959.

Discussion

Mr. FERRI: You go round the earth several times and you do not know if you can avoid the VAN ALLEN belts.

Mr. HILTON: This is true for the re-entry from 40,000 miles but for the hyperbolic re-entry having a negative apogee height, as I called it in the paper, that is to say coming from Mars on a transfer orbit governed by eq. (1), then it is essential that the vehicle enters the atmosphere at about the correct height.

In the paper I show an example of an unsuccessful re-entry with perigee at 60 miles, but by lowering the first perigee to 55 miles successful re-entry is achieved.

You are successful, for instance, if you come back from Mars to the perigee of 55 miles. You then get successive apogees of 36,000, 7,000 and 55, followed by landing.

Mr. FERRI: If you go from this orbit, do you transfer to an entering orbit?

Mr. HILTON: It is quite impossible, because you see, you are at 55 miles altitude and the retardation is so high.

Mr. FERRI: Would it not be useful if the pilot used more rocket braking?

Mr. HILTON: Yes, as I said in the paper, it is very useful if the pilot has enough rocket fuel in reserve to enable him to employ rocket braking, and to reduce his velocity from hyperbolic to elliptical velocity. However, we must consider the case of less successful missions, where the reserves of fuel have already been used in some other manoeuvre. The pilot may only have sufficient fuel to regain the earth's gravitational field, but not enough to carry out sufficient braking. The re-entry is very much easier and more certain if you can reduce velocity to an elliptical value.

Mr. FERRI: You cannot even drag in the first re-entry trajectory, you escape in an elliptic trajectory, you do not escape in a hyperbolic trajectory.

Mr. HILTON: You must lose enough velocity to reduce you to elliptical conditions, and the atmosphere has a very definite upper limit for this purpose. With regard to the pilot's comfort and heat transfer effects, 20 lb/sq.ft. is a reasonable mass loading, and the atmosphere has a top at about 55 miles for this loading. Anything about 60 miles will, as Dr. FERRI says, take a long time to re-enter, but if you make the first pass above 60 miles and at hyperbolic velocity you will be lost for ever.

Mr. FERRI: So you must transfer your trajectory to the right angle.

Mr. HILTON: I agree. I think it is the air density at perigee that is important.

Mr. FERRI: Not the re-entry angle?

Mr. HILTON: The angular re-entry is zero at perigee.

Mr. FERRI: I am talking of the angle at re-entry.

Mr. HILTON: Re-entry angle cannot be defined accurately, but perigee itself is quite precise defined as when the machine is flying tangentially.

Mr. FERRI: It is also very important to know the position in space of those radiation belts. Perhaps, sufficient information can be given in order to trace them, and do they change with the time. Then, of course it should be possible to detect movement of the belts. Perhaps this already has been done or it is going to be started.

Mr. HILTON: Naturally you are right, we know very little about the VAN ALLEN belts and they may assume a different formation at different times. They are very much affected by solar activity, particularly the electron belts, but I think by the time that man is returning from the Mars mission we are quite safe in assuming that the VAN ALLEN belts will be much more fully understood and can be forecast like the weather. This paper has shown how the pilot can avoid any known area of hazard by the use of aerodynamic lift.

Mr. STERN You should try to avoid it, because anyhow on the border the density of electrons and protons is not the same as it is in the centre. It is less dangerous on the border than it is in the centre.

Mr. HILTON: Frankly, I used this as an exercise to show how, by using aerodynamic lift, with a given distribution of VAN ALLEN belts, it is possible to use lift and navigate without the use of rocket fuel, to enable one to accomplish a satisfactory re-entry.

Aerodynamics of Trajectory Control for Re-Entry at Escape Speed

By

J. V. Becker[1], **D. L. Baradell**[2], and **E. B. Pritchard**[2]

(With 21 Figures)

Abstract — Zusammenfassung — Résumé

Aerodynamics of Trajectory Control for Re-Entry at Escape Speed. The problems of range control during re-entry of lifting manned vehicles are analyzed from the aerodynamic point of view. The ability to reach a given destination from the extremities of the entry corridor using wholly atmospheric modes of operation is found to require moderately higher lift/drag ratios than skip modes. However, the atmospheric cases do not suffer from the inherently sensitive guidance and control problems of the skip modes. An atmospheric technique is discussed whereby lateral and longitudinal range can be varied independently, using only roll control. For the longer re-entries the earth's sphericity is found to have major influences on attainable lateral range and on the shape of the maneuver envelope.

Several types of re-entry vehicle capable of meeting the aerodynamic requirements for range control are discussed briefly. From a fixed-base simulator study for one of these vehicles it is concluded that the typical maneuvers required for range control can be performed satisfactorily by a human pilot with the aid of artificial damping.

Aerodynamik der Bahnkontrolle beim Eintauchen in die Atmosphäre mit Fluchtgeschwindigkeit. Die mit der Bahnkontrolle bemannter Raumfahrzeuge zusammenhängenden Probleme werden von der aerodynamischen Seite untersucht. Es wird eine Methode beschrieben, bei der Breiten- und Längenwinkel voneinander unabhängig verändert werden können, wobei nur Rollen des Fahrzeuges zugelassen wird. Für längere Eintauchbahnen muß die Kugelgestalt der Erde berücksichtigt werden.

Verschiedene Typen von Eintauchflugkörpern werden untersucht. Auf Grund von Untersuchungen an einem Simulator wird geschlossen, daß die beim Eintauchen nötigen Manöver in hinreichender Weise von einem menschlichen Piloten bewältigt werden können, wenn eine künstliche Dämpfung vorgesehen ist.

Aérodynamique du contrôle de la trajectoire pour une rentrée à la vitesse de libération. Les buts du contrôle de la trajectoire sont, successivement, de réduire l'angle de rentrée à une faible valeur sans dépasser les limites de la décélération tolérable, d'éviter ou de contrôler un rebondissement hors de l'atmosphère et enfin, de contrôler longitudinalement et latéralement l'atterrissage dans la zone fixée. On décrit une façon d'opérer qui permet d'obtenir le contrôle du tangage et du lacet en agissant seulement sur la commande de roulis.

Les valeurs que doit prendre le rapport portance sur traînée du véhicule, pour réussir l'atterrissage dans la zone prescrite en commençant le pilotage à l'extrémité du corridor de rentrée, sont établies pour les diverses techniques de manoeuvre envisagées. Les dimensions de la zone d'atterrissage prédéterminée depuis de longues

[1] Chief, Aero-Physics Division, NASA-Langley Research Center, Langley Field, Virginia, U.S.A.

[2] Aero-Space Technologist, NASA-Langley Research Center, Langley Field, Virginia, U.S.A.

distances, supérieures au rayon de la terre, sont analysées en employant les équations de mouvement tridimensionnelles complètes, en coordonnées sphériques polaires. On trouve que les approximations communément employées pour obtenir le point d'impact du véhicule avec le sol sont très fausses si l'on part de longues distances.

Finalement, trois véhicules-type sont comparés pour leur comportement lors des manoeuvres de rentrée. On en conclut que dans beaucoup de cas, des rentrées pilotées peuvent être réussies sans même utiliser de systèmes artificiels amortissant les mouvements du véhicule autour de ses axes.

Notation

A	reference area
C_D	drag coefficient based on A
C_L	lift coefficient based on A
C_{L_1}, C_{L_3}	values of C_L at start and end of const. alt. maneuver (Fig. 4)
D	drag, component of resultant force along flight path
D_1	drag at start of range-control maneuver
g	acceleration due to gravity at earth's surface
G	resultant acceleration, $D/(W/g)\sqrt{1+(L/D)^2}$
h	altitude
l	lateral range measured normal to initial entry plane
L	lift, component of resultant force normal to flight path
m	mass
q	dynamic pressure, $\varrho V^2/2$
r_e	mean radius of earth
R	range
ΔR	range overlap for steep and shallow entries
$\Delta R_{E.P.}$	longitudinal range between overshoot and undershoot entry point for a given perigee location (Fig. 8)
t	time
T	kinetic energy
V	velocity
$\bar{V}$	$V\sqrt{g r_e}$
$\bar{V}_1$	V at the end of initial pull-up and start of range-control maneuver
$\bar{V}_2$, $\bar{V}_3$	values at start and end of constant q transition maneuver (defined on Fig. 4)
W	gross weight of vehicle
γ	path angle with respect to horizontal
γ_0	re-entry path angle at 400,000 ft altitude
γ_1	path angle at start of skip
λ	lateral displacement angle (Fig. 13)
ϱ	atmospheric density
ξ	heading angle (Fig. 13)
φ	roll angle (zero for unbanked vehicle)
φ_1	roll angle at start of range-control maneuver
ψ	range angle (Fig. 13)

I. Introduction

It is now apparent that re-entry from parabolic orbits is not only feasible but that considerable latitude exists in the choice of vehicle shape, heat protection methods, and other basic features of the re-entry system. There are, of course, many important problems remaining for research and development; the primitive solutions now envisioned lack refinement in many respects. But nevertheless the practical feasibility is clear. The researches supporting this position (some of which are covered by [1] through [14]) have naturally focused chiefly on the two prime problems of survival — deceleration and heating. Relatively little study has been devoted to the problems of successfully maneuvering the vehicle to a particular destination.

We have studied the maneuvers associated with lateral and longitudinal range control of manned lifting systems from the aerodynamic point of view. Some of the questions we have attempted to answer are: Is it possible within practical bounds of lift/drag ratio to achieve a point destination from any extremity of the re-entry corridor? Are the L/D requirements for point destination significantly reduced if a controlled skip out of the atmosphere is permitted, and to what extent is the range control problem aggravated by such a skip maneuver? How do the various possible types of atmospheric maneuver compare as to lateral and longitudinal range, control potential, vehicle L/D, attitude control, and heat-protection requirements? While we are not concerned specifically with guidance techniques, we will examine the piloting problems associated with typical entry maneuvers with the object of answering, in a preliminary way, the question: Can a human pilot perform the types of maneuver required for range control unaided by automatic control or artificial damping systems?

II. Uncontrolled Re-Entries

Severe restrictions are placed on the usable lift/drag ratio in trajectories which are controlled only in the sense that a fixed trim attitude is selected prior to contact with the atmosphere. A systematic computer study was made of such entries using the complete motion equations with the atmosphere specified in [15]. If a resultant deceleration of 10 g is chosen as a nominal limit for the tolerance of human pilots and a skip apogee of 400 miles is selected to avoid penetration of the inner radiation belt, it is seen in Fig. 1 that the L/D cannot exceed 0.48. The maximum corridor depth for such entries is about 0.8° (14 miles), and it occurs at $L/D \sim 0.2$. A still more serious limitation is the very large dispersion in longitudinal range for such entries, typically of the order of 1000 miles, that results from uncertainties in the atmosphere, the vehicle aerodynamics, and in the initial position in the entry corridor. It is readily concluded that control over the trajectory must be exercised during the re-entry if reasonable entry corridor depths and range accuracy are to be achieved.

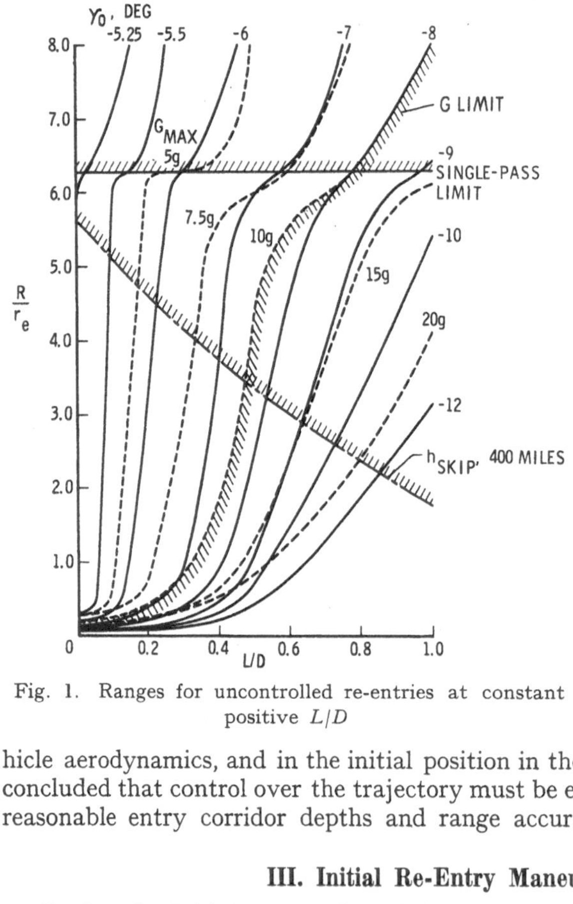

Fig. 1. Ranges for uncontrolled re-entries at constant positive L/D

III. Initial Re-Entry Maneuver

During the initial penetration of the atmosphere the primary objectives are to reduce the flight path angle to a value near zero without exceeding acceleration limits, and then to ascend or descend to the altitude for which the range-control maneuver is to be started. Range control is possible during the initial "pull-up" itself for noncritical entry angles. However, not much is lost in most cases by delaying until after the pull-up is completed. Obviously the character of the initial maneuver will have a strong influence on the subsequent range potential because it determines both the path angle limits (corridor boundaries) for which range control must be provided and the energy level available at the termination of the pull-up.

Control over the G schedule and reduction of peak G near the "undershoot" or G-limited boundary of the corridor can be accomplished either by variation in vehicle attitude or geometry or both. Large changes in geometry are required to achieve the same degree of modulation by geometry control as can be obtained by attitude control of a lifting vehicle (e.g. a 20 : 1 change in area for a drag-modulated device is roughly equivalent to a variable-attitude vehicle with L/D of about 0.6, [16]). Attitude modulation according to GRANT's method [7] is a simple technique for either human or automatic pilot, since it involves only a continuous reduction in pitch angle to hold the resultant deceleration fixed. Maximum modulation requires reducing the drag to $C_{D_{\min}}$ and C_L to zero. This is usually impractical from heating considerations [17] and a compromise schedule from $C_{D_{\max}}$ to $L/D_{\max}$ was selected for use in the present studies. The velocity at the end of this practical modulated maneuver is only slightly higher than for the corresponding fixed $(L/D)_{\max}$ case. Thus the subsequent range capability is about the same.

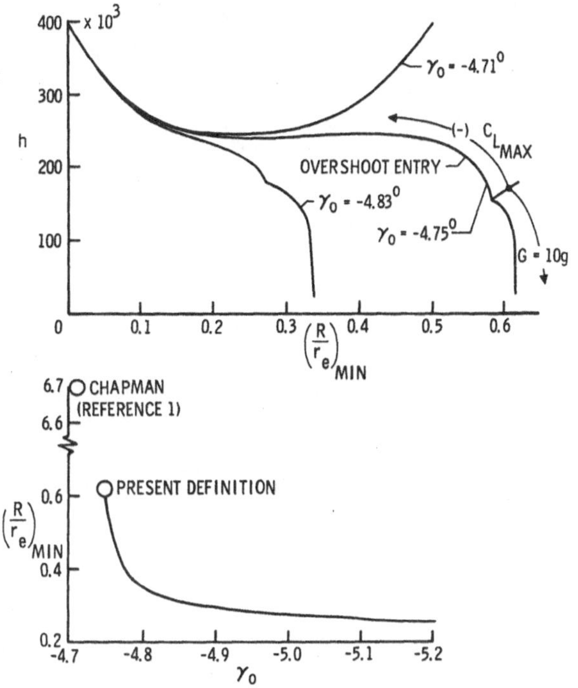

Fig. 2. Effect of entry angle on minimum range and definition of overshoot boundary used in the present study

The range achievable from the "overshoot" boundary of the corridor is extremely sensitive to the definition of this boundary (Fig. 2). CHAPMAN's definition (for $-C_{L_{\max}}$, $\bar{V}_{\text{exit}} = 1$) is impractical because it requires global range and because minute aberrations in the atmosphere or errors in entry angle will produce range variations of thousands of miles. From the piloting standpoint overshoot $-C_{L_{\max}}$ trajectories which never develop an ascending phase appear desirable. Furthermore they mark a borderline between very long range and short range trajectories, and they have the advantage of "digging in", i. e. developing high G which is desirable for minimizing range. This definition has accordingly been adopted for the present study. It will be noted that the practical overshoot entry angle so defined is only a minor fraction of a degree smaller than CHAPMAN's value.

Fig. 3 summarizes the corridor depths achievable by various initial entry maneuvers for lift/drag ratios ranging from 0 to 1. On the G-limited steep-entry boundary, with no modulation, most of the benefit of lift is realized with L/D of about one-half. Large increases in permissible entry angle result from attitude modulation. The benefits of the practical modulation technique (from $C_{L_{\max}}$ to $(L/D)_{\max}$) are particularly worthwhile for $L/D > 1/2$.

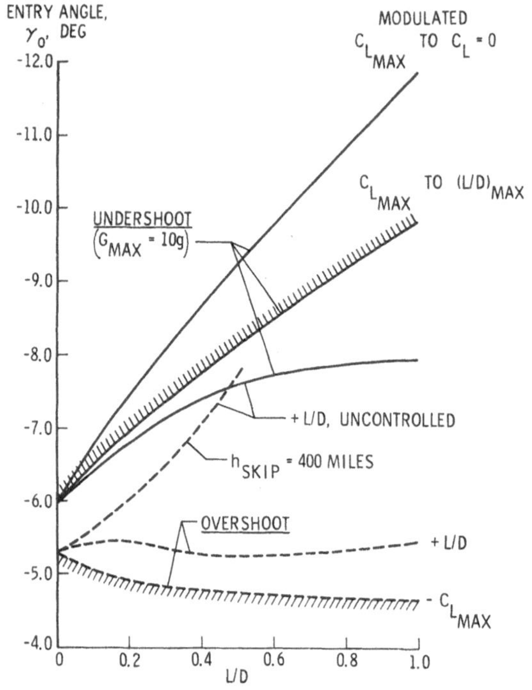

Fig. 3. Vertical boundaries of re-entry corridor as affected by L/D and type of initial maneuver. Cross-hatched boundaries were used in present range study

Without modulation and with the restriction of 400 miles for apogee of the skip only a very shallow corridor is available (the distance between the lowest solid curve and the uppermost dashed curve of Fig. 3). A much deeper corridor is achieved if, after an overshoot entry at fixed positive L/D, lift control is used or either avoid or control the skip. Still further expansion occurs if full negative lift is used in overshoot entries (defined as in Fig. 2) followed by lift control to avoid excessive G.

The corridor boundaries used in the subsequent range studies are indicated by the cross-hatching in Fig. 3. We will be interested in whether a given destination can be reached from both sides of the corridor or whether a reduced "usable" corridor depth is determined by this consideration.

IV. Aerodynamic Techniques for Range Control

The range covered during re-entry is a function of many variables: the entry velocity, entry path angle γ_0, L/D, $W/C_D A$, the orbit direction and inclination, the earth's rotation and oblateness, atmospheric density aberrations, and winds. Since we are concerned here with general comparisons rather than precise performance for a particular system, only the entry conditions and the vehicle parameters need be considered. It was assumed that the range control maneuver started immediately after the initial pull-up at the velocity, $\bar{V}_1$, existing at the end of pull-up to the desired maneuver flight path (Fig. 4). The range covered during pull-up and the flight variables at the end of pull-up were determined from digital computer results. Except where noted a value of $W/C_D A$ of 50 lb/sq ft, which is appropriate for manned vehicles in the $L/D \approx 1/2$ to 1 category, was used.

1. *Maximum Range*

The maneuver schedules which produce the true maximum longitudinal range for a practical set of constraints are not easily determined [18]. For present

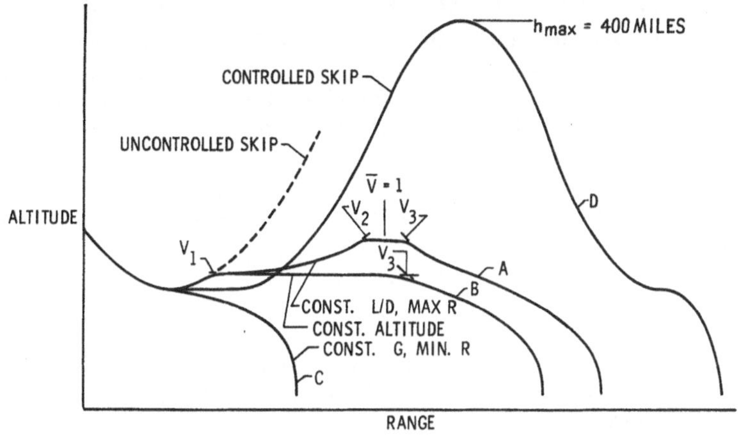

Fig. 4. Schematic comparison of trajectories used in range study

purposes, however, the constant-L/D glide, modified in the vicinity of $\bar{V} = 1$, is a satisfactory approximation. The differential equation for range is

$$\frac{dR}{r_e} = -\frac{\cos\gamma\,\bar{V}\,d\bar{V}}{D/W} \tag{1}$$

Since all of the paths to be considered have very small angles with respect to the local horizontal over most of the glide range after the initial pull-up has been completed it is assumed that $\cos\gamma = 1$. The balance between the vertical aerodynamic lift, gravity force, and centrifugal force then requires that

$$\frac{L\cos\varphi}{W} = (1 - \bar{V}^2) \tag{2}$$

whence

$$\frac{dR}{r_e} = \frac{(L/D)\cos\varphi\,\bar{V}\,d\bar{V}}{\bar{V}^2 - 1} \tag{3}$$

Integration of (3) for constant L/D results in infinite range if extended to $\bar{V} = 1$. This is avoided by joining the supercircular and subcircular glide paths by a fixed-deceleration path, Fig. 4 (path A). Since we are interested in producing the longest possible range in these glides we choose the highest feasible altitude for this transitional phase, limited by the requirement of adequate aerodynamic control which is assumed to be met if $q/(W/A) \gtrsim 0.2$. The range in the transitional maneuver itself is found to be independent of $q/(W/A)$ and is

$$\frac{R_{\text{transition}}}{(r_e)\,(L/D)} = \frac{\bar{V}_2^{\,2} - \bar{V}_3^{\,2}}{2(\bar{V}_2^{\,2} - 1)} \approx 1 \tag{4}$$

The L/D used in (4) is the constant value applying in the glide phases of the entry and at the start and end of the transition. D is assumed constant during transition and $L \cos \varphi$ is varied by rolling at fixed pitch from negative values for $\bar{V} > 1$ to positive values of $\bar{V} < 1$. The total range[1] for path A is then

$$\frac{R}{(r_e)(L/D)} = \frac{1}{2}\left[\ln\left(\frac{1-\bar{V}_1^2}{1-\bar{V}_2^2}\right) + \ln\left(\frac{1}{1-\bar{V}_3^2}\right) + 2\right] \tag{5}$$

Since $\bar{V}_2$ and $\bar{V}_3$ will depend on the value of $W/C_L A$ of the glide and thus on the design L/D, the value of eq. (5) will be independent of L/D only for sub-circular cases where $\bar{V}_1 = \bar{V}_3$. It is found that this equation is an adequate approximation for L/D values of about 1/2 or greater. For lower L/D values computer data were used.

The abrupt transitions in path angle at various points in the entries shown in Fig. 4 obviously represent an idealization which cannot actually be achieved. It remains for studies of the dynamic behavior of guided vehicles to determine how closely these ideal ranges can be approached.

2. Constant Altitude Mode

This mode is characterized by shorter maximum ranges than the constant L/D mode but it appears to offer possibly important advantages from the guidance and control standpoints. Obviously, the need for a transition maneuver in the

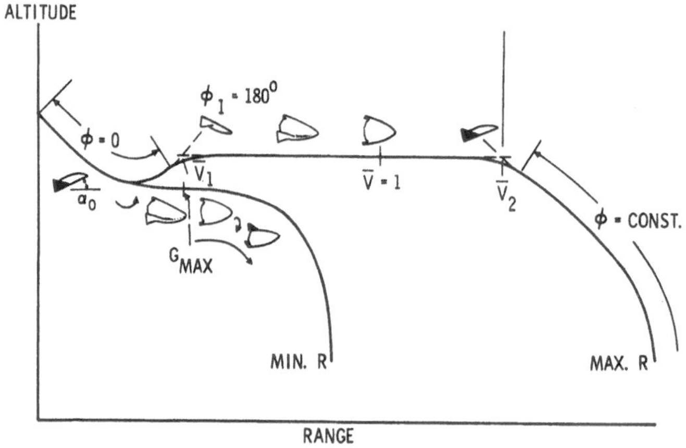

Fig. 5. Variable-roll control mode as applied to "constant altitude" trajectories

vicinity of $\bar{V} = 1$ is eliminated (Fig. 4). The rate of change in dynamic pressure dq/dt is smaller and constant in sign. The constant altitude may prove desirable from the guidance standpoint especially if the altitude can be measured directly by on-board radar. If we take the roll mode[2] as shown in Fig. 5 as an example we can see that if roll is applied in one direction only, lateral range would be developed. However, if the sign of the bank angle is periodically reversed, the lateral range can be reduced to zero. Since the pitch attitude and deceleration schedule are fixed regardless of the sign of the bank angle, the distance along

[1] The maximum range was computed for $\cos \varphi = 1$ in the glide phases preceding and following the transitional maneuver.

[2] The possible use of the fixed-pitch roll technique has been mentioned elsewhere (e.g. [21]) primarily from the point of view of simplifying the heat protection problem. We are concerned here with its use in range control.

the flight path, whatever its lateral characteristics, is essentially constant. Thus any desired lateral range from zero to the maximum for roll always in the same direction can be obtained with little effect on longitudinal range. Further details of the lateral range potentialities of this technique will be discussed later.

From the standpoint of attitude controls the roll mode is also attractive. Prior to entry a nominal pitch attitude would be selected and an aerodynamic flap type of pitch control would be set in the fixed position for this attitude. (If possible a center-of-gravity location requiring least possible control deflection would be selected.) Control of the vertical lift as required throughout the entry would be accomplished by rolling the vehicle. Longitudinal range would be controlled by initiating the constant altitude phase at the proper altitude (Fig. 5). Lateral range would be controlled as discussed previously by changing the sign of the roll angle at appropriate times during the entry.

Roll control can be accomplished economically by reaction jets because the rolling moments are low even in regions of high dynamic pressure, and the same reaction system employed for space attitude control can be used. (This is generally not possible foı the pitch-control mode because of the relatively large pitching moments required for trim for longitudinally stable aerodynamic vehicles.) Artificial roll damping can be applied by the same reaction devices.

The range equation for the constant altitude mode has no discontinuity at $\bar{V} = 1$. For the fixed-pitch variable-roll case the lift requirements are met if

$$\cos \varphi = \cos \varphi_1 \left(\frac{1 - \bar{V}^2}{1 - \bar{V}_1^{\,2}}\right)\left(\frac{\bar{V}_1}{\bar{V}}\right)^2 \tag{6}$$

Whence, from integration of eq. (3), in terms of L/D

$$\frac{R}{(r_e)\,(L/D)} = \frac{\cos \varphi_1\, \bar{V}_1^{\,2}}{(1 - \bar{V}_1^{\,2})} \ln \left(\frac{\bar{V}_1}{\bar{V}_3}\right) + \frac{1}{2} \ln \left(\frac{1}{1 - \bar{V}_3^{\,2}}\right) \tag{7}$$

or, in terms of the initial deceleration rate,

$$\frac{R}{(r_e)\,(W/D)_1} = \bar{V}_1^{\,2} \ln \left(\frac{\bar{V}_1}{\bar{V}_3}\right) + \frac{2}{1} \frac{L_1}{W} \ln \left(\frac{1}{1 - \bar{V}_3^{\,2}}\right) \tag{8}$$

The velocity ratio for which the constant altitude portion must terminate in the fixed-L/D glide, assuming $\varphi_1 = \varphi_3$, is given by

$$\frac{\bar{V}_1}{\bar{V}_3} = \sqrt{\bar{V}_1^{\,2}\left(1 - \frac{C_{L_3}}{C_{L_1}}\right) + \frac{C_{L_3}}{C_{L_1}}} = \sqrt{2\,\bar{V}_1^{\,2} - 1} \quad \text{for} \quad C_{L_1} = -\,C_{L_3} \tag{9}$$

If the constant-altitude mode uses fixed roll and variable pitch the deceleration and L/D schedules will depend both upon the force polar of the individual vehicle and the point on the polar where the maneuver is started. Flat-bottomed shapes were assumed in the present study and NEWTONian force relationships were used for vehicles having $(L/D)_{max}$ values of 1/2, 1, 1 — 1/2, and 2.

3. *Minimum Range*

In order to achieve the absolute minimum range for a given set of constraints complex modulated lift schedules are required. Minimum range was of interest in the present study primarily for the overshoot boundary condition. Our practical definition of the overshoot boundary involved $-\,C_{L_{max}}$ trajectories which tended to "dig in" to the atmosphere and develop high G values (Fig. 2). Minimum range could thus be approximated simply by assuming constant 10 g deceleration from the point where this first occurred to the lowest altitude where 10 g could still be maintained.

4. Controlled Skip

For comparison with the entirely atmospheric deceleration paths a family of controlled skips were evaluated. Control consisted of negative lift modulation by the variable roll technique, starting at the bottom of the pull-up to hold a fixed altitude until the velocity had diminished to the desired value. The vehicle was then rolled to $\varphi = 0$, and allowed to skip. Digital computer results were used to obtain the range for these partially controlled skipping trajectories.

V. Longitudinal Range

The maximum ranges attainable with the various atmospheric modes are compared in Fig. 6, starting in all cases at the conditions required for equilibrium

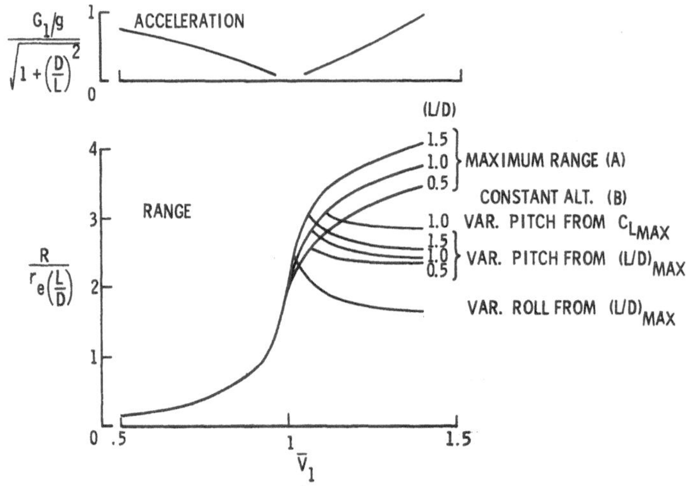

Fig. 6. Comparison of range for the various atmospheric modes starting at equal conditions at the end of the initial pull-up maneuver ($q/(W/A)_{min} = .2$)

glide at the starting velocity $\bar{V}_1$. Also shown is the starting value of the resultant deceleration, G_1. (This is also the actual G versus $\bar{V}$ schedule for the equilibrium glide cases.) The constant altitude cases increase in range capability as $\bar{V}_1$ diminishes because the reduced deceleration levels (corresponding to starting from the constant L/D glide condition) more than offset the reduced initial energy levels. Since the initial values of G are of the order of 1 g or less, minimum ranges of 1/10 or less of those shown could be achieved for $G_{1_{max}} = 10\,g$. Of the constant altitude cases, the pitch-control mode has inherently a somewhat larger range capability than the roll-control mode, especially if the maneuver is started near $C_{L_{max}}$ rather than L/D_{max}, but the differences are not large.

Fig. 6 can be used to obtain the total range achievable as a function of entry angle if the velocity ratio existing at termination of the pull-up maneuver is known together with the range covered during the pull-up phase. The latter quantities were obtained from the computer program for the modulated pull-ups described previously. The maximum atmospheric ranges are compared in Fig. 7 with the ranges attainable in partially controlled skipping entries for $L/D = 1$. Ranges approximately equal to those obtained by skipping could be attained if dynamic pressures as low as about 1 lb/sq ft were permissible. Control difficulties are encountered [19] at such low pressures, however, and a value of $q/(W/A)$ of the order of 0.2 is believed to be a reasonable approximate limit for satisfactory aerodynamic control. With this limit ranges of 1/2 to 3/4 of the ranges of controlled-skip entries can be achieved with the wholly atmospheric modes for $L/D = 1$.

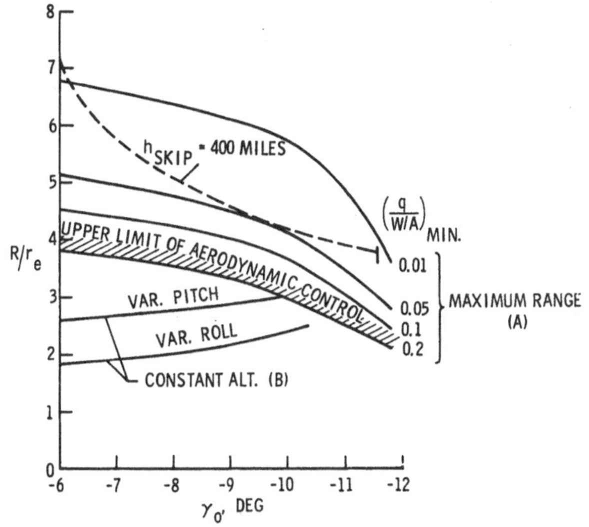

Fig. 7. Comparison of range for atmospheric and controlled skip modes. $L/D = 1$

We are now in a position to answer the question of whether the returning space vehicle can reach a given destination from any extremity of the re-entry corridor. The factors involved here are sketched in Fig. 8. Basically, the question is whether the least longitudinal range achievable for an overshoot entry will overlap the maximum range for an undershoot entry having the same perigee location. The most critical case is assumed to occur when the undershoot entry

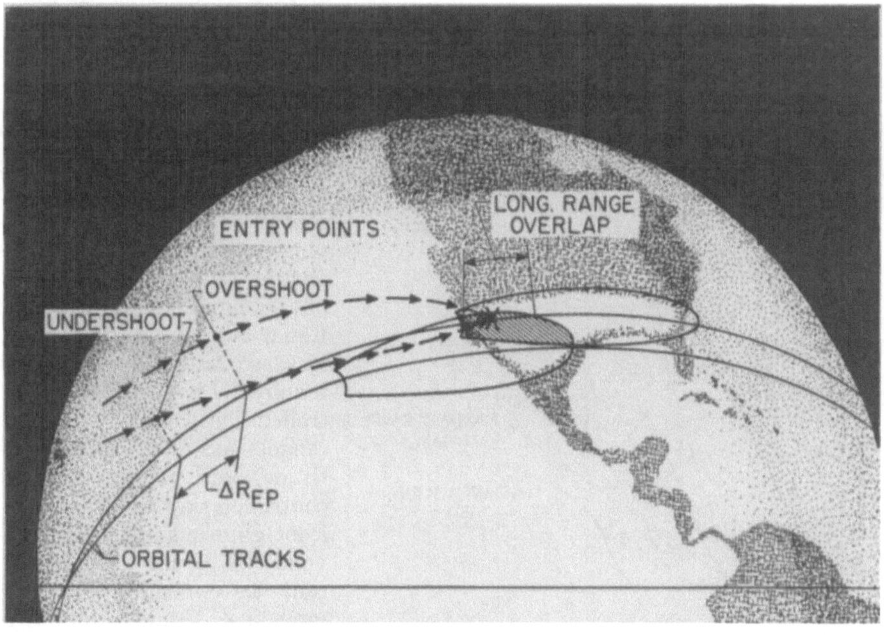

Fig. 8. Definition of longitudinal range overlap

plane does not coincide with that of the overshoot entry but is inclined so as to require a maximum lateral turn. Account is also taken of the fact that the entry point for a given perigee location lies farther down range for the overshoot case than does the entry point for the steep entry, the distance from entry point to perigee being $R/r_e = 2\gamma$. Fig. 9 presents the ranges achievable from overshoot and undershoot boundaries, both with and without the maximum lateral maneuver for undershoot, and for both the maximum-L/D glide and constant-altitude variable-roll modes. (The condition of $\varphi = 60°$ for the "variable-roll" case assumed that this value was used in the subcircular glide phase.) The cross-over points beyond which range overlap exists occur at $L/D = 0.36$ and 0.61, respectively, for the two types of trajectories for the condition of displaced orbital planes. If a range overlap of at least 2000 miles (i.e. $\pm$ 1000 miles) or more is considered desirable for manned re-entries from parabolic orbits the minimum L/D requirements become $L/D = 0.57$ and 0.71, respectively. A comparison with the overlap capabilities of the skip modes is made in Fig. 10. The controlled-skip maneuver provides overalps about 5000 miles greater than the best atmospheric mode, for $L/D = 0.5$. Equal overlaps of 2000 miles for zero lateral displacement require L/D values of 0.33 (controlled skip), 0.43 (maximum range), and 0.68 (constant altitude, variable φ). For the controlled skip mode no significant gain in range occurs beyond $L/D = 0.6$. It is important to remember that the foregoing results apply to entries from the extremes of the deepest practicable cor-

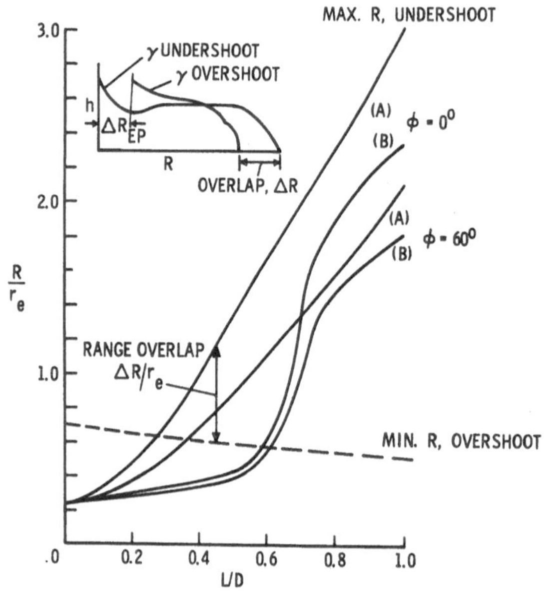

Fig. 9. Maximum and minimum ranges from corridor boundaries used to determine range overlap for atmospheric modes

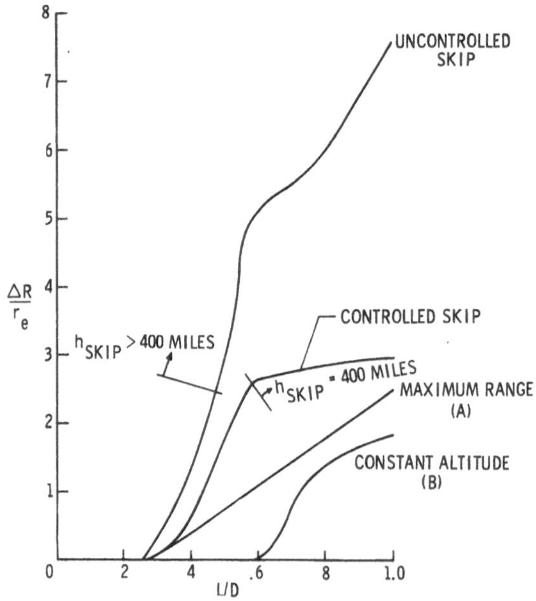

Fig. 10. Comparison of range overlap for atmospheric and skip modes

ridors, the corridors for the atmospheric cases being deeper than that for the controlled skip because the larger L/D values permit steeper entries (Fig. 3). If the condition of equal maximum corridor depth is imposed the L/D values for the aerodynamic modes can be reduced somewhat.

If we are willing to accept smaller-than-maximum corridor depths the L/D requirements for overlap can be considerably relaxed. For illustration consider the constant-altitude variable-roll mode for which an L/D of 0.58 is required to achieve overlap for an entry in which no lateral maneuver is involved (Fig. 9). The modulated entry corridor (Fig. 3) for this L/D extends from about $\gamma_0 = -4.7°$ to $\gamma_0 = -8.4°$, corresponding to a depth of about 60 miles. We now inquire how much the L/D requirement for overlap can be reduced if entry is made from more shallow corridors. Initially, we keep the overshoot boundary fixed and progressively reduce the value of $-\gamma_0$ for the undershoot boundary. Fig. 11 (solid lines) shows the effect of this on the L/D required for various overlap ranges. (The reference maximum corridor width on which the data of Fig. 11 are based is the value stated above for $L/D = 0.58$.) We note for this case that a large decrease in corridor width is required before any sizable reduction in L/D is possible. However, if we now hold the undershoot boundary fixed and relax the overshoot limit, large reductions in L/D can be obtained (dashed lines of Fig. 11) for the smaller range overlaps. Range overlaps significantly greater than about 1000 miles, however, can only be realized by relaxing the undershoot boundary either separately from or in conjunction with the change in the overshoot boundary.

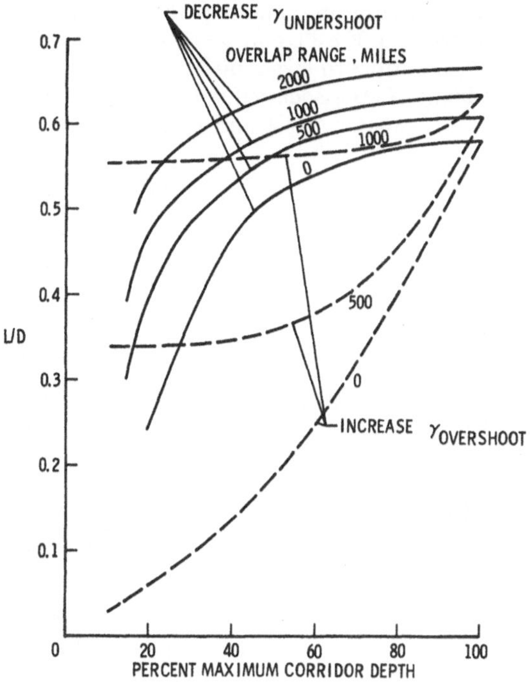

Fig. 11. Trade-off between corridor depth and L/D required for range overlap. Constant-altitude variable-roll mode

The choice of purely atmospheric versus controlled skipping modes is influenced strongly by guidance and control considerations. The present study revealed that the range for the skipping mode hinges delicately on the velocity and path angle for which the skip is initiated. Typical values of the rate of change of range with exit path angle and exit velocity for an exit angle of 1° and $\bar{V}_{exit}$ near unity were found to be

$$\frac{\partial R}{\partial \gamma_1} \approx 7500 \text{ miles/degree}$$

and

$$\frac{\partial R}{\partial V_1} \approx 50 \text{ miles/ft/sec}$$

Thus, although the skipping mode has a lower L/D requirement for range overlap, it has a serious guidance and control problem. From the same point of view it is not possible duringth e re-entry subsequent to a skip to correct the range for sizable guidance or control errors made in the initial entry because of the large magnitude of the range errors and the low L/D usually contemplated in this type of re-entry. For example, an angle error of $\pm$ 0.05° plus a velocity error of $\pm$ 10 ft/sec would cause a range error of some $\pm$ 900 miles which is nearly twice as great as the maximum possible range correction for an $L/D = 0.25$ vehicle reentering at $\bar{V} \sim 1$ and $\gamma \sim -1°$. In contrast, the atmospheric mode is both less susceptible to large initial errors and more tolerant of such errors because of the higher L/D associated with this type of re-entry and because of the capability for continuous control and correction.

VI. Lateral Range

Previous studies of lateral range ([20] and [21], for example) have been concerned primarily with satellite or subcircular-velocity entries at shallow angles,

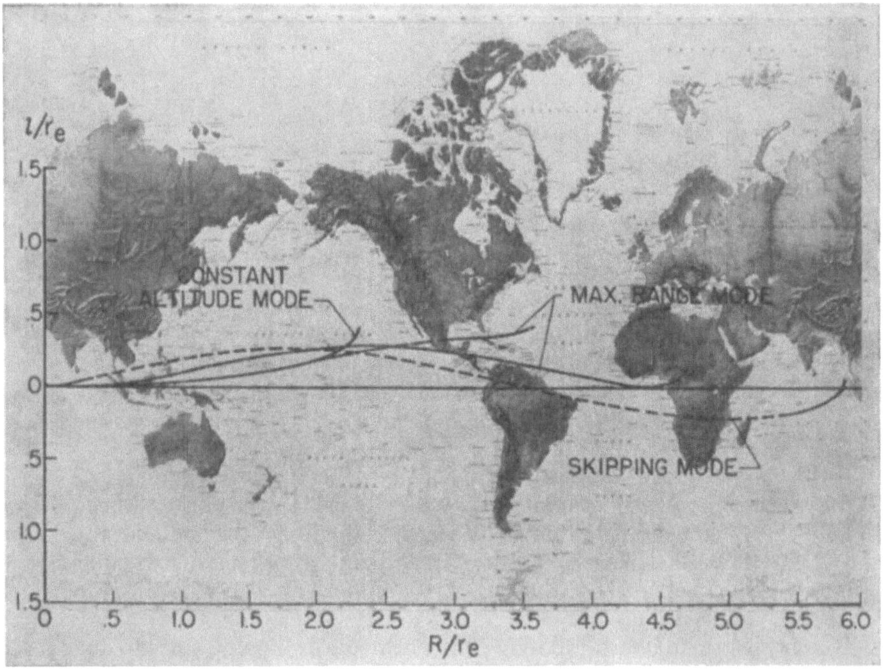

Fig. 12. Typical long-range re-entry paths for banked vehicles

usually along equilibrium fixed-L/D glide paths. The nature of such entries makes it impossible to develop large changes in heading in the initial phase of the glide and the major part of the lateral range develops later after much of the longitudinal range has been covered. For a subcircular glide at $L/D \approx 1$ or less the range covered during most of the turn is small enough that a simple approximation which neglects the spherical nature of the earth can be used for preliminary estimates [20].

Practical entry from a parabolic orbit, however, involves relatively steep initial penetration to a condition of high dynamic pressure where a significant change in heading can be accomplished early in the entry. Thus large lateral and longitudinal ranges can be developed for which it is clearly no longer permissible to neglect the earth's sphericity. An illustration of this is given in Fig. 12. The longest trajectory shown is that of a skipping vehicle which makes an uncontrolled banked pull-up (solid portion) a skip (dashed portion) followed by a second entry at the same bank angle and L/D. In the long-range skip the vehicle follows an orbital path inclined to the initial entry plane, which bends back toward the initial plane crossing it near $R/r_e = \pi$. The heading angle changes sign,

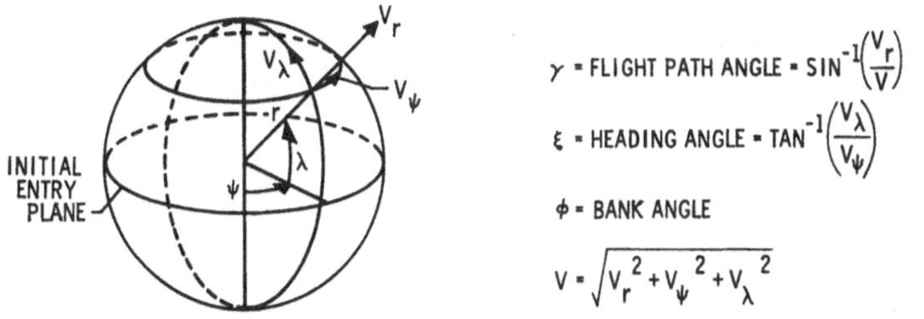

Fig. 13. Coordinate system used in particle motion analysis

twice in this particular case, before touchdown. The second entry is with a lateral displacement of opposite sign to that of the initial turn. Obviously, a greater final lateral range could have been achieved by eliminating the first turn. The constant L/D fixed-bank turns of Fig. 12 encounter a region of low dynamic pressure ($q_{\min} = 10$ lb/sq ft) over a large part of their range and they are therefore affected similarly by the earth's sphericity, but to a lesser extent. The heading angle undergoes one or more inflections and changes sign for the longer-range case. The shortest range case which operates at a higher q level appears to follow the simple pattern visualized in the aforementioned approximation; however, it also is actually influenced to some extent by sphericity.

In order to obtain accurate evaluations of lateral range, the equations of motion were set up in spherical coordinates (Fig. 13). The position of the vehicle is determined by the three coordinates r, ψ, and λ where r is the distance from the center of the earth to the vehicle, ψ is the angular distance from the re-entry point in the original entry plane, and λ is the angular distance measured normal to the original entry plane. The velocity of the vehicle at any point is composed of the components in these three directions: V_r, the velocity normal to the surface of the earth, V_ψ, the velocity along the surface of the earth parallel to the original entry plane, and V_λ, the velocity component along the surface of the earth and normal to V_r and V_ψ.

In addition we define γ, the flight path angle, and ξ the heading angle as indicated and use φ to denote the bank angle and V the total velocity of the vehicle. With these coordinates and definitions the kinetic energy of the system may be written

$$T = \frac{m}{2} (\dot{r}^2 + r^2 \cos^2 \lambda \, \dot{\psi}^2 + r^2 \, \dot{\lambda}^2) \tag{10}$$

and the LAGRANGE equations for the motion of the system, including aerodynamic forces, are

$$\frac{\partial T}{\partial r} - \frac{d}{dt}\left(\frac{\partial T}{\partial \dot{r}}\right) = -L\cos\varphi\cos\gamma + D\sin\gamma + m\,g$$

$$\frac{\partial T}{\partial \psi} - \frac{d}{dt}\left(\frac{\partial T}{\partial \dot{\psi}}\right) = -L(\cos\varphi\sin\gamma\cos\xi + \sin\varphi\sin\xi) - D\cos\gamma\cos\xi \tag{11}$$

$$\frac{\partial T}{\partial \lambda} - \frac{d}{dt}\left(\frac{\partial T}{\partial \dot{\lambda}}\right) = L(\sin\varphi\cos\xi - \cos\varphi\sin\xi\sin\gamma) - D\cos\gamma\sin\xi$$

From these relations, the equations of motion are obtained in the form

$$\frac{1}{g}\frac{dV}{dt} = -\frac{1}{2}\varrho V^2\left(\frac{C_D A}{W}\right) - \sin\gamma \tag{12}$$

$$\frac{1}{g}\frac{d\gamma}{dt} = \frac{1}{2}\varrho V\left(\frac{C_D A}{W}\right)\frac{L}{D}\cos\varphi - \frac{\cos\gamma}{V}\left[1 - \frac{V^2}{g\,r}\right] \tag{13}$$

$$\frac{1}{g}\frac{d\xi}{dt} = \frac{1}{2}\varrho V\left(\frac{C_D A}{W}\right)\frac{L}{D}\frac{\sin\varphi}{\cos\gamma} - \frac{V}{g\,r}\cos\gamma\cos\xi\tan\lambda \tag{14}$$

$$\frac{d\psi}{dt} = \frac{V\cos\gamma\cos\xi}{r\cos\lambda} \tag{15}$$

$$\frac{d\lambda}{dt} = \frac{V\cos\gamma\sin\xi}{r} \tag{16}$$

$$\frac{dr}{dt} = V\sin\gamma$$

The approximation often used in the calculation of lateral range is to neglect the second term in eq. (14). If this approximate relation is then combined with eq. (12) and small flight path angles are postulated, the resulting relation is

$$\frac{d\xi}{dV} = -\frac{L}{D}\frac{\sin\varphi}{V}$$

which is readily integrated for constant L/D and bank angle (φ). This approximation effectively considers a cylindrical earth and can lead to large errors in lateral range, especially when the longer ranges ($R/r_e > 1$) are considered.

Results of digital computer calculations of available landing area using both the complete (spherical earth) equations and the approximate (cylindrical earth) equations are presented in Fig. 14. A middle corridor re-entry ($\gamma = -6°$) was considered and vehicles of $W/C_D A = 50$ lb/sq ft performed unmodulated, unbanked pull-ups followed by the constant L/D mode of operation described previously with various constant bank angles during the glide portions. It is seen that the errors become very large for the longer ranges. The "approximate" lateral range is about 1.75 radii for the $L/D = 2$ re-entry as compared to about 0.75 radii for the complete equation. For $L/D = 0.5$ the errors involved are small because the ranges are small. If we consider controlled skip trajectories, however, where the distances traveled are on the order of several earth radii even for low L/D ratios, the approximate equation would again lead to large errors.

The landing area available to a vehicle with a maximum L/D of 0.5, following the controlled skip type of re-entry, initiated at a shallow entry angle ($\gamma_i = 5.5°$) is shown on Fig. 15. The vehicle maintains $(L/D)_{\max}$ during the entire re-entry as previously described. Some sample trajectory traces which went into the makeup of this footprint are also shown on the figures with the dashed portions of these traces representing the time the vehicle is outside of the atmosphere ($h > 400{,}000$ ft). The strong effect of the spherical curvature of the earth on these trajectories is obvious. Two interesting features of this footprint are the decrease

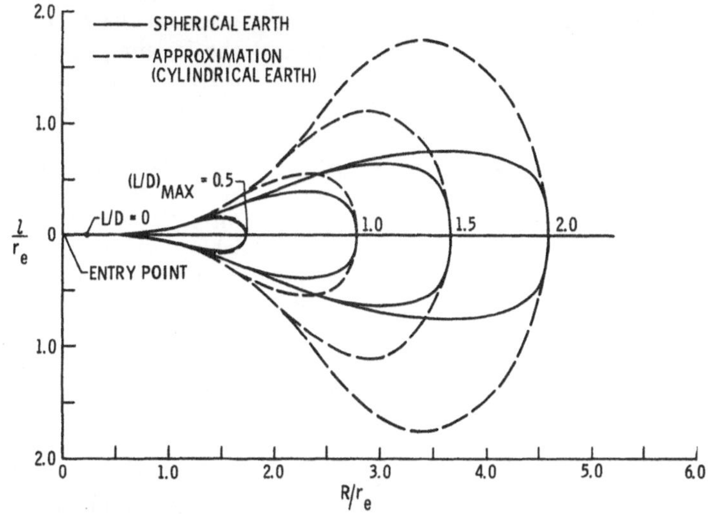

Fig. 14. Available landing area for maximum range mode as affected by earth's sphericity. $\gamma_0 = -6°$, $W/C_D A = 50$ lb/sq ft

in lateral range available for longitudinal ranges about half-way around the earth from the re-entry point, and the necessity of directing the vehicle initially

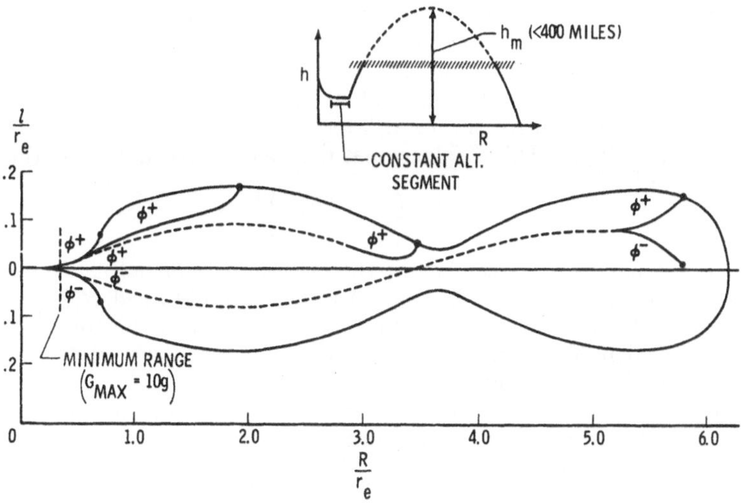

Fig. 15. Available landing area for controlled skip. $L/D = 0.5$, $\gamma_0 = -5.5°$, $W/C_D A = 50$ lb/sq ft

to the right of the entry plane in order to obtain sizable lateral displacements to the left of the entry plane for longitudinal distances greater than half-way around the earth.

For the atmospheric re-entry modes considered previously the lateral ranges available for vehicles with $L/D = 0.5$ and 1 are presented in Fig. 16 as a function of the velocity at the end of the pull-up. For the $L/D = 0.5$ vehicle a definite advantage is seen for the constant L/D mode over the constant altitude mode, while for the $L/D = 1$ vehicle, the advantage is small. In considering longitudinal ranges earlier in this paper it was seen that the constant L/D mode had a decided advantage over the constant altitude mode at all L/D values. The diminishing lateral range advantage is again due to the strong effects of the earth's spherical nature over the long ranges associated with the constant L/D mode, for higher L/D vehicles.

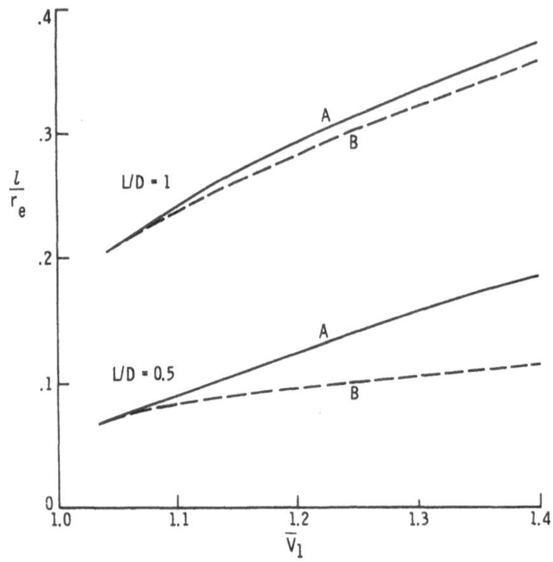

Fig. 16. Comparison of lateral range for atmospheric modes

During the constant altitude portion of flight, the direction and amount of displacement away from the original flight path can be controlled by varying the sign of the bank angle of the vehicle so as to direct the horizontal component of the lift vector to either the right or left of the flight path. If no effective lateral deviation is desired during all or part of the constant altitude flight, this may be accomplished by banking alternately to right and left, weaving around the original flight direction. By using this technique for only a portion of the total time at constant altitude and then banking in one direction for the rest of the time it is possible to control the total lateral displacement obtained duing the constant altitude flight. This displacement is given approximately by the relation

$$\frac{l_2}{l_{max}} = \left(1 - \frac{R_1}{R_{max}}\right)^2$$

where l_2 is the lateral displacement at the end of the constant altitude portion, l_{max} is the maximum lateral displacement that could be obtained, R_1 is the distance in the original flight direction (measured from the start of the constant altitude flight), over which no lateral displacement is affected, and R_{max} is the total distance traveled in the original flight direction.

Another scheme for controlling lateral displacement and heading angle during constant altitude flight, necessitating a minimum number of changes in direction of bank is illustrated on Fig. 17. On this sketch φ^+ signifies banking to the left and φ^- to the right. It is seen that only one change in direction of bank will enable the vehicle to achieve any lateral displacement up to the maximum value attainable, as shown by curves OBH and OCE and OD. This gives no control over heading angle at the end of the maneuver, however. The heading angle, which is zero at the end of maneuver OCE in the sketch will be positive for points between D and E and negative for points between E and H. If we allow a second change in the direction of bank we also have some control over the heading angle at the end of the maneuver. If we desire a zero heading angle at the end point, this can be achieved at any point between E and H by a two turn maneuver similar to curve $OBFG$. For points between E and D, however, a zero heading angle is not achievable at the end of the constant altitude phase. Heading angles other than zero can be achieved in a similar manner, with similar restrictions.

If no lateral displacement is desired during the constant altitude phase of flight, this may be achieved by either one change in direction of bank (OBH),

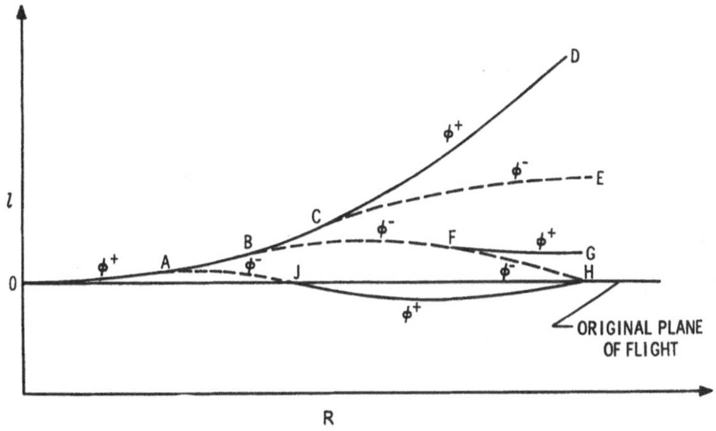

Fig. 17. Control of lateral range and final heading by change in sign of roll angle

two changes ($OAJH$), or any larger number. The larger the number of turns taken, the closer to the original flight path the vehicle remains.

In discussing the roll-control technique in this and in preceding sections we have considered its application to the constant-altitude mode. It is obvious, of course, that the roll-control technique can be equally well applied to other modes, for example, ones in which a G schedule may be specified. It can also be applied in the fixed-L/D mode in cases where a bank is permissible.

VII. Aerodynamic Requirements

The choice of lift/drag ratio will obviously depend on a number of other factors in addition to those so far considered in the discussion of corridors and range. The required entry corridor dimensions may in some cases be determined by the nominal accuracy of the guidance and control system used in the space phase. Studies such as [22], however, may be interpreted to indicate that practical guidance and control techniques can probably be devised which will require corridor depths and widths only a fraction of those attainable with an L/D of one-half. But invariably it is not the nominal design capability of such systems which determines the vehicle. Rather it is the off-design or emergency conditions which establish the requirements. From this point of view the maximum practicable corridor dimensions may be none too large. Unfortunately, these off-design requirements are impossible to determine in a general study because they depend upon details of each individual system. In the absence of any clear general requirements in these other important areas, we will base our choice of maximum L/D on the necessity for achieving range overlap from the corridor extremities for atmospheric modes. Accordingly from Figs. 9, 10, and 11, we choose maximum L/D values in the range 0.5 to 1.0 for vehicle design considerations.

Lift/drag values in this relatively low category can be achieved by a wide variety of hypersonic shapes. The chief problem for design is to develop configurations having high wave drag (as required by heating considerations) together with acceptable stability and control characteristics. Maximum flexibility in trajectory control requires operation of the vehicle in different regions of the lift-versus-drag force polar at different times during entry. Use of the low-drag portion of the polar is mandatory for initial G modulation, maximum range, and finally for tangential landing. Least heat rate is achieved by use of $C_{L_{max}}$, either in the initial pull-up or in the subsequent glide. Range permitting, the high-drag part of the polar would be used after the initial pull-up in heat-absorbing designs. It

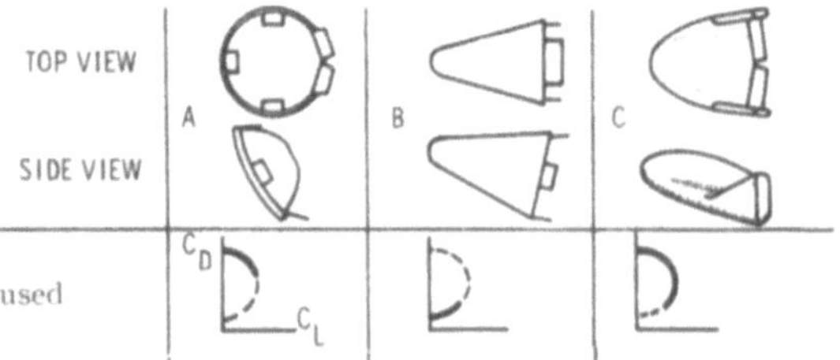

	A	B	C
Max. usable hyp. L/D	0.5	1.0	1.0
Lift curve slope, $dC_L/d\alpha$	—	+	+ and —
Corridor depth, miles	41	66 (mod.)	89 (mod.)
Range overlap, miles (const. alt. mode)	none	—	7200
Max. lateral range, miles	550	1650	1650
Relative heat load (middle of corridor)	1.0	2.1	2.1
Approx. relative weight	1.0	1.2	1.3

Fig. 18. Comparison of typical re-entry vehicle types

is not essential to achieve maximum capability in any one re-entry system; only parts of the polar are being used in a majority of current design studies. That is, limited performance and control are accepted in the interest of design simplifications or weight savings.

The blunt-nosed vehicle of Fig. 18 is similar in general concept to the orbital design for Project Mercury, the first manned orbital system developed by the United States. In this approach the brunt of the heating load is carried on a heat shield designed to absorb a major fraction of the heat load by the ablation process. The magnitude of the heat load is minimized by use of low L/D and high wave drag. The afterbody is not permitted to move out of the shadow of the heat shield so that it can be designed for relatively low heat rates for radiation cooling. The vehicle operates only in the high-drag, negative-C_{L_α} part of the force polar and it can be considered to be a minimum-weight solution with minimum acceptable trajectory control. A sizable overlap in range can be achieved only for a reduced corridor depth for the constant-altitude deceleration mode.

The cone-shaped design operates at the opposite extreme of the force polar and accepts a higher design heat load and higher weight in exchange for larger L/D and what are believed to be more favorable dynamic properties, i. e. a positive value of C_{L_α}, and a higher inherent damping factor than the minimal design. It is difficult to trim and stabilize at high angles of attack and for this reason it is usually considered only for the low-drag region of the polar, although there is no truly inherent restriction in this regard.

Vehicle C of Fig. 18 is intended to represent types capable of operation anywhere on the force polar in the re-entry environment and also capable of tangential landing. This type is somewhat heavier than the conical vehicle primarily because of the greater demands placed on its control system. In the region below $C_{L_{max}}$ the aerodynamics of the vehicle sketched are similar to those of the Dyna Soar type glider [23]. Our studies of vehicles of this type have indicated no important obstacle to successful operation over the polar from $(L/D)_{max}$ to $C_{D_{max}}$. The $C_{D_{min}}$ area, however, tends to be characterized by a loss of hypersonic longitudinal stability and increased heating problems of the upper surfaces. A study of a piloted vehicle of this kind in the high-drag region [24] concluded that for satellite entry, the piloting problems can be readily solved through use of combined aerodynamic and reaction controls designed for the high-drag attitude.

The relative heat loads shown in Fig. 18 include consideration of only the convective component for "cold" surfaces. Radiation from the hot gas to the body is an important additional heat load which must also be considered for re-entry at escape speed [25, 26]. In any case, however, these heat loads calculated for unheated surfaces are not a satisfactory index of relative heating. The surface temperature of a majority of the materials now contemplated will be very high over much of the entry so that a major fraction of the load will be disposed of by radiation away from the body. If we assume continued future growth in temperature capability [11], it appears likely that actual heat absorption by ablation or other forms of cooling may be required only during the initial pull-up maneuver of long-range re-entries. Subsequent maneuvers, whether of the A or B type, produce environmental conditions such that radiation cooling may dispose of virtually all of the applied heat. Minimum range entries, however, will require heat absorption, but here the total heat loads will be only a small fraction of the long-range values. Obviously, materials capable of both short-term heat absorption and long-term high surface temperatures would provide a desirable solution.

Our estimates of the weight penalties for increasing aerodynamic sophistication of these re-entry vehicle types (Fig. 18) may appear surprisingly small. It should be borne in mind, however, that the launch vehicle structure and control system may involve additional weight penalties not considered here. Furthermore, at the present stage of launch vehicle development, these re-entry vehicle weight penalties are by no means insignificant. Rather, they are large enough in some cases to dictate the use of the minimum vehicle.

VIII. Piloting Problems

The dynamic behavior of uncontrolled vehicles in the initial entry pull-up maneuver has been investigated theoretically ([27], for example). Prime factors in the motion are the large range of dynamic pressure, from zero to values comparable to those of current aircraft, and the low (or in some cases negative as for vehicle A) inherent dynamic damping factors of typical vehicles at re-entry speeds. The dominant effect of the increase in dynamic pressure causes the amplitude of the short period longitudinal oscillation (which develops if the vehicle is out of trim) to decrease in amplitude and increase in frequency as the bottom of the pull-up is approached. Typical peak frequencies of the short period motion of about 1 cycle per second occur for the postulated vehicle shapes for the steeper re-entries, and thus there is doubt whether a human pilot could control the motion, especially in the critical region where he must also reverse the direction of the lift. On the other hand it can logically be assumed that the reaction control system can be used by the pilot to reduce any initial out of trim attitude to a small value [28, 29]. Thus the subsequent high frequency oscillations would be inherently of very small amplitude and presumably would be tolerable even if uncontrolled or undamped by artificial damping devices.

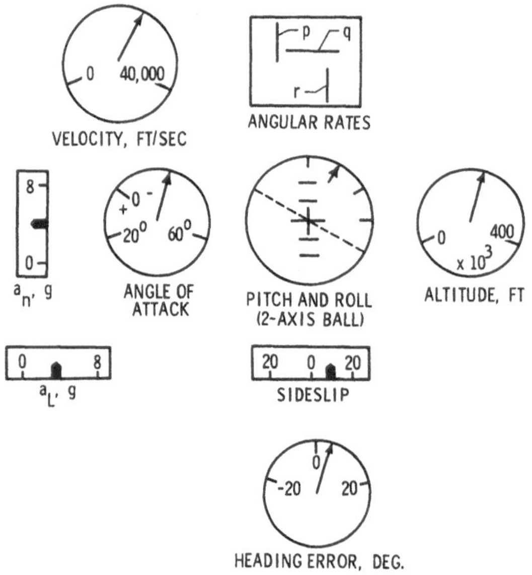

Fig. 19. Instrument panel used in simulator studies

The most difficult feature of the pull-up maneuver is the reversal of sign of the lift vector required to prevent a skip and to achieve a near horizontal glide path at the correct altitude. This must be accomplished in the presence of high but rapidly decreasing dynamic pressure, in most cases at supercircular velocity. Similar maneuvers are required later at $\bar{V}_2$ and $\bar{V}_3$ (Fig. 4). In the interim fixed L/D glide phase for $\bar{V} > 1$ the vehicle is subject to the unstable phugoid oscillation. That is, for fixed C_L any departure from the equilibrium glide altitude will tend to increase. The "period" is so long, however, that there is no question that the pilot can control the vehicle, except near $-C_{L_{max}}$ where any excess altitude cannot be corrected for because the lift cannot be increased.

A six-degree-of-freedom analog simulator has been developed at the Langley Research Center to permit investigation of a human pilot's ability to perform the basic pull-up and range-control maneuvers of re-entry from parabolic orbits. The preliminary results presented below have been obtained by Mr. A. Schy and Mr. M. Moul of the Langley staff.

The basic task set for the pilots was to enter at fixed C_L near L/D_{max} and to pull up to zero path angle at an altitude of 210,000 feet, starting alternately from a steep entry near undershoot, a middle corridor entry, and a shallow entry near the overshoot boundary for positive lift (Fig. 3).

A vehicle similar to A of Fig. 18 was used. The variable-pitch technique was employed to modulate lift. Negative lift was obtained by rolling 180°. Constant-gain automatic rate dampers were employed for all three axes. The pilot's display is shown in Fig. 19. A two-axis hand controller and rudder pedals were used.

The pilots found all three re-entry tasks simple to perform with dampers operating after some initial orientation practice runs (Figs. 20 and 21). Satisfactory entries could be made with either the yaw or the roll dampers inoperative.

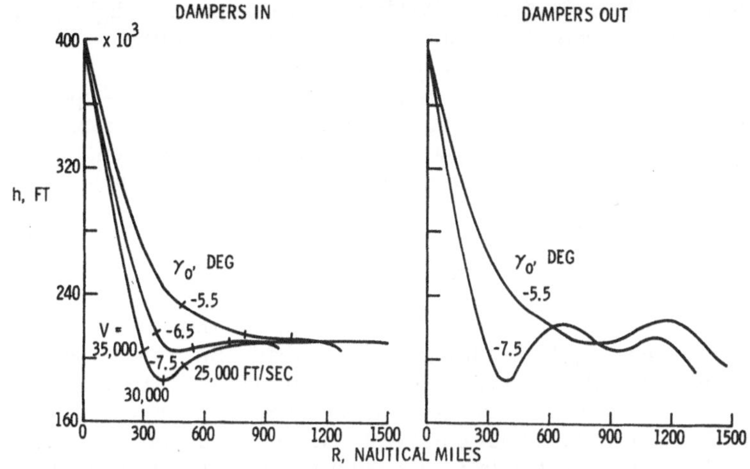

Fig. 20. Effects of dampers on piloted re-entry maneuvers for vehicle A of Fig. 18

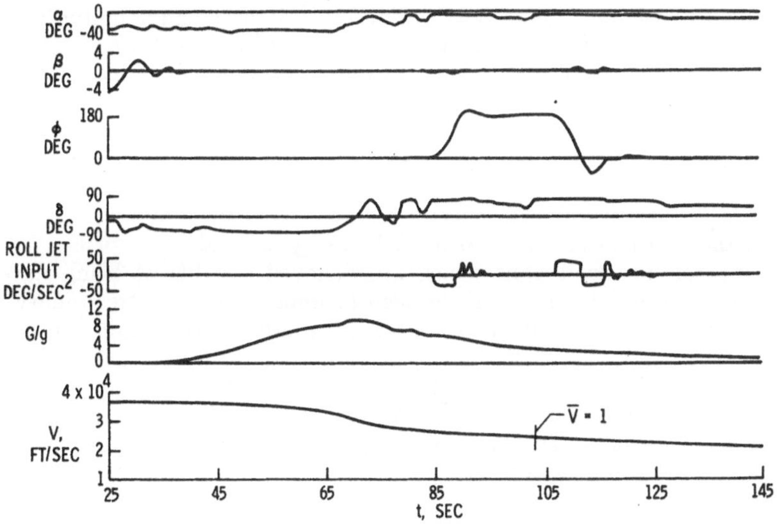

Fig. 21. Dynamic motions for steep entry of Fig. 20 a (dampers in, $\gamma_0 = -7.5°$)

Loss of pitch damping presented greater difficulty. The steep re-entry was difficult to control with any of the dampers out since the natural frequencies were high. With all rate dampers inoperative the behavior of the vehicle was generally unacceptable to the pilots. This type of vehicle is subject to aerodynamic control cross coupling and to inertial coupling as well. Thus, in the absence of damping, any roll control input, for example, will result in serious disturbances about the other axes. One possible mode of control is to pitch the vehicle to zero angle of attack before applying roll control, but this requires skillful action during the pull-up. Nevertheless, after considerable practice survivable entries could be achieved as shown by the typical example in Fig. 20.

In a similar series of entries using 235,000 feet as the pull-up altitude it was found that the shallow entry was difficult to control if positive flight path angles were allowed to develop. Control was somewhat more difficult for all entry angles because of the lower dynamic pressures.

No attempt was made in these preliminary tests to reach the specified glide altitude in the shortest time, as would be required for maximum range. Presumably, shorter times than those shown in Fig. 20 can be achieved. Preliminary simulated piloted entries using the fixed-pitch variable-roll technique have also been accomplished successfully.

IX. Conclusion

For equal range-control performance, atmospheric trajectories are found to require only moderately higher lift/drag ratios than skip trajectories which are limited to an apogee altitude below the inner radiation belt. In order to achieve a given destination from any extremity of the entry corridor with $\pm$ 1000 miles margin a minimum L/D of about 0.6 is required for a wholly atmospheric re-entry at fixed L/D. Constant-altitude modes require values up to about 0.7. These atmospheric modes do not suffer from the inherently sensitive guidance and control problems of the skip mode.

Certain atmospheric modes in which roll is the only control variable are capable of essentially independent variation of lateral and longitudinal range by changes in sign of the roll angle at appropriate points during the re-entry. Among other possible advantages this technique permits the use of the space reaction systems for control throughout re-entry. (Fixed trim at the desired pitch attitude must be provided aerodynamically.)

The sphericity of the earth has a marked influence on the achievable lateral range and on the shape of the maneuver envelope. Only for the shortest ranges of interest for re-entry at escape speed is it valid to use the familiar approximation in which the earth in effect is assumed to be cylindrical.

Three aerodynamic designs which are believed capable of being developed into practical re-entry vehicles have been examined briefly. Preliminary fixed-base simulator studies for one of these vehicles indicate that a human pilot can perform satisfactorily the basic maneuvers required for range control. In some cases, with special training, successful entries were achieved without the aid of artificial rate damping, but the pilots did not consider this an acceptable mode except for emergency. The presence of rate damping about three axes made the piloting problem relatively easy.

References

1. D. R. Chapman, An Analysis of the Corridor and Guidance Requirements for Supercircular Entry into Planetary Atmospheres. NASA TR R-55 (1959).
2. L. Lees, et al., Use of Aerodynamic Lift During Entry into the Earth's Atmosphere. ARS Journal **29**, No. 9 (1959).
3. Mac C. Adams, Recent Advances in Ablation. ARS Journal **29**, No. 9 (1959).
4. F. C. Grant, Importance of the Variation of Drag with Lift in Minimization of Satellite Entry Acceleration. NASA TN D-120 (Oct. 1959).
5. L. Broglio, Similar Solutions in Re-Entry Lifting Trajectories. Univ. of Rome, SIARgraph No. 54 (Dec. 1959).

6. K. WANG and L. TING, Analytic Solutions of Planar Re-Entry Trajectories with Lift and Drag. Polytechnic Institute of Brooklyn (PIBAL) Rep. No. 601 (April 1960).
7. F. C. GRANT, Analysis of Low-Acceleration Lifting Entry from Escape Speed. NASA TND-249 (June 1960).
8. J. D. C. CRISP and P. FEITIS, The Thermal Response of Heat-Sink Re-Entry Vehicles. PIBAL Rep. No. 576 (Juli 1960).
9. F. C. GRANT, Modulated Entry. NASA TN D-452 (Aug. 1960).
10. E. F. STYER, A Parametric Examination of Re-Entry Vehicle Size and Shape for Return at Escape Velocity. Paper presented at Third Annual Meeting, American Astronautical Society, Seattle, Washington, Aug. 8–11, 1960.
11. O. A. KELLY, JR., Parametric Study of a Manned Space Entry Vehicle. Aero-Space Engng. 19, No. 10 (1960).
12. B. A. GALMAN, Direct Re-Entry at Escape Velocity. Paper presented at Third Annual Meeting, American Astronautical Society, Seattle, Washington, Aug. 8–11, 1960.
13. R. TEAGUE, Flight Mechanics of Re-Entry After Circumlunar Flight by Means of Various Lifting Techniques. NASA-George C. Marshall Space Flight Center, MNN-M-Aero-4-60 (Sept. 15, 1960).
14. R. B. HILDEBRAND, Manned Re-Entry at Supersatellite Speeds. IAS Rep. No. 60–83 (Sept. 1960).
15. R. A. MINZNER, et al., ARDC Model Atmosphere, 1959 Geophysics Research Directorate, ARDC, U. S. Air Force (1959).
16. J. E. HAYS, et al., Analytical Study of Drag Brake Control System for Hypersonic Vehicles. Wright Air Develop. Div., U. S. Air Force Tech. Rep. 60-267 (Jan. 1960).
17. J. V. BECKER, Heating Penalty Associated with Modulated Entry into Earth's Atmosphere. ARS Journal 30, No. 5 (1960).
18. A. E. BRYSON, et al., Determination of the Lift or Drag Program that Minimizes Re-Entry Heating with Acceleration or Range Constraints Using a Steepest Descent Computation Procedure. IAS Paper, New York Meeting, Jan. 23–25, 1961.
19. R. C. WINGROVE and R. E. COATE, Piloted Simulator Tests of a Guidance System which Can Continuously Predict Landing Point of a Low L/D Vehicle During Atmospheric Re-Entry. NASA TN D-787.
20. R. E. SLYE, An Analytical Method for Studying the Lateral Motion of Atmosphere Entry Vehicles. NASA TN D-325, Sept. 1960.
21. D. S. MANDELL, A Study of the Maneuvering Performance of Lifting Re-Entry Vehicles. Paper presented at ARS 15th Annual Meeting, Washington, D. C., Dec. 5–8, 1960.
22. A. L. FRIEDLANDER and D. P. HARRY, III, Requirements of Trajectory Convective Impulses During the Approach Phase of an Interplanetary Mission. NASA TN D-255, 1960.
23. H. MULTHOPP, Design of Hypersonic Aircraft. Aero-Space Engng. 20, No. 2 (1961).
24. J. M. EGGLESTON, et al., Fixed-Base Simulator Study of a Pilot's Ability to Control a Winged Satellite Vehicle During High-Drag Variable-Lift Entries. NASA TN D-228, April 1960.
25. MAC C. ADAMS, A Look at the Heat Transfer Problem at Super-Satellite Speeds. Paper presented at ARS 15th Annual Meeting, Washington, D. C., Dec. 5–8, 1960.
26. M. J. BRUNNER, The Aerodynamic and Radiant Heat Input to Space Vehicles which Re-Enter at Satellite and Escape Velocity. Paper presented at ARS 15th Annual Meeting, Washington, D. C., Dec. 5–8, 1960.
27. H. J. ALLEN, Problems in Atmospheric Entry from Parabolic Orbits. Paper presented at Conference on Aeronautical and Space Engineering, Nagoya, Japan, Nov. 8–9, 1960.
28. W. H. STILLWELL and H. M. DRAKE, Simulator Studies of Jet Reaction Controls for Use at High Altitudes. NACA RM H58G18a, 1958.

29. G. W. FREEMAN, Reaction Controls for Re-entry Vehicles. Paper presented at Third Annual Meeting, American Astronautical Society, Seattle, Washington, Aug. 8–11, 1960.

Discussion

Dr. FERRI offered a substantiating comment on the part of Mr. BECKER's paper dealing with the pilot's ability to make a survivable entry in an emergency situation in which the position of the vehicle in the corridor was not known prior to entry. He displayed a slide based on his work showing that the rate of build-up of resultant deceleration could be used to identify the approximate position in the corridor. If the rate of onset was low for example, the pilot would know he was in the upper part of the corridor and could then follow an appropriate lift modulation schedule for that part of the corridor.

L'étagement optimum des groupes de fusées en fonctionnement parallèle

Par

B. M. Fraeijs de Veubeke[1]

(Avec 4 Figures)

Résumé — Zusammenfassung — Abstract

L'étagement optimum des groupes de fusées en fonctionnement parallèle. Etude théorique du cas limite où les moteurs et les réservoirs, indéfiniment fractionnés, sont largables de façon continue. Il est établi que dans un champ de gravitation uniforme les trajectoires optimales sont composées d'arcs à poussée constante, d'arcs à poussée monotone décroissante et éventuellement de discontinuités de poussée séparant deux arcs à poussée constante.

Pour une charge utile donnée ainsi qu'une vitesse terminale fixée en module et en inclinaison, on établit la synthèse des trajectoires minimales par rapport à la masse initiale ou à la poussée initiale. Les modifications dues à la présence d'un second étage sont discutées.

Bahn- und Stufenoptimierung mit gekoppelten Boostern. Es wird der Grenzfall untersucht, wo bei gekoppelten Boostern Tanks und Maschinen kontinuierlich abgeworfen werden können. Es zeigt sich, daß bei einem uniformen Gravitationsfeld die Bahnen aus mit konstantem Schub sowie mit abnehmendem Schub durchlaufenen Bögen bestehen, wobei endliche Unstetigkeiten zwischen den konstanten Schubabschnitten auftreten können.

Optimale Bahnen werden für gegebene Nutzlast sowie gegebene Größe und Richtung der Endgeschwindigkeit unter Berücksichtigung der Anfangsmasse oder des Anfangsschubes für ein- oder zweistufige Systeme angegeben.

Trajectory and Staging Optimization with Clustered Boosters. Theoretical analysis of the limiting case where engines and tanks in a clustered booster system can be jettisoned continuously. For a uniform gravitational field the trajectories are shown to consist of constant thrust and decreasing thrust arcs with possible finite thrust discontinuities between constant thrust segments.

Under given payload and given intensity and inclination of the terminal velocity vector, the minimal trajectories are determined with respect to initial mass or initial thrust, with or without second stage.

I. Introduction

Pour la mise en orbite de charges importantes ou la réalisation de trajectoires très énergétiques, la considération des pertes de gravitation domine celle de la trainée aérodynamique et la poussée requise au départ est très élevée.

Cette poussée peut s'obtenir en groupant un certain nombre de fusées en parallèle (*Saturne*). Une telle solution a l'avantage de se prêter en principe au largage successif de moteurs (*Atlas*) aussi bien que de réservoirs suivant les nécessités d'une optimisation de type donné.

[1] Professeur aux Universités de Liége et de Louvain, Belgique.

L'étude théorique entreprise ici considère le cas limite de l'étagement parallèle continu, c'est-à-dire le fractionnement de la poussée en une infinité de moteurs infinitésimaux alimentés par des réservoirs infinitésimaux largués au rythme de la consommation des ergols.

Les programmes de réduction de la poussée sont analysés pour les trajectoires naturellement incurvées dans un champ de gravitation uniforme, la vitesse et l'angle terminal d'injection sur orbite étant imposés.

De grandes différences dans les caractéristiques de la programmation optimale apparaissent suivant que l'on cherche à rendre la masse initiale ou la poussée initiale minimum, la charge utile étant fixe. Les modifications apportées par la présence d'un second étage sont aussi examinées.

Un grand nombre de considérations techniques et technologiques interviennent pour déterminer la mesure dans laquelle ces solutions idéales sont transposables en pratique. Elles sortent du cadre de cette étude mais l'importance des gains en jeu indique dans certains cas qu'elles méritent un examen approfondi.

II. Hypothèses et notations

a) Le champ de gravitation est uniforme;

b) la poussée est constamment tangente à la trajectoire;

c) la trainée aérodynamique est négligée;

d) le poids des moteurs à ergols liquides est à chaque instant proportionnel à la poussée F réalisée:

$$g\,M_e = \frac{F}{K} \tag{1}$$

Cette relation implique une correspondance entre la réduction de la poussée et une réduction du poids des moteurs. Ceux-ci sont donc fractionnés au point de pouvoir être largués de façon continue.

e) Le poids des réservoirs est à chaque instant proportionnel au poids d'ergols $g\,M_p$ restant à consommer

$$g\,M_s = \zeta\, g\,M_p \tag{2}$$

Cette relation implique également un fractionnement indéfini des réservoirs qui sont largués au rythme de la consommation en ergols.

f) Toutes les variables physiques sont considérées comme des fonctions d'une variable descriptive indépendante à croissance monotone entre une valeur initiale fixe σ_1 et une valeur fixe σ_3 correspondant à la consommation complète des ergols du premier étage. Les dérivées par rapport au paramètre descriptif sont notées

$$\frac{df(\sigma)}{d\sigma} = f^0$$

Le temps étant également à croissance monotone on aura en particulier

$$t^0 \geqslant 0$$

L'égalité à zéro sur un intervalle fini de variation de σ peut se produire. Elle correspond aux discontinuités dans le temps que peuvent subir certaines variables lors de la séparation d'une masse finie du système propulsif.

III. Equations fondamentales

M^* désignant la masse propulsée par le groupement en parallèle des fusées du premier étage, la masse totale à un instant quelconque est

$$M(\sigma) = M^* + M_p(\sigma) + M_s + M_e$$

soit, en vertu de (1) et (2)

$$M(\sigma) = M^* + (1 + \zeta)\, M_p(\sigma) + \frac{F(\sigma)}{K\, g} \tag{3}$$

Différentions cette relation par rapport à σ et éliminons le débit massique d'ergols par l'équation de la poussée

$$F = -\, c\, \frac{dM_p}{dt} = -\, c\, \frac{M^0}{t^0}$$

où c est la vitesse effective d'éjection, que nous supposerons constante; il vient

$$M^0 = -\, \frac{(I + \zeta)\, F}{c}\, t^0 + \frac{F^0}{K\, g} \tag{4}$$

Il est essentiel d'exprimer analytiquement le fait que la poussée ne peut que décroitre. L'équation

$$F^0 = -\, a^2 \frac{g}{c}\, g\, M\, t^0 \tag{5}$$

convient à cet effet. La variable a, dont la connaissance fixe la programmation de réduction de la poussée et qui n'interviendra pas dans sa dérivée, est notre variable de guidage, au sens général que nous avons donné à ce terme.

L'équation du mouvement le long de la tangente à la trajectoire est, suivant les hypothèses admises

$$M \frac{dV}{dt} = F - M\, g \sin\gamma \tag{6}$$

où γ est l'angle que fait le vecteur vitesse avec l'horizontale. L'équation correspondante suivant la normale est

$$V \frac{d\gamma}{dt} = -\, g \cos\gamma \tag{7}$$

Les quatre équations fondamentales du problème sont représentées par les équations (4) à (7). Elles gouvernent quatre fonctions inconnues du temps M, V, γ et F et forment un système qui requiert pour être intégré la connaissance de la fonction a, c'est-à-dire du programme de guidage. Celui-ci sera déterminé par le calcul des variations de façon à retirer certaines performances optimales du système. Auparavent il est avantageux pour la simplicité de l'écriture et pour le calcul numérique d'introduire des variables sans dimensions

$$\omega = \frac{V}{c} \qquad \tau = \frac{g}{c}\, t \qquad \beta = \frac{F}{M\, g} \qquad \mu = \frac{M}{M_u} \tag{8}$$

L'unité de masse M_u est la masse dite "utile" qui sera à définir dans chaque cas particulier.

Notant que, suivant la définition du facteur d'accélération instantanée β, on a

$$\frac{\beta^0}{\beta} = \frac{F^0}{F} - \frac{M^0}{M} = \frac{F^0}{F} - \frac{\mu^0}{\mu} \tag{9}$$

le système fondamental peut s'écrire

$$[G]_1 = \omega^0 + (\sin\gamma - \beta)\, \tau^0 = 0 \tag{10}$$

$$[G]_2 = \gamma^0 + \frac{\cos\gamma}{\omega}\, \tau^0 = 0 \tag{11}$$

$$[G]_3 = \frac{\mu^0}{\mu} - \frac{\beta^0}{K-\beta} + (1+\zeta)\frac{K\beta}{K-\beta}\tau^0 = 0 \tag{12}$$

$$[G]_4 = -\beta^0 - a^2\tau^0 + \frac{a^2}{K}\beta\tau^0 + \beta^2(1+\zeta)\tau^0 = 0 \tag{13}$$

L'avant dernière équation a été obtenue en éliminant a, la dernière en éliminant μ^0 entre les équations primitives.

IV. Equations d'Euler et conditions de transversalité

Désignons par $G(q, q^0, a)$ la fonction

$$G = \sum_1^4 \lambda_i [G]_i$$

construite à l'aide de quatre multiplicateurs lagrangiens λ_i; q désignant l'une des cinq fonctions $(\omega, \gamma, \mu, \beta, \tau)$. Les équations d'Euler du problème variationnel

$$\delta \int_{\sigma_1}^{\sigma_3} G\, d\sigma = 0$$

sont

$$[G]_q = -\frac{d}{d\sigma}\left(\frac{\partial G}{\partial q^0}\right) + \frac{\partial G}{\partial q} = 0$$

et

$$[G]_a = \frac{\partial G}{\partial a} = 0$$

c'est-à-dire explicitement

$$[G]_\omega = -\lambda_1{}^0 - \lambda_2 \frac{\cos\gamma}{\omega^2}\tau^0 = 0 \tag{14}$$

$$[G]_\gamma = -\lambda_2{}^0 + \lambda_1\tau^0\cos\gamma - \lambda_2\frac{\sin\gamma}{\omega}\tau^0 = 0 \tag{15}$$

$$[G]_\mu = -\frac{1}{\mu}\lambda_3{}^0 = 0 \tag{16}$$

$$[G]_\beta = \lambda_4{}^0 + (1+\zeta)\left(\frac{K}{K-\beta}\right)^2 \lambda_3\tau^0 - \lambda_1\tau^0 + \lambda_4\frac{a^2}{K}\tau^0 + 2\beta\lambda_4(1+\zeta)\tau^0 = 0 \tag{17}$$

$$[G]_\tau = U^0 = 0 \tag{18}$$

où

$$U = \lambda_1(\sin\gamma - \beta) + \lambda_2\frac{\cos\gamma}{\omega} + \lambda_3(1+\zeta)\frac{K\beta}{K-\beta} - \tag{19}$$
$$- \lambda_4 a^2 + \lambda_4\frac{a^2}{K}\beta + \lambda_4\beta^2(1+\zeta)$$

et

$$[G]_a = 2\lambda_4 a\tau^0\left(\frac{\beta}{K} - 1\right) = 0 \tag{20}$$

La condition de transversalité est

$$\left[\lambda_1 \,\delta\omega + \lambda_2 \,\delta\gamma + \lambda_3 \frac{\delta\mu}{\mu} - \left(\lambda_4 + \frac{\lambda_3}{K-\beta}\right)\delta\beta + U\,\delta\tau\right]_{\sigma_1}^{\sigma_3} = 0 \qquad (21)$$

On sait par la théorie générale du problème de MAYER sous forme paramétrique [2] que les équations d'EULER sont liées par une identité qui est ici

$$\omega^0[G]_\omega + \gamma^0[G]_\gamma + \mu^0[G]_\mu + \beta^0[G]_\beta + \tau^0[G]_\tau + \sum_1^4 \lambda_i^0[G]_i \equiv 0$$

Ceci montre que pour toute extrémale le long de laquelle une des variables q ne serait pas constante, l'équation d'EULER correspondante $[G]_q = 0$ est automatiquement vérifiée quand on satisfait aux autres et aux équations fondamentales (10) à (13).

V. Nature des extrémales

Les équations (16) et (18) montrent déjà que λ_3 et U sont des constantes le long de toute extrémale. De plus comme les conditions de raccord des extrémales (conditions de WEIERSTRASS et ERDMANN) demandent que les grandeurs $\partial G/\partial q^0$ soient continues, les multiplicateurs et la grandeur U sont des fonctions continues. Par conséquent λ_3 et U assument une valeur constante tout le long d'une trajectoire optimale.

Nous n'envisagerons que des problèmes d'optimum sans imposition de contraintes sur la durée de description de la trajectoire. Alors, comme la condition de transversalité comporte le terme $U\,\delta(\tau_3 - \tau_1)$ avec variation arbitraire sur la durée (on peut comparer des trajectoires de durées différentes), elle ne peut être satisfaite que par

$$U = 0. \qquad (22)$$

La valeur de la constante U sera donc nulle tout le long de la trajectoire ce qui, par (19), fournit une relation algébrique homogène entre les multiplicateurs. Ces derniers n'étant définis qu'à un facteur d'échelle près, la valeur effective de la constante λ_3 est sans importance; pour le calcul numérique il sera commode de lui assigner la valeur 1.

Les différentes extrémales possibles résultent des différentes façons de remplir la condition d'EULER (20) relative à la variable de guidage. La solution $\beta = K$ est à rejeter car cette valeur du facteur d'accélération ne peut être atteinte que par un moteur se propulsant lui-même à l'exclusion de toute autre masse, même du propergol qui lui est nécessaire. En fait on a toujours $K - \beta > 0$. Par contre trois autres possibilités conduisent à des solutions valables.

a) L'arc à poussée constante

Solution correspondant à

$$a \equiv 0 \qquad (23)$$

ce qui, eu égard à (5), revient à maintenir la poussée constante. Dans ces conditions (12) et (13) s'intègrent facilement en

$$\frac{1}{\beta} + (1 + \zeta)\,\tau = c\,t\,e \qquad (24)$$

$$\beta\mu = c\,t\,e \qquad (25)$$

Les équations (10) et (11) doivent être intégrées numériquement ainsi que (14) et (15) pour le calcul des multiplicateurs λ_1 et λ_2. Pour le calcul de λ_4 on peut, en vertu de l'identité entre équations d'Euler et du fait que β n'est pas constant, négliger (17) et se servir de (22) qui se réduit ici à

$$\lambda_1(\sin\gamma - \beta) + \lambda_2 \frac{\cos\gamma}{\omega} + \lambda_3 (1 + \zeta) \frac{K\beta}{K - \beta} + \lambda_4 (1 + \zeta)\beta^2 = 0. \tag{26}$$

b) L'arc à poussée décroissante

Solution de (20) correspondant à

$$\lambda_4 \equiv 0. \tag{27}$$

Dans ces conditions (17) se réduit à

$$(1 + \zeta) K^2 \lambda_3 = (K - \beta)^2 \lambda_1 \tag{28}$$

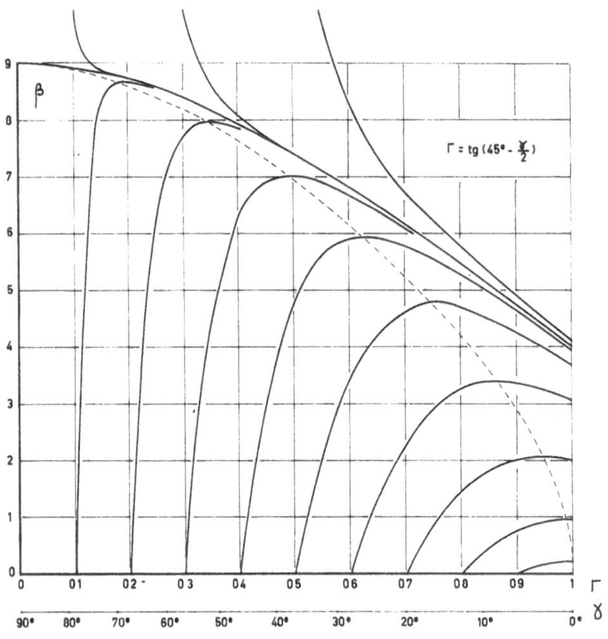

Fig. 1. Arcs à poussée décroissante. Famille de relations optimales entre le facteur d'accélération β et l'angle d'inclinaison γ de la trajectoire avec l'horizontale

Différentions cette relation en tenant compte de (16) et remplaçons λ_1^0 par sa valeur tirée de (14), il vient

$$2\lambda_1\beta^0 + \lambda_2 \cdot \frac{(K - \beta)\cos\gamma}{\omega^2}\tau^0 = 0. \tag{29}$$

Les équations (28) et (29) ainsi que (22) qui se réduit à

$$\lambda_1(\sin\gamma - \beta) + \lambda_2\frac{\cos\gamma}{\omega} + \lambda_3(1 + \zeta)\frac{K\beta}{K - \beta} = 0 \tag{30}$$

forment un système linéaire et homogène dans les multiplicateurs λ_1, λ_2 et λ_3. En annulant le déterminant du système on trouve pour condition de compatibilité

$$\beta^0 = -\frac{K - \beta}{2\omega}\left(\frac{\beta^2}{K} - \sin\gamma\right)\tau^0.$$

En y substituant β^0 par sa valeur tirée de (13) on trouve la loi de guidage explicite

$$a^2 = (1 + \zeta)\frac{K\beta^2}{K - \beta} + \frac{\beta^2 - K\sin\gamma}{2\omega} \tag{31}$$

La condition de compatibilité prend une forme remarquable quand on y élimine encore ω à l'aide de (11) et que l'on prend γ comme variable indépendante:

$$\frac{d\beta}{d\gamma} = \frac{K - \beta}{2\cos\gamma}\left(\frac{\beta^2}{K} - \sin\gamma\right). \tag{32}$$

Ainsi le facteur d'accélération β ne depend que de l'inclinaison de la trajectoire sur l'horizontale et des conditions initiales de description de l'arc. Une famille de courbes intégrales de (32) a été calculée pour la valeur $K = 81$ à l'ordinateur électronique. Le point ($\beta = \sqrt{K}$, $\gamma = \pi/2$) (ou encore $\Gamma = \tan(\pi/4 - \gamma/2) = 0$) est un point singulier (Fig. 1) d'où part une des courbes de la famille et au voisinage duquel les autres courbes sont violemment incurvées et tendent à rejoindre la première pour $\gamma = 0$ (ou $\Gamma = 1$).

Le long d'un tel arc de courbe on peut intégrer l'équation différentielle

$$\frac{d\omega}{\omega} = \frac{\sin\gamma - \beta}{\cos\gamma} d\gamma \tag{33}$$

déduite de (10) et (11) et l'on obtient alors le rapport entre les vitesses atteintes aux deux extrémités de l'arc. Par exemple

$$\frac{\omega_2}{\omega_3} = \exp\int_{\gamma_2}^{\gamma_3} \frac{\beta - \sin\gamma}{\cos\gamma} d\gamma \tag{34}$$

Au facteur d'échelle λ_3 près, les multiplicateurs sontcomplètement déterminés par les relations (27), (28) et (29).

c) L'arc à temps constant

Posons

$$a = \frac{A}{\sqrt{\tau^0}} \tag{35}$$

et faisons tendre maintenant τ^0 vers zéro. Comme

$$\lim a\,\tau^0 = 0$$

l'équation (20) est encore satisfaite. Comme d'autre part

$$\lim a^2\,\tau^0 = A^2$$

le système ne reste pas statique; on trouve

$$\beta^0 = A^2\left(\frac{\beta}{K} - 1\right) < 0$$

ce qui correspond, dans le temps, à une discontinuité possible du facteur d'accélération. D'autre part (12) s'intègre ici en

$$\mu\,(K - \beta) = c\,t\,e. \tag{36}$$

Rapprochant ce résultat de (3) qui peut s'écrire

$$\mu\,(K - \beta) = \frac{K}{M_u}\,[M^* + (1 + \zeta)\,M_p(\sigma)]$$

on en déduit qu'il n'y a pas de consommation d'ergols. Cette solution doit donc bien être interprétée comme une séparation à un instant donné d'une partie finie du système propulsif. Enfin les équations (10) et (11) se réduisent à

$$\omega = c\,t\,e \qquad \gamma = c\,t\,e. \tag{37}$$

En ce qui concerne les multiplicateurs on observera que a^2 devenant infini, l'équation (22) ne peut être satisfaite que pour

$$\lambda_4 \equiv 0. \tag{38}$$

Comme $\tau^0 = 0$, ce résultat est bien compatible avec (17) tandis que (14) et (15) donnent

$$\lambda_1 = c\,t\,e \qquad \lambda_2 = c\,t\,e. \tag{39}$$

VI. L'arc à temps constant comme arc intermédiaire

Sur la base des résultats précédents des conclusions importantes peuvent être tirées concernant l'existence possible d'un arc à temps constant comme arc intermédiaire. Soient

$$\beta = \alpha_1 \qquad \text{et} \qquad \beta = \alpha_2$$

les valeurs du facteur d'accélération aux extrémités d'un tel arc. En ces extrémités (38) impose la nullité de λ_4; il en découle l'identité des équations (26) et (30) qui demandent toutes deux de satisfaire aux extrémités les conditions

$$\lambda_1 (\sin\gamma - \alpha_1) + \lambda_2 \frac{\cos\gamma}{\omega} + \lambda_3 (1+\zeta) \frac{K\alpha_1}{K-\alpha_1} = 0 \tag{40}$$

$$\lambda_1 (\sin\gamma - \alpha_2) + \lambda_2 \frac{\cos\gamma}{\omega} + \lambda_3 (1+\zeta) \frac{K\alpha_2}{K-\alpha_2} = 0. \tag{41}$$

D'autre part il découle de (37) et (39) que les valeurs de ω, γ, λ_1 et λ_2 sont les mêmes dans ces deux conditions. Dès lors, après soustraction et simplification par $(\alpha_1 - \alpha_2)$ il vient

$$\lambda_1 = (1+\zeta) \frac{K^2}{(K-\alpha_1)(K-\alpha_2)} \lambda_3 \tag{42}$$

pour valeur constante du premier multiplicateur le long de l'arc intermédiaire à temps constant. Cette expression est valable en principe quelle que soit la nature de l'arc auquel on se raccorde (poussée constante ou décroissante) de part et d'autre. On constate cependant qu'elle est incompatible avec les valeurs

$$(1+\zeta) \frac{K^2}{(K-\alpha_1)^2} \lambda_3 \qquad (1+\zeta) \frac{K^2}{(K-\alpha_2)^2} \lambda_3$$

qui devraient être réalisées en vertu de (28) pour un raccord à gauche ou à droite avec un arc à poussée décroissante. On en déduit que "*un arc à temps constant ne peut constituer une transition qu'entre deux arcs à poussée constante*".

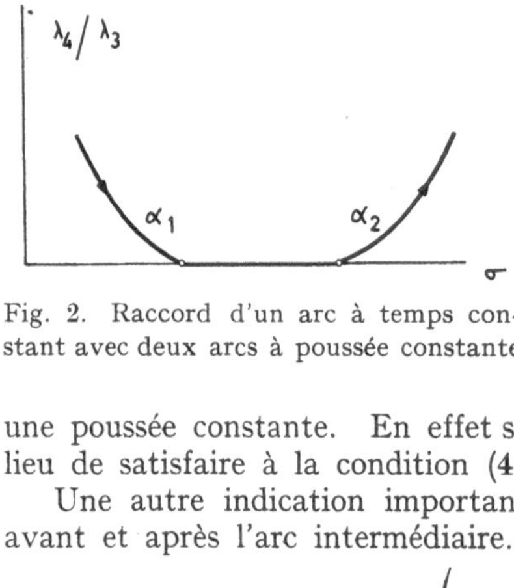

Fig. 2. Raccord d'un arc à temps constant avec deux arcs à poussée constante

Par contre un arc à temps constant peut terminer une trajectoire en faisant suite aussi bien à une poussée décroissante qu'à une poussée constante. En effet s'il n'est plus suivi d'un autre arc il n'y a plus lieu de satisfaire à la condition (41).

Une autre indication importante concerne le signe de λ_4 immédiatement avant et après l'arc intermédiaire. Suivant (17) on aura

$$\lambda_4{}^0 = \left(\lambda_1 - (1+\zeta) \frac{K^2}{(K-\alpha_1)^2}\right) \tau^0$$

pour valeur terminale de la dérivée le long de l'arc venant se raccorder à gauche à l'arc à temps constant. Remplaçant λ_1 par sa valeur (42) il vient

$$\lambda_4{}^0 = -(1+\zeta) \frac{K^2(\alpha_1 - \alpha_2)}{(K-\alpha_1)^2 (K-\alpha_2)} \lambda_3 \tau^0. \tag{43}$$

On trouve de même pour valeur initiale de la dérivée le long de l'arc à poussée constante quittant à droite l'arc intermédiaire

$$\lambda_4{}^0 = +(1+\zeta) \frac{K^2(\alpha_1 - \alpha_2)}{(K-\alpha_1)(K-\alpha_2)^2} \lambda_3 \tau^0. \tag{44}$$

Comme $(\alpha_1 - \alpha_2) > 0$ et que λ_4 est nul le long de l'arc intermédiaire on en déduit que le rapport λ_4/λ_3 conserve immédiatement après l'arc intermédiaire la valeur positive qu'il avait immédiatement avant de l'aborder (Fig. 2).

VII. Raccord entre un arc à poussée constante et un arc à poussée décroissante

Par (17) nous avons le long d'un arc à poussée constante ($a = 0$)

$$\frac{d\lambda_4}{d\tau} = -(1+\zeta)\frac{K^2}{(K-\beta)^2}\lambda_3 + \lambda_1 - 2(1+\zeta)\beta\lambda_4. \tag{45}$$

A hauteur du raccord avec l'arc à poussée décroissante λ_4 doit s'annuler et d'après (28) λ_1 a exactement la valeur requise pour que

$$\frac{d\lambda_4}{d\tau} = 0 \quad \text{au raccord.} \tag{46}$$

Pour décider du signe de λ_4 au voisinage du raccord nous différentions (45), remplaçons λ_1^0, λ_3^0 et β^0 par leurs valeurs générales tirées de (14), (16) et (13) (avec $a = 0$), puis évaluons l'expression au niveau du raccord à l'aide des résultats (27), (28), (29) et (46), il vient

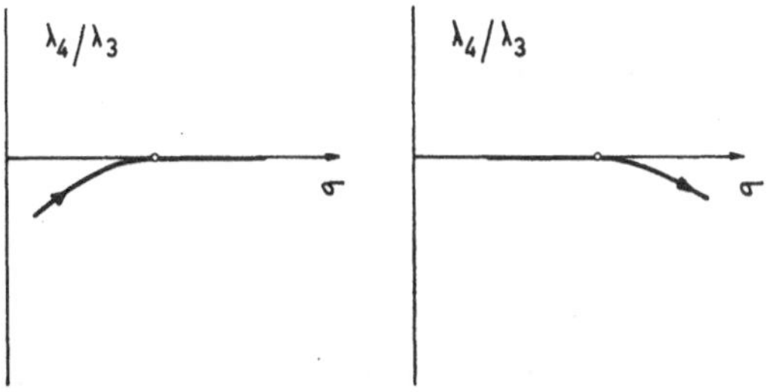

Fig. 3. Raccord entre un arc à poussée constante et un arc à poussée décroissante

$$\frac{d^2\lambda_4}{d\tau^2} = -(1+\zeta)\frac{2K}{(K-\beta)^2}\left[(1+\zeta)\frac{K\beta^2}{K-\beta} + \frac{\beta^2 - K\sin\gamma}{2\omega}\right]\lambda_3.$$

Le crochet est positif car, en vertu de (31), il correspond exactement à la valeur prise par a^2 à l'extrémité de l'arc à poussée décroissante (soulignons ici que, contrairement aux autres variables, la variable de guidage a subit des discontinuités). Par conséquent

$$\frac{1}{\lambda_3}\frac{d^2\lambda_4}{d\tau^2} < 0 \qquad \text{au raccord.}$$

Il en résulte que le rapport λ_4/λ_3 est négatif au voisinage immédiat du raccord comme l'indique la Fig. 3.

VIII. Le critère de variation forte

Une discontinuité de la variable de guidage, faisant sauter sa valeur de a à $\bar{a}$, reste compatible avec la continuité des variables q du système fondamental. Comme le montrent les équations (7) à (10), seules les dérivées q^0 subissent des discontinuités qui font passer leur valeur à $\bar{q}^0$.

Comme la fonction G est homogène dans les dérivées q^0, la fonction excès E de WEIERSTRASS peut être calculée par l'expression simplifiée

$$E = G(q, \bar{q}^0, \bar{a}) - G(q, \bar{q}^0, a).$$

Dans le cas présent on trouve

$$E = \lambda_4 \bar{\tau}^0 (\bar{a}^2 - a^2)\left(\frac{\beta}{K} - 1\right). \tag{47}$$

Si une grandeur, dont la variation figure dans la condition de transversalité, est à rendre minimum, E doit être nul ou d'un signe opposé au coefficient de cette variation; nul ou de même signe pour le cas d'un maximum. C'est la forme prise ici par le critère de variation forte.

IX. Fusée de poids initial minimum

Cherchons à rendre $M_1 = M(\sigma_1)$ minimum pour une masse utile M_u donnée, c'est-à-dire $\mu(\sigma_1)$ minimum, dans les conditions suivantes

$$\omega(\sigma_1) = \omega_1 \qquad \omega(\sigma_3) = \omega_3 \qquad \gamma(\sigma_3) = \gamma_3$$

toutes valeurs imposées. Les variations de ces grandeurs sont donc nulles a priori et, eu égard à (22), la condition (20) de transversalité se réduit à

$$- \lambda_2(\sigma_1)\, \delta\gamma_1 + \lambda_3 \frac{\delta\mu_3}{\mu_3} - \lambda_3 \frac{\delta\mu_1}{\mu_1} -$$
$$\left(\lambda_4(\sigma_3) + \frac{\lambda_3}{K - \beta_3}\right) \delta\beta_3 + \left(\lambda_4(\sigma_1) + \frac{\lambda_3}{K - \beta_1}\right) \delta\beta_1 = 0$$

Les variations qui subsistent aux limites ne sont pas indépendantes car l'équation (3), appliquée en σ_3 quand les ergols sont épuisés, fournit

$$M(\sigma_3) = M^* + \frac{F(\sigma_3)}{K g}\,. \tag{48}$$

Nous commencerons par supposer qu'en σ_3 la trajectoire ne se prolonge pas en réalité par la mise à feu d'un second étage. Dans ces conditions nous pouvons assimiler M^* à M_u. D'où, divisant (48) par $M(\sigma_3)$ et résolvant pour μ_3

$$\mu_3 = \frac{K}{K - \beta_3}\,. \tag{49}$$

Cette relation permet d'exprimer la variation de β_3 en fonction de celle de μ_3. Après substitution dans la condition de transversalité, nous pouvons annuler les coefficients des variations arbitraires sur γ_1, μ_3 et β_1 et obtenons les conditions aux limites naturelles du problème, qui sont

$$\lambda_2(\sigma_1) = 0 \tag{50}$$

$$\lambda_4(\sigma_3) = 0 \tag{51}$$

$$\lambda_4(\sigma_1) = - \frac{\lambda_3}{K - \beta_1}\,. \tag{52}$$

Il ne subsiste de la condition de transversalité que le terme

$$- \lambda_3 \frac{\delta\mu_1}{\mu_1}\,.$$

Si le coefficient de la variation est différent de zéro, c'est bien la condition nécessaire pour que μ_1 soit minimum. En fixant l'échelle des multiplicateurs par le choix

$$\lambda_3 = 1 \tag{53}$$

le coefficient en question est négatif et le critère de variation forte s'écrit

$$E \geqslant 0 \tag{54}$$

Il reste à déterminer la synthèse des arcs constituant une trajectoire minimale. Elle doit respecter la continuité des variables physiques et des multiplicateurs.

Puisque l'on a toujours $K - \beta > 0$, la condition (52) demande une valeur initiale négative de λ_4. Par conséquent le premier arc est à poussée constante ($a = 0$) et, par (48), on constate que le critère (54) est bien vérifié. D'autre part la condition terminale (51) doit être satisfaite et ceci peut avoir lieu de deux façons distinctes:

a) λ_4 s'annule pour la première fois en fin de trajectoire (σ_3). Celle-ci est alors simplement constituée de l'arc à poussée constante.

b) λ_4 s'annule en un point intermédiaire $\sigma = \sigma_2$. On ne pourra remplir la condition terminale (51) qu'en maintenant λ_4 nul. Car, en prolongeant encore l'arc initial, ou λ_4 deviendra positif et le critère de variation forte sera violé, ou il deviendra négatif, supprimant toute possibilité ultérieure de raccorder un nouvel arc et de remplir la condition (51).

On peut écarter le cas $\tau_2 = \tau_3$, car alors l'arc initial à poussée constante est simplement prolongé par un arc à temps constant. Ceci veut dire qu'en fin de propulsion, une fois la vitesse et l'angle d'injection atteints, on sépare la masse utile de la totalité ou d'une partie du système propulsif restant. Ce n'est qu'une variante technique secondaire du cas a).

Pour $\tau_2 < \tau_3$, il n'est pas question de prolonger l'arc initial par un arc à temps constant qui serait nécessairement de caractère intermédiaire car, suivant l'analyse faite à la section VI, ceci exigerait une valeur positive de λ_4 antérieure au raccord. Il ne subsiste en définitive que la possibilité de raccorder directement un arc à poussée décroissante et ceci apparait compatible avec les conclusions tirées à la section VII. Encore une fois une variante technique secondaire consiste à terminer la trajectoire par un arc à temps constant.

Ce qui décide en pratique de l'occurence du cas a) de celle du cas b) est l'importance du gain de vitesse $\omega_3 - \omega_1$ requis. Ceci apparaitra clairement sur l'exemple de la fusée sonde, traité à la section suivante.

Quelle que soit la procédure suivie, le problème de l'intégration complète d'une trajectoire comporte un double jeu d'approximations successives. La procédure que nous allons décrire et qui s'applique au cas b), est simplement destinée à montrer le caractère cohérent du problème. Supposons que nous disposions en σ_2 d'une première approximation aux valeurs prises par β et γ. Par intégration de (32) jusqu'à la valeur γ_3 imposée, nous sommes en mesure de calculer (34) et, comme ω_3 est connu, nous en déduisons la valeur de ω_2. Les valeurs prises en σ_2 par les multiplicateurs découlent alors de (27), (28), (30) et (53). Nous disposons alors des valeurs initiales nous permettant d'intégrer à rebours le long de l'arc à poussée constante jusqu'à ce que, par exemple, la condition initiale (50) soit remplie. En général la vitesse initiale n'aura pas la valeur ω_1 requise et la condition (52) ne sera pas remplie. Il faudra procéder par perturbation des deux estimations de β_2 et γ_2 pour arriver à remplir ces deux dernières conditions.

L'intégration de (12) peut être différée jusqu'à établissement de la trajectoire exacte. Elle fournira le rapport μ_1/μ_3 et le minimum μ_1 sera connu par utilisation de (49).

Une procédure analogue mais évidemment simplifiée s'applique au cas a).

X. Fusée de poids initial minimum en vol vertical (fusée-sonde)

Pour $\gamma = \pi/2, \cos\gamma = 0, \sin\gamma = 1$, le problème précédent s'intègre sous forme finie, fournissant un guide précieux pour le cas des trajectoires incurvées.

On peut éliminer l'équation (11) du système fondamental et faire $\lambda_2 \equiv 0$, supprimant de ce fait la nécessité de se préoccuper de la condition (50). Il en découle de plus que (28) et (30) forment déjà un système homogène aux inconnues λ_1 et λ_3, dont la condition de compatibilité fournit

$$\beta^2 = K \qquad \text{ou} \qquad \beta = \sqrt{K} \tag{55}$$

Ici l'arc à poussée décroissante est aussi un *arc à accélération constante*. Le point représentatif dans le diagramme de la Fig. 1 se maintient au point singulier.

a) Le long de l'arc à poussée constante les équations fondamentales ont les intégrales suivantes

$$\mu \beta = \mu_1 \beta_1 \tag{56}$$

$$(1 + \zeta)(\omega - \omega_1) = \ln \frac{\beta}{\beta_1} - \frac{1}{\beta_1} + \frac{1}{\beta} \tag{57}$$

$$(1 + \zeta)\tau = \frac{1}{\beta_1} - \frac{1}{\beta} \qquad (\text{avec } \tau_1 = 0). \tag{58}$$

Il découle de (14) que λ_1 reste constant, sa valeur étant fixée par le raccord à l'arc à accélération constante. Celle-ci se déduit de (28) et (55)

$$\lambda_1 = (1 + \zeta) \frac{K}{(\sqrt{K} - 1)^2} \qquad (\text{avec } \lambda_3 = 1) \tag{59}$$

On en déduit par (26) la valeur initiale de λ_4

$$\lambda_4(\sigma_1) = \frac{K}{\beta_1{}^2} \left[\frac{\beta_1 - 1}{(\sqrt{K} - 1)^2} - \frac{\beta_1}{K - \beta_1} \right]. \tag{60}$$

Dès lors la condition aux limites (52) fournit la valeur initiale du facteur d'accélération qui, fait remarquable, ne dépend pas de la performance demandée mais uniquement de la constante K

$$\beta_1 = \frac{K}{2\sqrt{K} - 1}. \tag{61}$$

Notons que cette valeur est relativement élevée; pour $K = 81$, par exemple, on trouve $\beta_1 = 4.765$.

b) A l'aide de (56), (57) et (58) on obtient les valeurs suivantes au début de la phase d'accélération constante

$$\mu_2 = \mu_1 \frac{\sqrt{K}}{2\sqrt{K} - 1}$$

$$(1 + \zeta)(\omega_2 - \omega_1) = \ln \frac{2\sqrt{K} - 1}{\sqrt{K}} - \frac{\sqrt{K} - 1}{K}$$

$$(1 + \zeta)\tau_2 = \frac{\sqrt{K} - 1}{K}$$

Pour la même valeur de K les seconds membres valent respectivement 0.529, 0.537 et 0.0988.

Le gain de vitesse est à ce moment de l'ordre de la moitié de la vitesse effective d'éjection. Ce n'est que pour des gains plus faibles qu'il y a lieu d'étudier les trajectoires qui sont simplement à poussée constante et la valeur initiale du facteur d'accélération dépend alors du gain demandé.

La durée de la phase à poussée constante est relativement courte; de l'ordre du dixième de l'impulsion spécifique c/g.

c) Le long du segment à accélération constante λ_4 reste nul; le facteur d'accélération conserve sa valeur (55) et les équations fondamentales s'intègrent sans difficultés

$$\mu_2 = \mu_3 \exp\left\{(1 + \zeta) \frac{K}{\sqrt{K} - 1} (\tau_3 - \tau_2)\right\} \tag{62}$$

$$\omega_3 - \omega_2 = (\sqrt{K} - 1)(\tau_3 - \tau_2) \tag{63}$$

Tandis que (49) devient

$$\mu_3 = \frac{\sqrt{K}}{\sqrt{K} - 1} \tag{64}$$

et permet de remonter jusqu'à la valeur de μ_1.

XI. Fusée de poussée initiale minimum

Le problème de rendre minimum

$$\frac{F(\sigma_1)}{g M_u} = \mu_1 \beta_1$$

est probablement plus significatif du point de vue technique. En effet la conception et la mise au point de moteurs-fusée entre pour une part très importante dans le coût total et dans les délais de réalisation d'un projet.

On substituera alors

$$\delta\beta_1 = \frac{1}{\mu_1} \delta(\beta_1 \mu_1) - \frac{\beta_1}{\mu_1} \delta\mu_1$$

dans la condition de transversalité (20) en même temps que l'on continue à se servir de (49) pour éliminer la variation de β_3. L'annulation des variations arbitraires sur γ_1, μ_3 et μ_1 fournit le nouveau jeu de conditions naturelles (50), (51) et

$$\lambda_4(\sigma_1) = -\frac{K}{\beta_1(K - \beta_1)} \tag{52'}$$

qui remplace maintenant (52). La condition de transversalité exprime maintenant la condition de minimum nécessaire

$$\delta(\beta_1 \mu_1) = 0$$

pour autant que le coefficient de cette variation qui est $-\lambda_3/(\beta_1 \mu_1)$ soit réellement différent de zéro.

Le choix $\lambda_3 = 1$ donne à nouveau une valeur négative à ce coefficient et conserve la validité du critère (54). La nouvelle condition (52′) exige encore une valeur initiale négative de λ_4. Par conséquent rien n'est à changer qualitativement aux discussions de la section IX. La section suivante montrera l'étendue des modifications quantitatives en jeu.

XII. Fusée de poussée initiale minimum en vol vertical

Les résultats (55), (56), (57), (58), (59) et (60) restent valables. La nouvelle condition (52′) fournit maintenant

$$\beta_1 = 1 \tag{61'}$$

si bien que l'accélération nette de la fusée au départ est maintenant nulle. Les valeurs en début de phase d'accélération constante deviennent

$$\mu_2 = \mu_1 \frac{1}{\sqrt{K}}$$

$$(1 + \zeta)(\omega_2 - \omega_1) = \ln \sqrt{K} - \frac{\sqrt{K} - 1}{\sqrt{K}}$$

$$(1 + \zeta)\, \tau_2 = \frac{\sqrt{K} - 1}{\sqrt{K}}$$

soit respectivement 0.111, 1.308 et 0.889 pour $K = 81$. La phase de poussée constante est donc beaucoup plus longue, étant de l'ordre de 85% de l'impulsion spécifique et à ce moment le gain de vitesse dépasse d'environ 30% la vitesse effective d'éjection. Les résultats (62), (63) et (64) restent valables.

XIII. Modifications introduites par l'addition d'un second étage

La masse M^* devient celle d'un second étage

$$M^* = M_u + M_s^* + M_e^* + M_p^*$$

Pour simplifier le calcul nous adoptons des indices constructifs

$$\sigma = \frac{M_s^*}{M^*} \qquad \varepsilon = \frac{M_e^*}{M^*}.$$

Autrement dit nous maintenons la masse structure (y compris les réservoirs) et la masse du moteur proportionnelles à la masse totale de l'étage. Dans ces conditions

$$M^*(1 - \varepsilon - \sigma) = M_u + M_p^*.$$

L'accroissement de vitesse donné par le dernier étage est évalué sans tenir compte des pertes de gravitation, ceci étant d'autant plus exact que l'angle γ_3 se rapproche de zéro:

$$\varDelta\omega = \frac{c^*}{c} \ln \frac{M^*}{M^* - M_p^*} = \frac{c^*}{c} \ln \frac{\mu^*}{1 + (\varepsilon + \sigma)\,\mu^*}$$

où

$$\mu^* = M^*/M_u.$$

L'équation (49) est maintenant à remplacer par la suivante

$$\mu^* = \mu_3 \frac{K - \beta_3}{K}$$

toujours déduite de (48). Par conséquent

$$\varDelta\omega = \frac{c^*}{c} \ln \frac{\mu_3(K - \beta_3)}{K + (\varepsilon + \sigma)\,\mu_3\,(K - \beta_3)}. \tag{65}$$

Pour une vitesse terminale ω_3 imposée on aura maintenant pour valeur de la vitesse en fin de combustion du premier étage

$$\omega(\sigma_3) = \omega_3 - \varDelta\omega.$$

Nous en prenons la variation que nous substituons dans la condition de transversalité

$$\delta\omega(\sigma_3) = K\,\frac{c^*}{c}\,\frac{1}{K + (\varepsilon + \sigma)\,\mu_3\,(K - \beta_3)} \left[\frac{\delta\beta_3}{K - \beta_3} - \frac{\delta\mu_3}{\mu_3}\right]$$

a) Dans le problème du minimum de μ_1

Les conditions (50) à (52) sont conservées. La condition supplémentaire

$$K\,\frac{c^*}{c}\,\lambda_1(\sigma_3) = \lambda_3\,[K + (\varepsilon + \sigma)\,\mu_3\,(K - \beta_3)] \tag{66}$$

remplace (49) pour le calcul de μ_3 et par conséquent de μ_1. Dans le cas de la fusée-sonde par exemple

$$\mu_3 = \frac{1}{\varepsilon + \sigma}\,\frac{\sqrt{K}}{\sqrt{K} - 1} \left[\frac{c^*}{c}\,(1 + \zeta)\,\frac{K}{(\sqrt{K} - 1)^2} - 1\right] \tag{64'}$$

remplace (64) avec dès lors

$$\Delta\omega = \frac{c^*}{c}\left[\ln\frac{1}{\varepsilon+\sigma} - \ln\left(1 + \frac{1}{\frac{c^*}{c}(1+\zeta)\frac{K}{(\sqrt{K}-1)^2} - 1}\right)\right].$$

Pour qu'un second étage améliore les performances ($\Delta\,\omega > 0$) il faut que

$$\varepsilon + \sigma < 1 - \frac{c(\sqrt{K}-1)^2}{c^* K(1+\zeta)}.$$

Cette condition devient difficile à tenir pour les valeurs élevées de K et faibles de ζ, à moins que la vitesse effective d'éjection du second étage ne soit nettement supérieure à c.

b) Dans le problème du minimum de $\beta_1\mu_1$

Les conditions (50), (51) et (52') de la section XI sont conservées et la condition additionnelle (66) s'applique également. Nous pouvons en tirer les mêmes conclusions quant à l'efficacité d'un second étage.

XIV. Conclusions et remarques finales

Les exemples illustrés à la Fig. 4 et relatifs au cas d'une fusée-sonde, permettent peut être de tirer certaines conclusions ayant une valeur générale. Pour les caractéristiques

$$K = 81 \qquad \zeta = 0.05 \qquad \omega_1 = 0 \qquad \text{et} \qquad \omega_3 = 4.7$$

la courbe *1* représente en fonction du temps réduit la poussée $\beta\,\mu$ par unité de poids utile, dans le cas où le minimum cherché est celui (μ_1) de la masse initiale par unité de masse utile. La courbe *2* représente la masse réduite μ en fonction du temps dans les mêmes conditions.

Les courbes *3* et *4* correspondent respectivement à *1* et *2* quand le minimum cherché est $\beta_1\mu_1$, la poussée initiale par unité de poids utile.

Les courbes *5*, *6*, *7* et *8* correspondent respectivement à *1*, *2*, *3* et *4* lorsqu'on ajoute un second étage ayant les caractéristiques

$$\varepsilon + \sigma = 0.1 \qquad c^* = 1.5\,c$$

(environ le rapport entre vitesses effectives d'éjection pour la combinaison Lox-Hydrogène liquide contre Lox-Kérosène) donnant un accroissement de vitesse réduite

$$\Delta\,\omega = 2.414$$

(soit environ la moitié de la performance totale requise).

On observera combien dans le cas *7*, la présence d'un second étage noble réduit aussi bien la durée que l'importance de la phase de réduction de poussée du premier étage. On observera aussi l'écart considérable qui se manifeste entre les poussées initiales, les masses initiales et par conséquent les consommations d'ergols, suivant que l'on adopte l'un ou l'autre type de minimum. Ceci suffit à souligner l'importance des études conduisant à une définition correcte du type de minimum à adopter.

Enfin il faut bien reconnaitre que dans la solution idéale pour une trajectoire incurvée, telle qu'elle a été exposée, on n'est pas maître de l'altitude terminale. Pour certaines applications il suffit que cette altitude soit supérieure à un minimum. Cette condition a plus de chances d'être réalisée pour un minimum de poussée initiale que pour un minimum de masse initiale. Si elle n'était pas réalisée, le traitement du problème devrait être élargi en incorporant l'équation du gain d'altitude avec un cinquième multiplicateur, ce qui compliquerait fort la nature de l'arc à poussée décroissante. La distance couverte horizontalement peut aussi devenir telle qu'il faille tenir compte de la courbure du champ de gravitation.

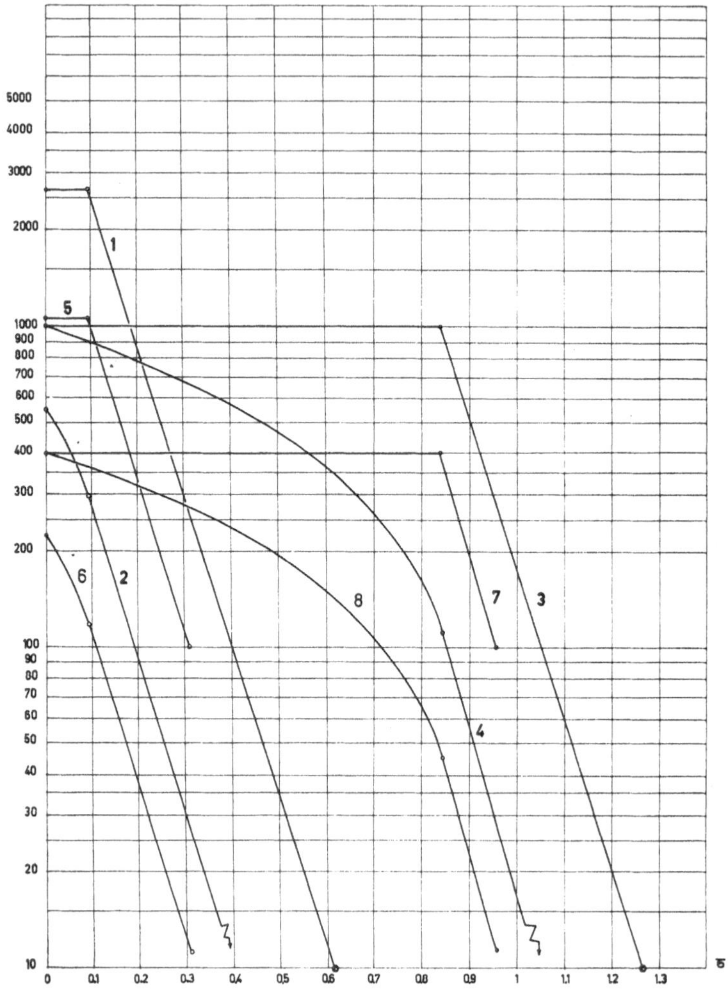

Fig. 4. Programmes optimums pour une fusée-sonde en fonction du temps réduit $\tau = tg/c$. *1*. Le rapport poussée/charge utile pour une fusée de poids minimum à l'envol. *3*. Pour une fusée de poussée minimum à l'envol. *5*. Comme sub *1*, mais avec second étage. *7*. Comme sub *3*, mais avec second étage. Les courbes *2*, *4*, *6* et *8* donnent dans les mêmes conditions le rapport de la masse instantanée à la masse utile

Références

1. M. Barrère, A. Jaumotte, B. Fraeijs de Veubeke et J. Vandenkerckhove, Rocket Propulsion, Chapter 12. Amsterdam: Elsevier, 1960.
2. B. Fraeijs de Veubeke, Méthodes variationnelles et performances optimales en aéronautique. Bull. soc. math. Belg. **2** (1957).

Discussion

M. Roy: Je vous remercie pour cette présentation volontairement raccourcie et qui me paraît parfaitement claire. Je voudrais vous poser une petite question. Vous avez dit qu'on détermine les deux termes fondamentaux par intégration en remontant à l'origine et à l'aide de deux conditions, dont l'une est une inégalité.

M. Fraeijs de Veubeke: Non, l'inégalité était là pour montrer que la valeur demandée au multiplicateur est négative, ce qui explique la constitution qualitative de la trajectoire, mais la valeur demandée au multiplicateur est de plus déterminée.

M. Crocco: Je voudrais demander ceci: vous avez, si je ne me trompe pas, employé la poussée dans la direction tangentielle. Evidemment, ceci constitue une limitation, peut-être pas pour le cas de fusée-sonde mais bien pour les autres cas.

M. Fraeijs de Veubeke: Le Professeur Lawden a étudié le problème de l'optimisation de l'orientation d'une poussée constante pour obtenir le maximum de rayon d'action d'un engin balistique. Je crois qu'il confirmera l'affirmation que par rapport à une poussée tangentielle les gains sont faibles. Je pense que ce sera le même cas ici.

(s'adressant à M. Lawden) You optimized the range with respect to thrust orientation for a ballistic missile and you found gains which are small compared to the case of tangential thrust.

M. Lawden: That is true. The gain in range was about 7%.

M. Fraeijs de Veubeke: If my memory is correct, you find that the thrust must be held constant in orientation in space, but the velocity vector's inclination changes with respect to the thrust orientation and at the end of the flight both are coincident. The gain on range is very small. Presumably this will also be true in the present case.

M. Lawden: No. The two vectors are coincident at the termination of the powered flight only when the final energy of the vehicle is being maximized, not the total range.

Aspects analytiques du problème des perturbations d'un satellite artificiel

Par

Jean Kovalevsky[1]

(Avec 1 Figure)

Résumé — Zusammenfassung — Abstract

Aspects analytiques du problème des perturbations d'un satellite artificiel. L'auteur décrit les effets d'origine non gravitationnelle, (principalement frottement atmosphérique et pression de radiation). Il semble cependant possible, de généraliser la méthode de von Zeipel jusqu'à inclure ces effets. Bien que la solution obtenue n'ait plus les caractéristiques d'une théorie générale au sens de la Mécanique Céleste classique, on peut néanmoins conduire les calculs de manière parallèle et obtenir des solutions formelles. Le problème de la convergence des séries obtenues reste entier.

En conclusion, il semble qu'on puisse d'ores et déjà considérer que le problème principal du mouvement d'un satellite autour d'une planète non sphérique ait reçu une solution satisfaisante. La famille d'orbites ainsi obtenue pourrait maintenant être considérée comme une famille d'orbites intermédiaires, chaque orbite étant caractérisée par ses éléments moyens. Les effets des forces d'origines différentes devraient être calculés à partir de ces orbites intermédiaires et il serait avantageux que les efforts soient maintenant concentrés sur les perturbations subies par ces orbites.

Über die Anwendung der Störungsrechnung bei Satelliten. Der Autor beschreibt die Effekte, die nicht von der Gravitation bewirkt werden (Widerstand, Strahlungsdruck). Es scheint möglich zu sein, die von-Zeipel-Methode auch auf diese Effekte anzuwenden. Allerdings gestatten diese Lösungen dann nicht mehr so allgemeine Aussagen wie diejenigen der klassischen Himmelsmechanik. Nichtsdestoweniger kann man aber analoge Überlegungen anstellen und zumindest formale Lösungen erhalten. Das Konvergenzproblem bleibt jedoch ungelöst.

Im Zusammenhang damit kann geschlossen werden, daß das Problem der Bewegung eines Satelliten um einen nicht kugelförmigen Planeten hinreichend genau gelöst werden kann. Die Klasse der so erhaltenen Bahnen kann als eine Familie von Zwischenbahnen aufgefaßt werden, wobei jede Bahn durch Mittelwerte ihrer Elemente bestimmt ist. Die Kraftwirkungen verschiedenen Ursprungs können dann mit Hilfe dieser Bahnen bestimmt werden und es erscheint möglich, damit Aussagen über die Störungen dieser Bahnen zu gewinnen.

Analytical Approach of the Perturbation Problem for an Artificial Satellite. The author describes the effects of non-gravitational origin (chiefly atmospheric drag and radiation pressure). It seems possible, however, to generalize the von Zeipel method to include these effects. Although the solution attained no longer has the characteristics of a general theory in the sense of classic celestial mechanics, one can nevertheless make parallel calculations and obtain formal solutions. The problem of the convergence of the series obtained remains unsolved.

[1] Bureau des Longitudes, 3, rue Mazarine Paris 6e, France.

In conclusion, it seems that one can now consider the principal problem of the movement of a satellite around a non-spherical planet as having been satisfactorily solved. The family of orbits thus obtained could be considered as a family of intermediate orbits, each orbit having the characteristics of its mean elements. The effects of forces of different origins would have to be calculated on the basis of these intermediate orbits and it would be advantageous to concentrate efforts on the perturbations undergone by these orbits.

I. Introduction

Le problème du mouvement d'un satellite artificiel autour de la Terre a, au cours des quatre dernières années, été l'objet de très nombreuses études. Il n'est sans doute pas exagéré d'évaluer à plus de mille, le nombre de rapports, articles, etc. — distribués ou publiés à ce sujet depuis le lancement du premier objet dans l'espace.

C'est en effet le premier problème à résoudre que de trouver, à partir des observations visuelles, photographiques, radioélectriques (radar ou effet DOPPLER), une trajectoire dont les particularités permettent de déterminer le système de forces auquel a été soumis l'objet étudié. La connaissance de ce système de forces permet alors de calculer la trajectoire précise et fournit des prévisions sous forme d'éphémérides.

Par suite de la complexité de ces forces, d'origines très diverses, on a été conduit souvent à fractionner le problème en isolant les difficultés et à étudier séparément les effets des forces d'origines différentes. Cette procédure fournit une première approximation des perturbations. Elle ne peut, cependant, s'appliquer indéfiniment, dès que les effets secondaires (ou mixtes) provenant de l'effet de l'une des forces sur les perturbations produites par une autre, ne sont plus négligeables. C'est ce qui se produit notamment dans la conjonction du frottement atmosphérique et de l'aplatissement terrestre, et ceci complique singulièrement le problème.

Dès lors, si l'on désire produire une théorie de toutes ces perturbations, avec une précision de l'ordre de celle des meilleures observations photographiques, on est conduit à écrire des équations très complètes, donc très compliquées, dont la solution analytique est longue et difficile. A vrai dire, cette solution n'existe pas encore. Pressés par le temps, devant fournir très rapidement des résultats, la plupart des auteurs se sont contentés de théories simplifiées, lorsqu'ils n'ont pas, tout simplement, intégré numériquement les équations différentielles du mouvement, au risque de ne plus distinguer les effets suivant leurs causes.

D'ailleurs, on a pu objecter que la recherche d'une solution analytique du problème du mouvement d'un satellite artificiel n'est qu'un exercice purement académique, inutile, puisqu'une fois les équations écrites, leur intégration numérique fournit les éphémérides et même peut améliorer les valeurs numériques des forces en présence. Cependant, tout effet de force dont on n'aurait pas tenu compte, ne peut être isolé et reconnu que grâce à la connaissance analytique de sa forme possible. La découverte des termes du troisième ordre dans le potentiel terrestre n'est qu'un exemple de ceci. La solution analytique est donc, en fait, un instrument essentiel de recherche dans l'étude dynamique de l'espace environnant. Seule (à condition d'être complète dans ses hypothèses), elle peut fournir une preuve irréfutable de l'origine d'un effet, en excluant les coïncidences numériques, d'autant plus probables que les effets à discerner sont faibles.

Il semble bien maintenant que, grâce aux très nombreuses tentatives faites de traiter analytiquement le problème principal du mouvement d'un satellite artificiel (perturbations dues à l'aplatissement de la Terre), et aux quelques solutions analytiques partielles publiées sur les autres causes de perturbations, il soit possible de discerner un caractère commun à ces méthodes et de distinguer celles qui sont les plus puissantes et sont susceptibles, sinon de fournir une solution analytique du problème complet, du moins de tenir compte du plus grand nombre possible de forces simultanément.

Il apparaît en effet nettement que toutes les théories n'ont pas la même puissance. De nombreuses solutions proposées, si elles permettent d'atteindre rapidement l'objectif poursuivi (en général une théorie du premier ordre et souvent avec l'hypothèse d'une faible excentricité), deviennent d'un maniement très délicat sinon difficile dès que l'on désire poursuivre les calculs au delà. Les difficultés proviennent soit d'une définition peu adéquate des constantes d'intégration, dépendant parfois de l'ordre de la solution, soit d'hypothèses simplificatrices portant sur certains effets, sinon sur le champ de forces lui-même. En fait la poursuite de ces théories jusqu'à la limite de précision des observations ne serait possible qu'avec des modifications plus ou moins grandes du mode de résolution proposé.

II. Perturbations dues à l'aplatissement terrestre

a) Le potentiel terrestre

Le développement en harmoniques sphériques du potentiel terrestre est le suivant:

1) Si on suppose que le potentiel est de révolution autour de l'axe des pôles, on a:

$$V = \frac{\mu}{a_e}\left[\frac{a_e}{r} + \sum_{n=2}^{\infty} \frac{J_n P_n(\sin\varphi)\, a_e^n}{r^n}\right] \tag{1}$$

où a_e est le rayon équatorial terrestre, μ est le produit de la constante de la gravitation par la masse de la Terre, P_n est le polynome de LEGENDRE d'ordre n et φ la latitude du point considéré, situé à une distance r du centre de la Terre ($\sin\varphi = z/r$). Les coefficients J_n sont sans dimension et décroissent rapidement lorsque n croît.

Les déterminations récentes des coefficients J_n sont nombreuses, mais souvent incompatibles si l'on compare les erreurs probables données par leurs auteurs. Les deux plus récentes, basées chacune sur les mouvements de 3 satellites: celle de KOZAÏ [1] à l'aide de Vanguard 1, Explorer 7 et Vanguard 3 et celle de KING-HELE [2] à l'aide de Vanguard 1, Spoutnik 2 et Explorer 7 sont résumées dans le Tableau 1.

Tableau 1

Coefficient	Résultats de KOZAÏ	Résultats de KING-HELE
$10^6 J_2$	$1082{,}49 \pm 0{,}024$	$1082{,}79 \pm 0{,}15$
$10^6 J_3$	$-\ 2{,}29 \pm 0{,}02$	$-\ 2{,}4 \pm 0{,}3$
$10^6 J_4$	$-\ 2{,}12 \pm 0{,}04$	$-\ 1{,}4 \pm 0{,}2$
$10^6 J_5$	$-\ 0{,}23 \pm 0{,}02$	$-\ 0{,}1 \pm 0{,}1$
$10^6 J_6$	0	$+\ 0{,}9 \pm 0{,}8$

Il est probable que les erreurs moyennes indiquées — qui sont des erreurs internes provenant des résidus — ne représentent pas, surtout dans le premier cas, la précision avec laquelle on connaît ces constantes. Certes il est souhaitable d'utiliser plus complètement et sur un nombre plus grand de satellites les observations de haute précision (notamment les photographies à l'aide de chambres de BAKER-NUNN) pour améliorer ces erreurs externes. Il ne faut cependant pas négliger la part que les théories peuvent avoir dans ces différences. Nous avons déjà indiqué autre part (3) pourquoi l'interprétation de certaines constantes d'intégration pour certaines théories pouvait être délicate et peut-être à l'origine d'erreurs systématiques. La facilité d'interprétation des observations en fonction des constantes d'intégration doit être un critère fondamental dans le choix des théories.

2) Si l'on ne suppose plus que l'équateur soit circulaire, le potentiel dépend alors de la longitude géographique λ du point étudié. Au potentiel donné par (1), il faut encore ajouter les termes:

$$\mu \sum_{n=2}^{\infty} \sum_{q=1}^{n} J_{nq} \frac{a_e^r}{r^{n+1}} P_{nq} (\sin \varphi) \sin q (\lambda + \beta_{nq}) \tag{2}$$

P_{nq} (sin φ) est une fonction de LEGENDRE associée, J_{nq} et β_{nq} sont des constantes donnant la magnitude et le déphasage par rapport au méridien, origine des perturbations.

On a détecté jusqu'à présent J_{22} à l'aide des observations de Vanguard 2 et Vanguard 3 [4, 5] et J_{41} [5].

b) *Équations dans le problème gravitationnel principal*

Par suite de la faiblesse des corrections au potentiel introduites par les termes de la formule (2), il suffit de ne considérer, comme on l'a fait d'ailleurs, dans tous les travaux relatifs à ce problème, que le potentiel (1) limité de J_2 à J_5.

Les équations du mouvement en coordonnées rectangulaires à direction fixe, centrées au centre de la Terre supposée fixe (ce qui revient à négliger les perturbations luni-solaires) sont:

$$\frac{d^2x}{dt^2} = \frac{\partial V}{\partial x}; \quad \frac{d^2y}{dt^2} = \frac{\partial V}{\partial y}; \quad \frac{d^2z}{dt^2} = \frac{\partial V}{\partial z} \tag{3}$$

soit encore, tout système équivalent à celui-ci, utilisant tel ou tel système classique de variables utilisé en Mécanique Céleste: voir par exemple MOULTON [6], TISSERAND [7] ou tout autre ouvrage de Mécanique Céleste.

On aura ainsi le système d'équations de LAGRANGE, où les inconnues sont les six éléments osculateurs classiques:

a: demi-grand axe
e: excentricité
i: inclinaison
Ω: longitude du noeud
ω: argument du périgée
M: anomalie moyenne.

R est la fonction perturbatrice, $R = V - \mu/r$ et n, le moyen mouvement est une écriture abrégée de $\sqrt{\mu} a^{-3/2}$

$$
\left\{
\begin{aligned}
\frac{da}{dt} &= \frac{2}{n\,a}\frac{\partial R}{\partial M}\\
\frac{de}{dt} &= \frac{1-e^2}{n\,a^2\,e}\frac{\partial R}{\partial M} - \frac{\sqrt{1-e^2}}{n\,a^2\,e}\frac{\partial R}{\partial \omega}\\
\frac{di}{dt} &= \frac{1}{n\,a^2\sqrt{1-e^2}\,\mathrm{tg}\,i}\frac{\partial R}{\partial \omega} - \frac{1}{n\,a^2\sqrt{1-e^2}\sin i}\frac{\partial R}{\partial \Omega}\\
\frac{d\Omega}{dt} &= \frac{1}{n\,a^2\sqrt{1-e^2}\sin i}\frac{\partial R}{\partial i}\\
\frac{d\omega}{dt} &= -\frac{1}{n\,a^2\sqrt{1-e^2}\,\mathrm{tg}\,i}\frac{\partial R}{\partial i} + \frac{\sqrt{1-e^2}}{n\,a^2\,e}\frac{\partial R}{\partial e}\\
\frac{dM}{dt} &= n - \frac{1-e^2}{n\,a^2\,e}\frac{\partial R}{\partial i} - \frac{2}{n\,a}\frac{\partial R}{\partial a}
\end{aligned}
\right. \qquad (4)
$$

On aura aussi le système classique d'équations canoniques de Delaunay, si l'on pose:

$$
\begin{aligned}
L &= \sqrt{\mu\,a}\,; & l &= M\\
G &= \sqrt{\mu\,a\,(1-e^2)}\,; & g &= \omega\\
H &= \sqrt{\mu\,a\,(1-e^2)}\cos i\,; & h &= \Omega
\end{aligned}
$$

$$
\text{et si } F = \frac{\mu^2}{2\,L^2} + R \qquad \text{on a:}
$$

$$
\left\{
\begin{aligned}
\frac{dL}{dt} &= \frac{\partial F}{\partial l}\,; & \frac{dG}{dt} &= \frac{\partial F}{\partial g}\,; & \frac{dH}{dt} &= \frac{\partial F}{\partial h}\\
\frac{dl}{dt} &= -\frac{\partial F}{\partial L}\,; & \frac{dg}{dt} &= -\frac{\partial F}{\partial G}\,; & \frac{dh}{dt} &= -\frac{\partial F}{\partial H}
\end{aligned}
\right. \qquad (5)
$$

c) *Analogie avec les problèmes de Mécanique Céleste classique*

Dans les systèmes d'équations (3), (4) ou (5), les seconds membres sont des fonctions des coordonnées x, y, z seulement. Si l'on tient compte de (2), il s'y introduit λ, qui est une fonction linéaire du temps.

Cette fonction perturbatrice R peut s'exprimer en fonction des six variables choisies (canoniques, de Lagrange etc.) à l'aide de séries de Fourier de la forme:

$$
\sum A_{jk} \cos (j\,M + k\,\omega) \qquad (6)
$$

où A_{jk} dépend des variables $a\,e\,i$ (ou $L\,G\,H$) et j et k sont des nombres entiers. Si l'on tient compte des termes (2), un troisième argument linéaire, contenant $\Omega + \lambda$ s'introduit dans le cosinus.

Or, il se trouve que telle est exactement la forme des équations que l'on rencontre dans les divers types de problèmes de 3 corps ou plus qui sont l'objet, depuis plus de 200 ans, de l'étude des astronomes. Que ce soit la théorie des planètes ou celle des satellites, les équations du mouvement ont une des formes (3), (4) ou (5) et la fonction perturbatrice R prend la forme générale:

$$
\sum A_{j\,k\,l\,m_1\,m_2} \cos (j\,M + k\,\omega + l\,\Omega + m_1\,\lambda_1 + m_2\,\lambda_2 + \dots) \qquad (7)
$$

où λ_1, λ_2 sont des fonctions linéaires du temps, très analogues, mais en général plus complexes que la forme (6).

A titre d'exemple, pour la théorie de la Lune, cette fonction perturbatrice se développe sous la forme (7) avec une seule fonction linéaire du temps supplémentaire, l'anomalie moyenne du Soleil.

Il s'ensuit que toutes les méthodes ayant été essayées avec succès en Mécanique Céleste, soit pour la Lune, soit dans d'autres cas, seront valables pour le problème gravitationnel principal d'un satellite artificiel. Un certain nombre de ces méthodes sont décrites par BROWN [8] et toutes peuvent convenir pour traiter le problème qui nous concerne. Pratiquement d'ailleurs, la plupart de ces méthodes ont été essayées.

La méthode de DELAUNAY par BROUWER [9]. La lourdeur du procédé paraît inutile pour ce problème simple. Le mouvement d'un satellite artificiel offre cependant un exemple permettant d'étudier à fond le mécanisme de cette méthode, ce qu'est en train de faire MORANDO à Paris.

La méthode de HILL-BROWN par BROUWER [9] ou GREBENIKOV [10];

La méthode de HANSEN, par MUSEN [11];

La méthode des équations de LAGRANGE a été souvent utilisée. Elle m'a permis de retrouver, jusqu'au second ordre les résultats de BROUWER [3]. KOZAÏ l'a utilisée aussi sous une forme évitant l'emploi de l'anomalie moyenne afin d'obtenir des formules finies [12]. Citons encore MERSON et PLIMMER [13].

Parmi les méthodes utilisant une orbite intermédiaire citons celles de STERNE [14] ou de GARFINKEL [15] qui recherchent des Hamiltoniens partiels donnant une solution sous forme finie.

Enfin, appliquant une méthode proposée par VON ZEIPEL en 1916 [16], D. BROUWER a fourni une remarquable solution de ce problème [17][1].

En conséquence, l'énoncé du problème étant similaire à l'énoncé des principaux problèmes classiques de Mécanique Céleste, les méthodes mises au point dans la résolution de ces problèmes, devaient donner pour le mouvement d'un satellite artificiel une solution. Cette solution, en général, apparaît beaucoup plus rapidement et à la suite de calculs moins nombreux, puisque le problème gravitationnel principal dépend essentiellement de deux paramètres angulaires (M et ω) seulement.

d) *Analogie avec la théorie de la Lune*

L'analogie, en fait, est encore beaucoup plus précise entre le problème principal de la théorie de la Lune et le problème qui nous préoccupe. Elle ne repose pas seulement sur l'analogie entre les formes (6) et (7) de la fonction perturbatrice, mais aussi entre les moyens et le formulaire permettant d'arriver à ces développements.

En effet, dans le problème principal de la théorie de la Lune (perturbations par le Soleil supposé se mouvoir suivant les lois de KEPLER), la fonction perturbatrice R est:

$$R = k^2 m' \left[\frac{1}{\Delta} - \frac{x x' + y y' + z z'}{(x'^2 + y'^2 + z'^2)^{3/2}}\right] \tag{8}$$

où x', y', z' sont les coordonnées géocentriques du Soleil (masse m')

x, y, z sont les coordonnées géocentriques de la Lune,

Δ est la distance Lune-Soleil.

---

[1] Je dois signaler que pendant que je rédigeais cette communication, j'ai reçu un exemplaire avant tirage d'un excellent travail de W. M. KAULA, discutant de façon beaucoup plus détaillée qu'il ne m'a été possible de le faire ici, les diverses questions relatives aux aspects géodésiques du problème des satellites artificiels et décrivant en particulier la plupart des théories que je cite. "Celestial Geodesy" paraitra dans le volume 9 de "Advances in Geophysics".

Si nous appelons r' et r les distances de la Terre respectivement au Soleil et à la Lune (Fig. 1) et si S est l'angle $(\overrightarrow{TL}, \overrightarrow{TS})$ on a:

$$\Delta^2 = r^2 + r'^2 - 2\, r\, r' \cos S$$

Or r' est beaucoup plus grand que r, et on peut écrire:

$$\frac{1}{\Delta} = \frac{1}{r'}\left[1 - 2\frac{r}{r'}\cos S + \left(\frac{r}{r'}\right)^2\right]^{-1/2} \tag{9}$$

$$= \frac{1}{r'}\left[1 - \frac{r}{r'} P_1(\cos S) + \left(\frac{r}{r'}\right)^2 P_2(\cos S) + \ldots + \left(\frac{r}{r'}\right)^n P_n(\cos S)_j \ldots\right]$$

où P_n représente le polynome de Legendre d'ordre n. Si on remarque encore que $P_1(\cos S) = \cos S = (x\,x' + y\,y' + z\,z')/r\,r'$ et qu'on n'utilise que les dérivées partielles de R par rapport aux quantités relatives à la Lune, on peut omettre le terme $1/r'$ et écrire

$$R = V - \frac{\mu}{r} = k^2 m' \sum_{n=2}^{\infty} \frac{r^n}{r'^{n+1}} P_n(\cos S) \tag{10}$$

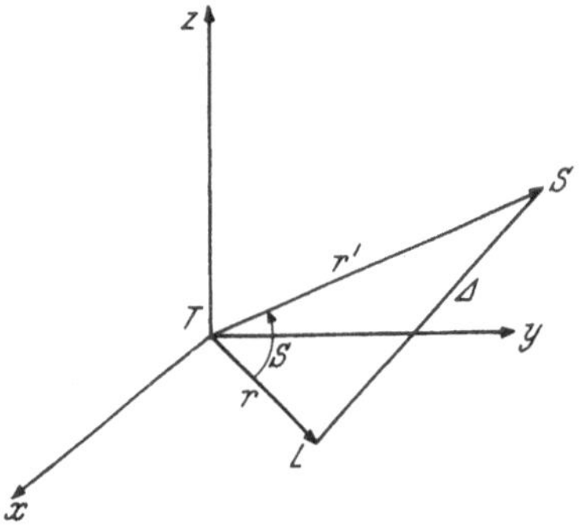

Fig. 1

formule qui ressemble à la formule (1) où les petites quantités J_n seraient remplacées par $A_n = (k^2 m')/a'^{n+1}$ où a' est le demi-grand axe fixe de l'orbite solaire, qui est grand par rapport à z. Les quantités A_n diminuent donc rapidement avec n. Mise sous la forme suivante:

$$R = \sum \frac{k^2 m'}{a'^{n+1}} \times \frac{r^n}{\left(\frac{r'}{a'}\right)^{n+1}} P_n(\cos S) \tag{11}$$

nous voyons que dans la théorie du satellite artificiel la latitude φ, qui ne dépend que de ω et de M, joue le role de S qui, lui, dépend aussi de la longitude du noeud de la Lune et de l'anomalie moyenne du Soleil. De même le coefficient $1/r^n$ joue le rôle de r^n/r'^{n+1} qui, en plus de M, dépend aussi de l'anomalie moyenne du Soleil.

Ainsi, les développements (1) et (11) sont-ils très voisins et peuvent-ils *essentiellement* se traiter de la même façon, bien que formellement, les expressions de l'angle ou du coefficient soient différents. En particulier, dans les deux cas on envisagera, dans une première approximation, de négliger tous les termes de R, sauf le premier, puis, à chaque approximation, on tiendra compte des puissances successives du petit paramètre.

III. Autres perturbations d'origine gravitationnelle

Comme pour la théorie de la Lune où, en plus des perturbations solaires, il faut également tenir compte de perturbations d'origine planétaire, ou encore dues à l'aplatissement de la Terre, un satellite artificiel subit d'autres perturbations d'origine gravitationnelle ou non.

Les perturbations d'origine gravitationnelle, comme toutes les perturbations rencontrées en Mécanique Céleste, proviendront d'une fonction perturbatrice. Dans le cas qui nous occupe, il s'agira principalement des perturbations luni-solaires, et aussi les perturbations dues à l'ellipticité de l'équateur terrestre.

Soit R_0 la fonction perturbatrice du problème principal et R, la fonction perturbatrice provenant de la considération d'une autre force. Les systèmes 3, 4 ou 5 restent vrais; il suffit d'écrire: $R = R_0 + R_1$.

Rigoureusement, on est donc conduit à résoudre un problème analogue au problème principal, mais avec une fonction perturbatrice R plus compliquée, pouvant dépendre du Temps et de la longitude du noeud du satellite (par l'intermédiaire des coordonnées de la Lune et du Soleil et la distance Satellite-Lune ou Satellite-Soleil par exemple). Ici encore, essentiellement nous retrouvons un problème typique de Mécanique Céleste qu'on peut attaquer par les mêmes méthodes. Le problème des perturbations luni-solaires d'un satellite artificiel est le même que celui des perturbations par un autre satellite et par le Soleil de tout satellite naturel comme il s'en trouve dans les systèmes de Saturne ou de Jupiter.

Toutefois il est, dans certains cas, possible de linéaniser le problème, ce qui permet d'étudier séparément les perturbations d'origines différentes. Raisonnons sur le système (4) qu'on écrira sous la forme:

$$\frac{dE_i}{dt} = \sum_j A_{ij} \frac{\partial R}{\partial E_j} \qquad 1 \leqslant i, j \leqslant 6 \tag{12}$$

où E_i est un élément, A_{ij} un coefficient dépendant des éléments.

Si $R = R_0 + R_1$, on a:

$$\frac{dE_i}{dt} = \sum_j A_{ij} \frac{\partial R_0}{\partial E_j} + \sum_j A_{ij} \frac{\partial R_1}{\partial E_j} \tag{13}$$

Si E_{0j} sont des éléments elliptiques fixes dans le temps (par exemple les éléments moyens) et si nous écrivons que la solution est

$$E_j = E_{0j} + \delta E_j \tag{14}$$

(13) peut s'écrire:

$$\frac{d\delta E_i}{dt} = \sum_j A_{ij} \frac{\partial R_0}{\partial E_j} + \sum_j A_{ij} (E_{0k} + \delta E_k) \frac{\partial R_1}{\partial E_j} (E_{0k} + \delta E_k) \tag{15}$$

Si la contribution à δE_i de la variation du second membre provenant des variations δE_k est négligeable, c'est-à-dire si l'intégrale

$$\sum_j \int_0^t \sum_k \left(\frac{\partial A_{ij}}{\partial E_k} \delta E_k \frac{\partial R_1}{\partial E_j} + A_{ij} \frac{\partial^2 R_1}{\partial E_j \, \partial E_k} \delta E_k \right) dt \tag{16}$$

est bornée par une quantité faible par rapport aux erreurs d'observations, on peut remplacer le système (13) par:

$$\frac{dE_i}{dt} = \sum_j A_{ij} \frac{\partial R_0}{\partial E_j} + \sum_j A_{ij} (E_{0k}) \frac{\partial R_1 (E_{0k})}{\partial E_j} \quad . \tag{17}$$

Les coefficients de la 2ème partie étant indépendants de E_i et ne dépendant, par conséquent que du temps, on a leur contribution ΔE_k *directe* à E_i en intégrant par rapport au temps la deuxième expression.

Si cette contribution est telle que l'addition de ΔE_k à E_k dans l'expression $\Sigma A_{ij} (\partial R_0)/(\partial E_j)$ introduit une variation des éléments E_i non négligeable, nous dirons qu'on a ainsi une *perturbation indirecte*. Si, par contre, cette variation est négligeable, les perturbations dues à R_1 seront entièrement connues en intégrant le système:

$$\frac{dE_i}{dt} = \sum_j A_{ij}(E_{0k}) \frac{\partial R_1(E_{0k})}{\partial E_j} \tag{18}$$

C'est la méthode employée pour le calcul des perturbations planétaires de la Lune et pour la plupart des théories des planètes. Elle est valable pour un satellite artificiel en ce qui concerne les perturbations dues à l'ellipticité de l'équateur terrestre et, à l'intérieur même du problème principal, étant donnée la précision actuelle des observations, pour les perturbations provenant de tous les termes de (1) sauf J_2, pour lequel il est indispensable de poursuivre les calculs au second ordre.

En ce qui concerne les perturbations luni-solaires, la séparation est légitime tant que ces perturbations ne sont pas trop grandes. Il n'en est pas toujours ainsi. MUSEN [18] a montré que pour certains satellites à forte excentricité, l'effet pouvait être très grand, au point de modifier entièrement l'aspect du mouvement.

IV. Perturbations d'origine non gravitationnelle

Les perturbations d'origines diverses, introduisent dans les équations, des termes parfois prédominants ayant une structure essentiellement différente des termes d'origine gravitationnelle. Les deux causes les plus importantes sont le frottement atmosphérique et la pression de radiation, sans que l'on puisse considérer comme négligeable dans *tous les cas* d'autres effets possibles.

1. Frottement atmosphérique

De façon générale, on convient que l'accélération de frottement

$$\vec{\gamma} = - K S \varrho |v| \vec{v} \tag{19}$$

où S est la surface de la maitresse section du corps, perpendiculaire à la vitesse $\vec{v}$ prise par rapport à l'air supposé fixe, ϱ est la densité et K un coefficient dépendant de la forme du satellite. En fait K dépend de la nature des chocs (élastiques, avec diffusion, etc.) et par suite, dans une certaine mesure de ϱ. Pratiquement, les divers auteurs considèrent K comme une constante, ce qui revient à reporter sur ϱ les variations de K et considérer une densité dynamique qui peut être différente de la densité physique du milieu.

De plus, il est nécessaire d'avoir une loi pour ϱ. La loi exponentielle est choisie en général:

$$\varrho = \varrho_0 \, e^{(r \cdot r_0)/H} \tag{20}$$

où ϱ_0 est la densité à l'altitude r_0 prise souvent proche du rayon vecteur au périgée et H étant un certain facteur de réduction ayant la dimension d'une altitude. Cependant JACCHIA [19] a été conduit à considérer que H lui-même dépendait du $r - r_0$

$$H = H_0 + \beta(r - r_0) \tag{21}$$

β étant de l'ordre de 0,2. De plus, pour tenir compte de l'aplatissement de l'atmosphère, r_0 devrait dépendre de la latitude φ

$$r_0 = r_0' - s \cos \varphi \tag{22}$$

s étant de l'ordre de 20 kilomètres.

Enfin, la rotation de l'atmosphère a pour effet que la vitesse $\vec{v}$ dans l'équation (19) n'est pas exactement celle du satellite par rapport aux axes de la théorie du mouvement. Une manière approchée d'en tenir compte c'est de multiplier la force par un facteur dépendant de l'inclinaison du satellite et de la vitesse de rotation de la Terre [20].

Enfin, on a pu montrer déjà l'existence d'un effet de l'éclairement solaire sur le frottement atmosphérique [21], effet qui est compliqué par des variations aléatoires dont certaines sont des conséquences des grosses éruptions solaires [22].

La complexité de ces forces, leur aspect parfois aléatoire, rend pratiquement impossible tout traitement analytique complet du problème. Les très nombreux travaux s'y rapportant traitent le problème sous une forme semi-empirique, seule la forme des termes perturbateurs étant déterminée par des considérations analytiques. L'analyse et l'intégration numérique se montrent les seuls instruments capables de traiter ce problème avec la précision requise par les observations. Cependant pour les théories de haute précision, dans les travaux d'ordre géodésique ou d'analyse gravitationnelle, il est indispensable d'avoir une possibilité d'exprimer analytiquement ces perturbations afin d'en tenir compte sur tout un long intervalle de temps. Ceci exclut évidemment toute étude de ce genre sur des satellites bas, mais pour les satellites élevés, lorsque les effets atmosphériques sont faibles, la recherche d'une théorie analytique est indispensable pour tout travail de précision. Les travaux de Brouwer et Hori [23] sont un premier pas vers une telle théorie.

2. *Pression de radiation*

En principe, le problème est plus simple que celui du frottement atmosphérique, par suite de la constance de l'énergie de radiation. La pression dépend évidemment de la section maitresse éclairée, mais, comme pour la résistance de l'air, c'est moyenné par la rotation rapide du satellite, à moins que, pour la théorie, on ne considère un satellite sphérique. Dans ce dernier cas, la pression est une force constante en valeur absolue. Ceci conduit à une théorie du type de ceux que traite la Mécanique Céleste [24] où l'on peut introduire une fonction perturbatrice. Lorsque le satellite n'a pas un rapport surface sur masse trop grand, un calcul de perturbations du type de celui décrit à la fin du paragraphe III est suffisant. Pour les satellites ayant une faible masse pour un gros volume, l'effet est prépondérant et peut être mélangé à l'effet gravitationnel dans les équations.

Cependant, une discontinuité dans l'expression de la force est introduite par les éclipses de ces satellites. Pendant le passage du satellite dans l'ombre de la Terre, cette force est réduite à zéro. Cette discontinuité rend impossible tout traitement purement analytique du problème. Encore une fois, seule une solution numérique, comme celle de Kozaï [25] peut donner une solution précise dans les cas pratiques.

3. *Autres effets; traitement des faibles perturbations*

Les forces d'origine électro-magnétiques ont été évaluées, mais jamais explicitement décelées, surtout par suite de l'incertitude où l'on est sur la magnitude du frottement atmosphérique.

Pour le traitement de tous ces effets, la méthode analytique ou numérique qui s'impose est la considération séparée des perturbations et l'addition des variations des éléments ainsi calculées à la solution du problème principal. La même méthode s'appliquera en fait pour la pression de radiation sur les satellites denses, sauf les cas de résonnance du mouvement du noeud ou du périgée avec celui du Soleil. Enfin, pour les satellites denses de très haute altitude (périgée au delà de 1500 kilomètres), les seuls qui aient un intérêt astrométrique ou géodésique, la résistance de l'air pourra être traitée analytiquement de cette façon, à condition que l'effet de l'activité solaire ne soit pas prédominant, ce qui reste encore à démontrer. Pour ce type de satellite, il est donc de la plus haute importance de concevoir une méthode unique permettant de traiter les perturbations d'origines les plus diverses.

V. Possibilité d'une théorie unifiée

En ce qui concerne le problème gravitationnel principal, les diverses théories présentent chacune certains avantages. Néanmoins le volume de calculs nécessaires pour atteindre la solution varie dans une grande mesure d'une méthode à l'autre. J'ai personnellement essayé, en relation avec la préparation de cours à la Faculté des Sciences de Paris, de comparer le travail nécessaire pour arriver à une théorie du second ordre, en utilisant successivement la méthode des équations de LAGRANGE [3], la méthode de DELAUNAY [9] et celle de VON ZEIPEL [16].

La première méthode est la plus rapide et peut-être la plus facilement mise en oeuvre, si l'on dispose d'une calculatrice électronique pour traiter les séries de FOURIER à coefficients littéraux que l'on rencontre. Conduisant à des expressions identiques à celles de la méthode de DELAUNAY, elle est beaucoup plus simple à mettre en oeuvre et exige bien moins de calculs. Ces deux méthodes conduisent à exprimer les éléments osculateurs (ou les coordonnées) sous forme de séries de FOURIER, dont les coefficients sont des séries dépendant des éléments moyens. Le désavantage de ces méthodes réside dans le fait que les expressions trouvées sont infinies. Cependant le traitement des équations de LAGRANGE, de KOZAÏ évite ce dernier inconvénient [12].

Une autre méthode, permettant d'arriver aux mêmes résultats avec les éléments moyens est celle de BROUWER [17]. La somme de calculs demandée est équivalente à la méthode de KOZAÏ. Mais, comme l'a montré G. HORI [26] la méthode de VON ZEIPEL conduit à une solution valable dans le cas de l'inclinaison critique i (telle que $\cos^2 i = 1/5$; $i = 62° 26'$ environ), pour laquelle la méthode de LAGRANGE et la solution de DELAUNAY ne sont plus valables. La solution ainsi proposée est d'ailleurs dans la ligne des théories de la libration en Mécanique Céleste (théorie du satellite de Saturne, Hyperion, par exemple). En fait d'autres théories, comme celle de GARFINKEL [15] permettent d'attaquer le problème de l'inclinaison critique [27] et si ce critère élimine certaines méthodes, pour le choix de la théorie la plus complète, il ne peut être seul. Il est d'ailleurs probable que plusieurs théories (celles de BROUWER et de GARFINKEL en particulier) soient de puissance analogue pour attaquer les perturbations autres que celles du problème principal.

Cependant, il faut, dans la mesure du possible, utiliser une théorie où les constantes d'intégration ont une signification intrinsèque, indépendante de la notion d'origine de Temps. Seules de telles constantes pourront être ensuite aisément transformables en variables par la méthode de la variation des constantes par exemple, pour tenir compte d'autres perturbations. A puissance égale pour le problème principal, il faut adopter les théories utilisant les éléments moyens (valeur moyennée sur un intervalle de temps infini) de préférence aux éléments à un instant donné. Ce sont elles qui permettent d'inclure le plus directement les effets gravitationnels ou non qui nécessitent un traitement plus complet que l'isolement pur et simple que nous avons décrit.

A l'heure actuelle, la méthode de VON ZEIPEL est celle qui répond le mieux à ces critères. C'est aussi celle qui a été le plus travaillée dans le sens indiqué ici et pour laquelle on a pu montrer qu'il est possible de traiter d'autres effets.

En particulier, BROUWER et HORI ont montré [23] qu'il est possible d'inclure le frottement atmosphérique dans les équations avec les variables de DELAUNAY. Etant donnée la solution du problème principal, on peut par la méthode des substitutions dans les seconds membres et approximations successives, mettre la solution sous forme d'une série de FOURIER avec termes séculaires et mixtes en plus des termes périodiques classiques. En particulier, dépendront du coefficient K de frottement, seuls des termes périodiques et séculaires aussi bien dans les variables angulaires, que dans les variables métriques (a, e, i), contrairement aux resultats des solutions générales de tous les problèmes gravitationnels. Il s'introduira aussi des termes mixtes (périodiques dont les coefficients dépendent du Temps), qui seront proportionnels à K et à J_r. On arrive ainsi à une généralisation de la forme des séries dans une théorie générale. Les calculs sur ces séries peuvent se faire suivant les mêmes méthodes (approximations successives) que pour les séries classiques en Mécanique Céleste. Elles s'apparentent d'ailleurs du point de vue forme, à la théorie des planètes, quoique l'origine des termes séculaires soit ici physique et ne provienne pas seulement d'un artifice de calcul.

Ce problème de la mise sous forme analytique des perturbations dues au frottement atmosphérique n'est pas encore entièrement résolu. Il est probable — comme c'est le cas dans la théorie des planètes — que les séries obtenues ne soient pas convergentes pour de grandes valeurs du temps. BROUWER et HORI n'ont encore donné la solution que dans l'hypothèse d'une atmosphère sphérique avec une densité répartie suivant une loi exponentielle. Cependant, le fait même qu'ils aient pu mettre les équations sous une forme qui peut se traiter comme des équations canoniques est un grand pas vers la solution analytique du problème complet du mouvement d'un satellite artificiel. Peut-être des méthodes de Mécanique Céleste autres que celle de VON ZEIPEL se prêtent-elles à la solution de ce problème. Celle-ci, en tous cas, a tous les avantages requis pour être adoptée comme base de départ aux futures théories du mouvement.

VI. Conclusion

Le nombre de solutions concordantes au problème principal de la théorie d'un satellite artificiel est tel qu'on peut considérer ce problème comme entièrement résolu. Les solutions ont été publiées sous formes finies, qui suffisent à la quasi-totalité des besoins. Il semble que ces orbites (dépendant seulement de deux paramètres angulaires) doivent maintenant être considérés comme des données de base au même titre que le mouvement elliptique dans un stade moins avancé de la théorie. Il est maintenant temps de s'attaquer au problème fondamental de la solution analytique de toutes les perturbations non gravitationnels, de caractère non aléatoire. Cette solution sera indispensable dans un court laps de temps pour les satellites géodésiques. Elle le sera davantage encore lors des lancements de satellites destinés à vérifier la relation qui existe entre le temps atomique et le temps gravitationnel. Lancés à une altitude de 40.000 kilomètres, quoique très massifs, ces satellites subissent des forces non gravitationnelles sensibles. Il faudra en faire une théorie analytique valable avec une précision de quelques millièmes de seconde par an. Les difficultés d'élaboration de telles théories, provenant du caractère très divers des forces en présence, doivent être attaquées dès maintenant, profitant de l'acquit en Mécanique Céleste et plus particulièrement dans l'étude du problème principal. Il n'est pas encore certain que toutes ces difficultés soient vaincues avant que les progrès techniques rendent possibles ces réalisations.

Références

1. Y. Kozaï, The Gravitational Field of the Earth Derived from the Motion of Three Satellites. Astronom. J. **66**, 8 (1961).
2. D. G. King-Hele, The Earth's Gravitational Potential, Deduced from the Orbits of Artificial Satellites. Geophysic. J. **4**, 3 (1961).
3. J. Kovalevsky, Influence des termes du second ordre sur la théorie du mouvement d'un satellite artificiel. Space Research, Proceedings of the First International Space Science Symposium, Nice, p. 458 (1960).
4. I. G. Izsak, A Determination of the Ellipticity of the Earth's Equator from the Motion of Two Satellite Observations. Smithsonian Institution Astrophysical Observatory Special Report no. 56, p. 11 (1961).
5. W. M. Kaula, Analysis of Satellite Observations for Longitudinal Variations of the Gravitational Field. Proceedings of the Second International Space Science Symposium, Florence (1961) — sous presse.
6. F. R. Moulton, An Introduction to Celestial Mechanics. New York: Mac Millan Co., 1947.
7. F. Tisserand, Traité de Mécanique Céleste, Tome I. Paris: Gauthier-Villars, 1960.
8. E. W. Brown, An Introductory Treatise on the Lunar Theory. New York: Dover Publications Inc., 1960.
9. D. Brouwer, Outlines of General Theories of the Hill-Brown and Delaunay Types of Orbits of Artificial Satellites. Astronom. J. **63**, 433 (1958).
10. E. A. Grebenikov, Sur l'application de la méthode de Hill dans l'étude du mouvement d'un satellite artificiel de la Terre. Bull. Inst. Astronomie théor. Leningrad **7**, 811 (1960).
11. P. Musen, Application of Hansen's Theory to the Motion of an Artificial Satellite in the Gravitational Field of the Earth. J. Geophysic. Res. **64**, 2271 (1959).
12. Y. Kozaï, The Motion of a Close Earth Satellite. Astronom. J. **64**, 367 (1959).
13. R. H. Merson, The Motion of a Satellite in an Axi-symmetric Gravitational Field. Geophysic. J. **4**, 17 (1961).
14. T. E. Sterne, The Gravitational Orbit of a Satellite of an Oblate Planet. Astronom. J. **63**, 28 (1958).
15. B. Garfinkel, The Orbit of a Satellite of an Oblate Planet. Astronom. J. **64**, 353 (1959).
16. H. von Zeipel, Arkiv för Matematik, Astronomi och Fysik **11**, 1 (1916).
17. D. Brouwer, Solution of the Problem of Artificial Satellite Theory Without Drag. Astronom. J. **64**, 378 (1959).
18. P. Musen, Contributions to the Theory of Satellite Orbits. Proceedings of the First International Space Science Symposium, Nice, p. 437 (1960).
19. L. G. Jacchia, The Effect of a Variable Scale Height on Determination of Atmospheric Density from Satellite Acceleration. Smithsonian Institution Astrophysical Observatory Special Report no. 46, p. 1 (1960).
20. T. E. Sterne, Effect of the Rotation of a Planetary Atmosphere Upon the Orbit of a Close Satellite. ARS Journal **29**, 777 (1959).
21. L. G. Jacchia, A Variable Atmospheric Density Model from Satellite Accelerations. Smithsonian Institution Astrophysical Observatory Special Report no. 39, p. 1 (1960).
22. L. G. Jacchia, The Atmospheric Drag of Artificial Satellites During the November 1960 Events. Proceedings of the Second International Space Science Symposium, Florence (1961) — sous presse.
23. D. Brouwer et G. Hori, The Drag Problem Treated with Delaunay Variables. Notes of the Summer Institute in Dynamical Astronomy. Edité par Mac Donnel Aircraft Co., Saint Louis, p. 247 (1961).
24. P. Musen, The Influence of the Solar Radiation Pressure on the Motion of an Artificial Satellite. J. Geophysic. Res. **65**, 1391 (1960).

25. Y. KOZAÏ, Effects of Solar Radiation Pressure on the Motion of an Artificial Satellite. Smithsonian Institution Astrophysical Observatory Special Report no. 56, p. 25.
26. G. HORI, The Motion of an Artificial Satellite in the Vicinity of the Critical Inclination. Astronom. J. **65**, **291** (1960).
27. B. GARFINKEL, On the Motion of a Satellite in the Vicinity of the Critical Inclination. Astronom. J. **65**, **624** (1960).

Discussion

M. MORANDO demande quelle est celle, parmi les théories citées, qui serait la plus aisée à programmer sur des machines électroniques. M. KOVALEVSKY pense que c'est celle de DELAUNAY dont la procédure se répète indéfiniment à chaque pas, contrairement à celle de VON ZEIPEL qui se traite différemment à chaque approximation. Il souligne le fait qu'on peut, ainsi qu'il l'a fait sur des machines moyennes comme l'IBM **650**, programmer les calculs algébriques littéraux qui entrent dans la résolution des équations de la mécanique et qu'il serait bon qu'un grand calculateur soit employé à de telles recherches.

M. MACÉ demande des précisions sur l'importance du frottement atmosphérique dans l'ensemble de la théorie d'un satellite artificiel. Il apparaît, dans la discussion qui suit, qu'elle devient prépondérante en deçà de **300** kilomètres. M. KOVALEVSKY fait remarquer qu'on utilise une formule de frottement en V^2, ce qui conduit à la définition d'une densité dynamique qui, suivant les conditions où se font les chocs avec les molécules, peut ne pas correspondre à la densité physique.

D'une discussion engagée par M. DÉSVEAUX, il se dégage l'idée générale que dans les problèmes de satellites — contrairement aux problèmes de rentrée — on ne tient pas compte de la portance. Il est probable que cet effet soit difficile à dégager des autres effets secondaires.

The Effect of Atmospheric Oblateness on a Satellite Orbit

By

D. G. King-Hele[1]

(With 9 Figures)

Abstract — Zusammenfassung — Résumé

The Effect of Atmospheric Oblateness on a Satellite Orbit. The most serious perturbations to the orbit of an earth satellite are usually those caused by air drag. The simplest equations to specify the evolution of an orbit under the action of air drag are given; then, after a survey of the upper atmosphere which indicates that the atmosphere shares the ellipticity of the earth beneath it, the corresponding equations for an oblate atmosphere are formulated. Finally, numerical results are given which show that the effect of atmospheric oblateness can alter the spherical-atmosphere results by up to 30% for some orbits, though for the majority of orbits the effect is much smaller, and 5% would be a more representative figure.

Der Einfluß der atmosphärischen Abplattung auf Satellitenbahnen. Die bekannteste Bahnstörung wird durch den Luftwiderstand hervorgerufen. Stark vereinfachte Gleichungen für die Bahn eines Satelliten unter Berücksichtigung des Luftwiderstandes werden angegeben, wobei nachher auch die atmosphärische Abplattung berücksichtigt wird. Schließlich werden numerische Resultate angegeben, die zeigen, daß der Effekt der atmosphärischen Abplattung Änderungen der Bahn gegenüber der kugelsymmetrischen Atmosphäre bis zu 30% bewirken kann. Allerdings ist für den Großteil der Bahnen der Effekt viel kleiner und kann größenordnungsmäßig etwa mit 5% angenommen werden.

Effet de l'aplatissement de l'atmosphère sur les orbites des satellites. Les perturbations les plus sérieuses du mouvement orbital ont été celles causées par la résistance de l'air, qui enlève au satellite un peu d'énergie à chaque révolution, pour éventuellement anéantir le mouvement orbital. Du fait que la résistance de l'air est plus grande dans la section de l'orbite proche du périgée, son effet principal est de retarder le satellite chaque fois qu'il passe au périgée, avec comme résultat que le satellite ne va pas être lancé aussi loin dans le côté opposé de la terre, qu'à sa révolution précédente. Des résultats numériques sont fournis montrant comment la contraction de l'orbite sous l'influence du frottement de l'air peut être affectée par l'aplatissement de l'atmosphère. Les résultats obtenus avec l'hypothèse d'une atmosphère sphérique sont peu modifiés par l'introduction de l'hypothèse de son aplatissement; toutefois des changements atteignant 30% peuvent survenir pour des orbites particulièrement sensibles à ce phénomène.

I. Introduction

For the majority of artificial satellites so far launched, the most serious orbital perturbations have been those caused by air drag, which deprive the satellite of a little energy on each revolution and eventually destroy its orbit altogether. The effects of air drag are most important for satellites with perigee

[1] Royal Aircraft Establishment, Farnborough, Hants, England.

heights lower than 300 km, but still need to be taken into account for perigee heights up to about 1,000 km; at heights above 1,000 km solar radiation pressure is greater than air drag. The results given in this paper therefore apply primarily to orbits with perigee heights of less than, say, 600 km; and the method is also limited to orbital eccentricities less than 0.2. Most of the satellites so far launched comply with these two conditions.

The other major perturbation to the orbit of a near satellite [1] is the effect of the earth's oblateness, which causes the orbital plane of the satellite to rotate about the earth's axis and makes the major axis of the orbit rotate in its own plane so that the perigee latitude changes. Neither of these two motions has any first-order effect on the orbital changes caused by air drag in a spherically-symmetrical atmosphere; but the change in the perigee latitude must be taken into account when the atmosphere is assumed to be oblate. The changes in perigee distance from the earth's centre caused by the odd harmonics in the earth's gravitational field amount to less than 10 km and can usually be allowed for separately.

Luni-solar perturbations [2] are very small for near-satellites with orbital eccentricity less than 0.2, and can also be considered separately.

In this paper the equations specifying the contraction of a satellite orbit under the influence of air drag are first given in their simplest form, in section II. Then, in section III, the properties of the upper atmosphere are briefly surveyed. In section IV the equations for an orbit contracting in an oblate atmosphere are stated, and section V gives numerical results.

II. General Effect of the Atmosphere on Elliptic Orbit, with Simplest Results

Since the air drag is greatest over the section of the orbit near perigee, where the ambient air is most dense, the chief effect of drag is to retard the satellite each time it passes perigee, with the result that the satellite does not swing out so far on the opposite side of the earth as it did on its previous revolution. The apogee height is therefore steadily reduced while the perigee height remains almost constant: the orbit contracts and becomes more nearly circular.

In deriving the simplest mathematical formulae to describe this contraction, we assume that the atmosphere is spherically symmetrical and does not vary with time. We also assume that the air drag D acts in the direction opposite to the satellite's velocity V relative to the ambient air and is expressible as

$$D = \tfrac{1}{2} \varrho V^2 S C_D, \tag{1}$$

where ϱ is the air density, S the mean cross-sectional area and C_D a drag coefficient. An uncontrolled satellite tends to rotate about an axis of maximum moment of inertia, and the direction in space of this axis changes only slowly. S is therefore taken as the mean of the cross-sectional areas in the two extreme modes of motion, which, for a long thin satellite, are tumbling end-over-end and rotating exactly like an aeroplane propeller. C_D remains almost constant during a satellite's life, its numerical value [3] being near 2.2. $S C_D$ is not likely to vary by more than about 10% even for satellites of large length/diameter ratio, and in developing the theory it is assumed constant.

We further assume that the air density ϱ varies exponentially with height y in the height-band of perhaps 100 km above perigee where drag is important, so that

$$\varrho = \varrho_p \exp\left(-\frac{y - y_p}{H}\right), \tag{2}$$

where suffix p denotes perigee and H is a constant, representing the height in which density falls off by a factor of 2.718.

Under these assumptions, it can be shown that, if the eccentricity e lies between 0.02 and 0.2, the variation of perigee distance r_p from the earth's centre with e is given [4], to the first order, by

$$r_p \simeq r_{p_0} - \frac{H}{2} \ln \frac{e_0}{e}, \tag{3}$$

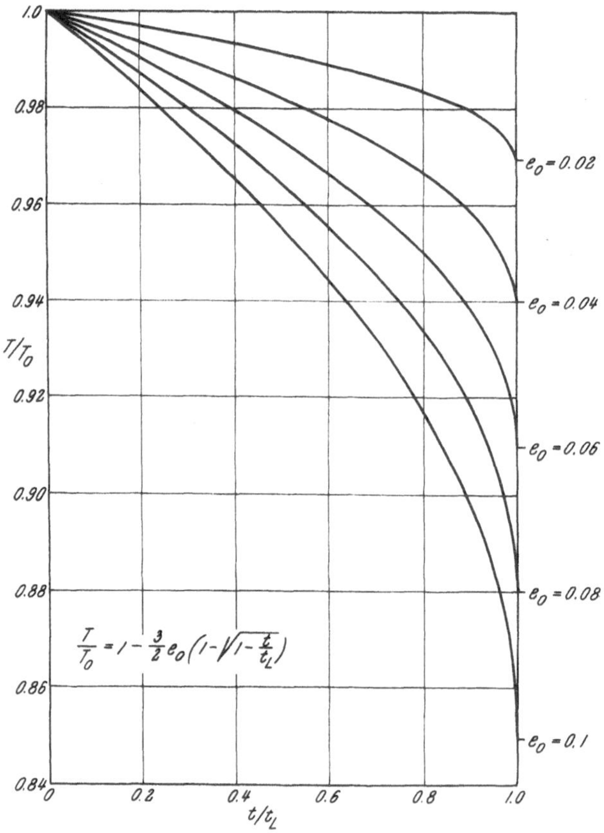

Fig. 1. Variation of orbital period with time as given by eq. (8). T period of revolution, T_0 initial value of T, e_0 initial eccentricity, t time, t_L life-time of satellite

where suffix o denotes initial values. Orbital period T can then be expressed in terms of e as

$$T \simeq T_0 \left(\frac{1 - e_0}{1 - e}\right)^{3/2}. \tag{4}$$

The variation of eccentricity with time t is given by

$$e \simeq e_0 \sqrt{1 - t/t_L}, \tag{5}$$

where t_L is a constant which represents, very nearly, the lifetime of the satellite. Hence the equation for period in terms of time is

$$T \simeq T_0 \left\{1 - \frac{3}{2} e_0 \left(1 - \sqrt{1 - t/t_L}\right)\right\}. \tag{6}$$

On differentiating (6) with respect to t and putting $t = 0$, we obtain a formula for lifetime in terms of the initial rate of change of T:

$$t_L \simeq -\frac{3\, e_0\, T_0}{4\, \dot{T}_0}. \tag{7}$$

These five equations provide the simplest mathematical description of how an elliptic satellite orbit ($e < 0.2$) contracts under the influence of air drag until the time, usually very near the end of its life, when e falls to 0.02.

Forgetting, for the moment, that these equations are approximate rather than exact, we see from eq. (5) that e^2 varies linearly with time, decreasing towards zero at the end of the satellite's life. Eq. (6) gives the characteristic form of a satellite's period-versus-time curve, with increasing slope towards the end, as shown in Fig. 1. Eq. (7) allows the lifetime to be found from the initial elements e_0, T_0 and the rate of change of period $\dot{T}_0$, which can usually be measured after a few days in orbit. The orbital period T of a satellite can be found simply and accurately, merely by timing its transits; eq. (4) is therefore particularly powerful because, if the initial orbit is known, the eccentricity e at any later time can be found from T. Since the semi major axis can also be found directly from T, the size and shape of the orbit are directly determinable from T. Once e is known, the change in perigee height can be found from (3), with error less than 10 km, since H is quite small, of order 50 km.

These simplest equations suffer from various inaccuracies, which can be eliminated by including smaller terms of order e, $H/a\,e$, etc. [5]. The assumption of a spherically symmetrical atmosphere is, however, questionable, and, in the next section, the properties of the upper atmosphere are briefly surveyed, with a view to specifying more realistic assumptions for the theory.

III. The Upper Atmosphere, at Heights from 200 to 700 km

It is now established [6—11] that the atmosphere at heights above 200 km responds vigorously to solar disturbances: there is a tendency towards the 27/28-day periodicity characteristic of solar influence [7]; day-to-day fluctuations in density, of up to $\pm$ 20% at 200 km height, are found to be linked with solar activity [6, 8, 9] (the density being highest when the sun is most active); exceptional solar outbursts lead to very great temporary increases in density [10]; and at heights above 400 km the density decreased considerably [11] (by a factor of $2\frac{1}{2}$ at 500 km) between 1958 (sunspot maximum) and 1960.

Furthermore, the air density is higher by day than by night. At heights of 200—300 km this effect is small, not more than 10%, but at heights above 400 km it is the dominant feature of the atmosphere. The maximum daytime density, attained 2 to 3 hours after noon, is 4 to 8 times higher than the minimum night-time density, attained between midnight and dawn, at heights of 500—600 km. These results are summed up in Fig. 2, which presents data obtained [11] from the orbits of 29 satellites.

The slopes of the curves in Fig. 2 give values of the quantity H defined in eq. (2). It is found [11] that H increases from about 45 km at a height of 200 km to about 90 km (by day) and 55 km (by night) at a height of 600 km. Since there is some scope for artistic licence in drawing the curves of Fig. 2, the value

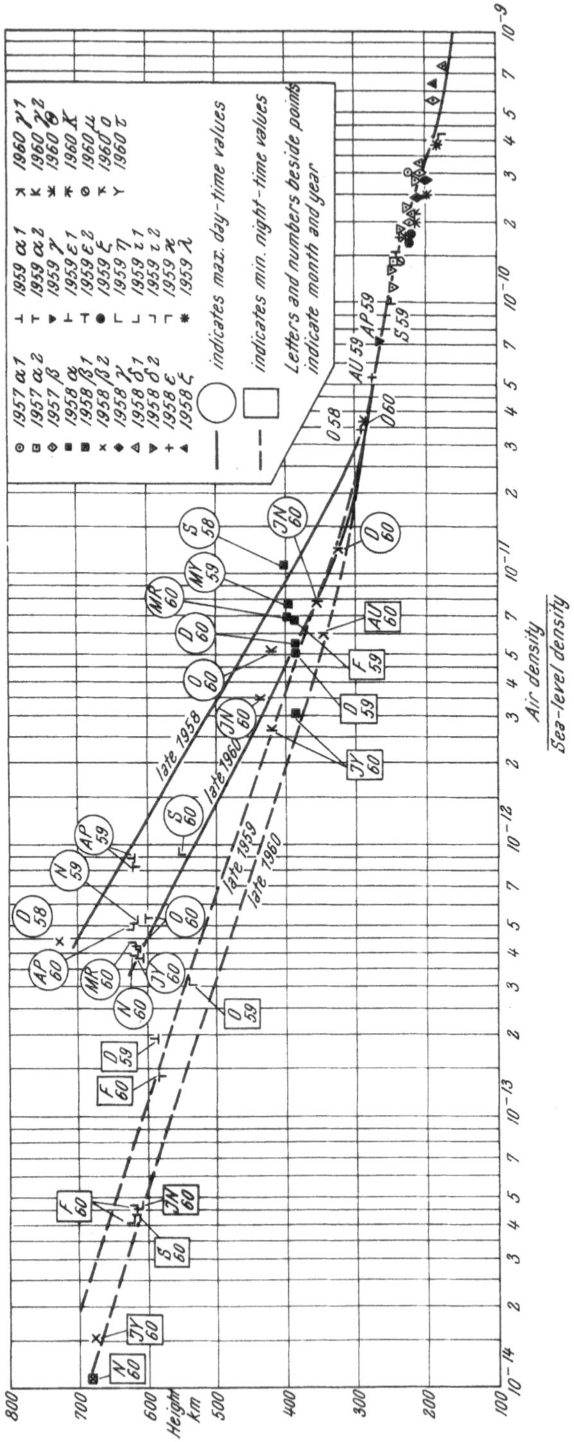

Fig. 2. Upper-atmosphere density obtained from the orbits of 29 satellites launched before January 1961

of H at any particular height is subject to an error (s. d.) of perhaps 15%. Consequently the variation of H with height cannot be exactly specified, and it is therefore not yet worth considering any more accurate assumption about the variation of density with height than eq. (2). The average value of H over a height-band of 50—100 km, and hence the value of H for use in (2), should, however, not be in error by more than about 10%.

The air density certainly varies between day and night, and fluctuates in response to solar activity: does it also vary systematically with latitude? The best evidence on this question has come from Sputnik 3, whose perigee moved slowly from 50° N to 65° S and back again to north of the equator. Fig. 3 shows the air density at a height of 235 km above the earth's surface as found from the orbit of this satellite [11]: the full line gives the calculated values, but a

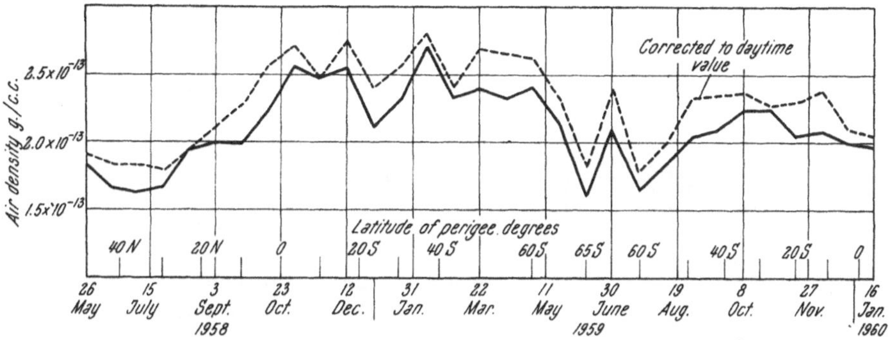

Fig. 3. Air density at height of 235 km, as given by Sputnik 3

better comparison is obtained from the broken line, for which the values are corrected to daytime conditions. From Fig. 3 we can conclude that the density at a height of 235 km did not depart from its average value by a factor of more than 1.25 in 1958—9, and most of this variation could well be ascribed to the influence of solar activity. Studies of Discoverer satellites [11—14] with perigees at latitudes up to 84°, have also shown no sign of any significant variation of density with latitude, for heights of 200—300 km.

In view of these results it is appropriate to assume that the air density does not vary with latitude, so that the ellipticity of the atmosphere, ε, may be taken equal to that of the earth [15], namely 0.00335.

The difference of density between day and night, and the variation in the course of the sunspot cycle, will inevitably affect satellite orbits. The variations of the orbital elements with time as given by theory, assuming density to be independent of time, are smooth curves; but in practice there will be oscillations superposed on these curves as the perigee experiences the day-to-night cycle and time progresses through a solar cycle. The theoretical curves are still necessary, however, to define the shape of the mean curves (i. e. when stripped of the oscillations), and to provide a measure of how much the air density has varied in the course of the fluctuations. The theoretical equations which do not involve time, such as (3) and (4), are of course unaffected by the time-variations in density.

IV. Theory of the Contraction of a Satellite Orbit in an Oblate Atmosphere

1. Introduction

If the atmosphere is assumed oblate, the effect of air drag on an orbit will vary according to the perigee latitude, since a given perigee distance from the earth's centre corresponds to a greater air density if perigee is at the equator than if perigee is near the poles. It is therefore necessary to specify the latitude of perigee, or, more precisely, the argument of perigee, ω, defined as the angle from the ascending node N to the perigee point P, measured along the orbit, as shown in Fig. 4. Under the influence of the earth's gravitational field, ω changes at a rate given by [1]

$$\dot{\omega} = 4.98(R/a)^{3.5}\,(5\cos^2 i - 1)\,\{1 + 0(e^2)\} \quad \text{degrees per day,} \tag{8}$$

where R is the earth's equatorial radius, a the semi major axis, and i the inclination of the orbit to the equator. Eq. (8) shows that ω is positive if $i < 63.4°$ and negative if $i > 63.4°$. Since a varies only slightly during a satellite's life, ω varies almost linearly with t; while e^2, and hence $(a\,e)^2 = x^2$, also varies almost linearly with t. One of the best simple approximations to ω is therefore

$$2\,\omega = A \pm G\frac{x^2}{x_0^{\,2}} \tag{9}$$

where A and G are constant, and the sign is chosen so that G is always positive: in practice this means that the plus sign is to be taken when $i > 63.4°$ and the minus sign when $i < 63.4°$.

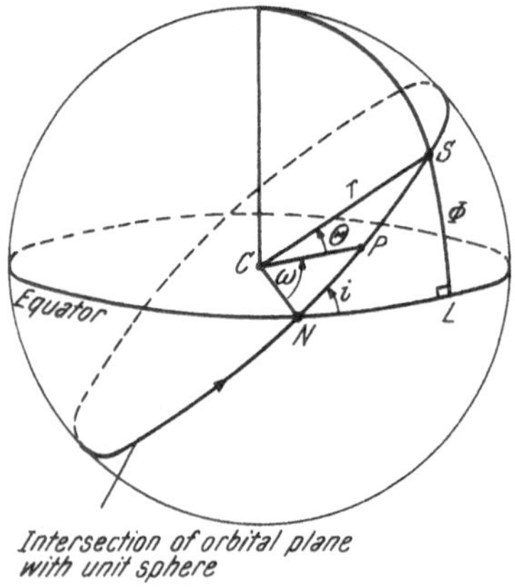

Fig. 4. Diagram of satellite orbit projected on unit sphere. P perigee point, S satellite position

Analytical formulae to specify the contraction of a satellite orbit under the action of air drag have been obtained [16] under the assumptions that:

(i) air density varies exponentially with height above perigee, as in eq. (2);

(ii) the surfaces of constant density are spheroids of fixed ellipticity ε, and density does not vary with time;

(iii) ω is given by eq. (9);

(iv) $e < 0.2$.

The equations take different forms according as e is greater or less than 0.02 (or, more strictly, according as $a\,e/H$ is greater or less than 3). The results for Phase 1, $0.02 < e < 0.2$, are given in section IV.2 and those for Phase 2, when $e < 0.02$ (or, strictly, $a\,e/H < 3$) are given in section IV.3.

2. Equations for Phase 1 $(0.02 < e < 0.2)$

During Phase 1, under the assumptions stated above, the equation for perigee distance in terms of eccentricity, given in its simplest form in eq. (3), is found to be

$$\frac{r_{p_0} - r_p}{H} = \frac{1}{2}\left\{\left(1 - \frac{5\,H}{2\,a_0}\right)\ln\frac{e_0}{e} - (e_0 - e)\left(1 - \frac{3\,H}{4\,a_0}\cdot\frac{1+e_0}{e\,e_0} - \frac{e+e_0}{2}\right) + \right.$$
$$\left. + \zeta + 0\left(\frac{a\,e^5}{H}, \frac{H^2}{a^2\,e}; \frac{a\,e\,c^4}{H}, \frac{c\,H^2}{a^2\,e^2}\right)\right\} \tag{10}$$

Here ζ represents the contribution of atmospheric oblateness and is given by

$$\zeta = \frac{2\,c\,H}{a_0\,e_0}\left[\cos 2\,\omega_0 - \frac{u_0}{u}\cos 2\omega + \pi\,u_0[\cos A\{S(u_0) - S(u)\} \pm \sin A\{C(u_0) - C(u)\}]\right], \tag{11}$$

where

$$c = \frac{\varepsilon\,r_{p_0}}{2\,H}\sin^2 i \tag{12}$$

$$\frac{u}{u_0} = \frac{e}{e_0}\{1 - e_0 + e + 0(e^2)\} \tag{13}$$

$$u_0 = \sqrt{2\,G/\pi} \tag{14}$$

and $C(u)$ and $S(u)$ are the FRESNEL integrals. The quantity c, which is a measure of the effect of atmospheric oblateness on the orbit, usually does not exceed 0.2 and can be regarded as of the same order of magnitude as e. If ω is constant, $G = 0$ and (11) can be written as

$$(\zeta)_{\omega\ \mathrm{const}} = \left(\frac{1}{e_0} - \frac{1}{e}\right)\varepsilon\sin^2 i\cos 2\,\omega. \tag{15}$$

The variation of T/T_0 with e is then given by

$$\frac{T}{T_0} = \left(\frac{1-e_0}{1-e}\cdot\frac{r_p}{r_{p_0}}\right)^{3/2}. \tag{16}$$

The equation expressing t/t_L in terms of e is extremely lengthy and here we only quote the simpler form for ω constant, which is

$$\left(\frac{t}{t_L}\right)_{\omega\ \mathrm{const}} = 1 - \lambda^2 - \frac{e_0\,\lambda^2}{3}(1-\lambda) + \frac{e_0^{\,2}\,\lambda^2}{144}(1-\lambda)(119 + 135\,\lambda) - \\ - \frac{3\,H\,\lambda^2}{4\,a_0}\ln\frac{1}{\lambda} - \xi + 0\left(\frac{a\,e^5}{H}, \frac{a\,e\,c^4}{H}; e^3, c^3; \frac{H\,e}{a}, \frac{H\,c^2}{a\,e}; \frac{H^2}{a^2\,e}\right), \tag{17}$$

where $\lambda = e/e_0$ and

$$\xi = \left\{8\,\lambda\,(1-\lambda) + \frac{6\,H}{a_0\,e_0}\ln\frac{1}{\lambda}\right\}\frac{\varepsilon}{2\,e_0}\sin^2 i\cos 2\,\omega. \tag{18}$$

The lifetime formula (7) becomes, for an oblate atmosphere, again with ω constant,

$$(t_L)_{\omega\ \mathrm{const.}} = -\frac{3\,e_0\,T_0}{4\,\dot{T}_0}\left\{1 + \frac{7}{6}e_0 + \frac{5}{16}e_0^{\,2} + \frac{H}{2\,a_0\,e_0}\left(1 + \frac{11}{12}e_0 + \frac{3\,H}{4\,a_0\,e_0} + \frac{3\,H^2}{4\,a_0^{\,2}\,e_0^{\,2}}\right) + \right. \\ \left. + \frac{2\,\varepsilon}{e_0}\sin^2 i\cos 2\,\omega + 0\left(\frac{a\,e^5}{H}, \frac{a\,e\,c^4}{H}; e^3, c^3; \frac{H\,e}{a}, \frac{H\,c^2}{a\,e}; \frac{H^2}{a^2\,e}, \frac{c\,H^2}{a^2\,e^2}\right)\right\}. \tag{19}$$

The air density at perigee, ϱ_p, is connected with the orbital elements and their rates of change by the equation

$$\varrho_p = -\frac{\dot{T}}{3\,\delta}\sqrt{\frac{2\,e}{\pi\,a\,H}}\left\{1 - 2\,e + \frac{5}{2}e^2 - \frac{H}{8\,a\,e}\left(1 - 10\,e + \frac{7\,H}{16\,a\,e}\right) + \right. \\ \left. + \frac{\varepsilon}{e}\sin^2 i\cos 2\,\omega + 0(e^3, c^3; H\,e/a, H\,c^2/a\,e; H^2/a^2\,e, c\,H^2/a^2\,e^2)\right\}, \tag{20}$$

where $\delta = F\,S\,C_D/m$, m being the mass of the satellite and F a factor which allows for the rotation of the atmosphere and usually [5] lies between 0.9 and 1.

3. *Equations for Phase 2* $(0 < e < 0.02)$

During Phase 2, when $a\,e/H$ is written as z and suffix 1 denotes initial values, the equation for r_p, with constant ω, is

$$\frac{r_{p_1} - r_p}{H} = \left[\left(1 - \frac{3H}{a_1}\right)\ln\frac{z_1 I_1(z_1)}{z\,I_1(z)} + \frac{2H}{a_1}\left\{\frac{z_1 I_0(z_1)}{I_1(z_1)} - \frac{z\,I_0(z)}{I_1(z)}\right\} - \right.$$

$$\left. - (z_1 - z) - \{\psi(z) - \psi(z_1)\}\right][1 + 0(c^4)], \tag{21}$$

where

$$\psi(z) = b\left\{\ln\frac{I_1(z)}{z^3} - \frac{z\,I_0(z)}{6\,I_1(z)} - \frac{z^2}{12}\right\}. \tag{22}$$

Here $I_n(z)$ is the BESSEL function of the first kind and imaginary argument, of order n and argument z, and

$$b = \frac{c}{2}\cos 2\,\omega. \tag{23}$$

T/T_1 can be found from (16), with suffix o replaced by suffix 1.

The variation of $t - t_1 = \tau$ with z, again with ω constant, is given by

$$\frac{\tau}{\tau_L} = \left\{1 - \frac{z^2}{z_1{}^2} - \frac{4\,b}{z_1 + 2\,b}\cdot\frac{z}{z_1}\left(1 - \frac{z}{z_1}\right)\right\}\{1 + 0(0.03)\}, \tag{24}$$

for $1 \leqslant z \leqslant 3$. For $0 \leqslant z \leqslant 1$, with suffix 2 denoting initial values and $\tau' = z - z_2$, we find

$$\frac{\tau'}{\tau_L{}'} = \left[1 - \left(\frac{z}{z_2}\right)^{2-2b}\left\{1 + \frac{28\,H}{3\,a_2}\left(1 - \frac{z}{z_2}\right)\right\}\right][1 + 0(0.02]). \tag{25}$$

The 0 (..) terms in these equations represent the maximum errors. The appropriate formula for lifetime, with ω constant or variable, is

$$\tau_L = -\frac{3\,e_1\,T_1}{4\,\dot{T}_1}\cdot\frac{I_0(z_1)}{I_1(z_1)}\left\{1 + 2\,e_1\frac{I_1(z_1)}{I_0(z_1)} - \frac{5\,e_1}{12} + \frac{H}{2\,a_1} + c\frac{I_2(z_1)}{I_0(z_1)}\cos 2\,\omega_1 + 0(e^2, c^2/2)\right\}, \tag{26}$$

and the equation for air density at perigee is

$$\varrho_p = -\frac{\dot{T}}{3\pi\,a\,\delta}\,\frac{\exp(z + c\cos 2\,\omega)}{I_0 + 2\,e\,I_1 + c\,I_2\cos 2\,\omega + \tfrac{1}{4}c^2 I_0 + 0(e^2, c^4)}. \tag{27}$$

4. *Circular Orbits*

For circular orbits the argument of perigee ω becomes meaningless, and there can be no first-order effects of atmospheric oblateness because these always enter the equations via the parameter $c\cos 2\,\omega$. There are some second-order effects [16], but these are always very small.

V. The Effect of Atmospheric Oblateness: Numerical Results

1. *Introduction*

Eqs. (10)—(27) show that the effects of atmospheric oblateness usually depend upon several orbital parameters — e, i, ω and sometimes H/a — as well as the ellipticity of the atmosphere, ε. For Phase 1 the effects are usually best measured by the quantity $q = (\varepsilon/e)\sin^2 i\cos 2\,\omega$, and are therefore usually greatest when $i = 90°$, $\omega = 0, 90°, \ldots$, and e has its minimum value of 0.02, so that $q_{max} = 50\,\varepsilon$. The "average" value of q, when, say, $i = 45°$, $\cos 2\,\omega = 2/\pi$

and $e = 0.1$, is however much less, $q_{av.} = 3\,\varepsilon$. This wide disparity between the maximum and "average" values makes it difficult to generalize about the results, and in this section both maximum and average values will be indicated, the average values being obtained by assigning each relevant parameter approximately its arithmetic mean value.

2. *Results for Phase 1* $(0.02 < e < 0.2)$

The variation of $(r_{p_0} - r_p)/H$ with e/e_0 for a spherical atmosphere in Phase 1, as given by (10) with $\zeta = 0$, is shown in Fig. 5. The correction to be applied to allow for atmospheric oblateness depends on a large number of parameters,

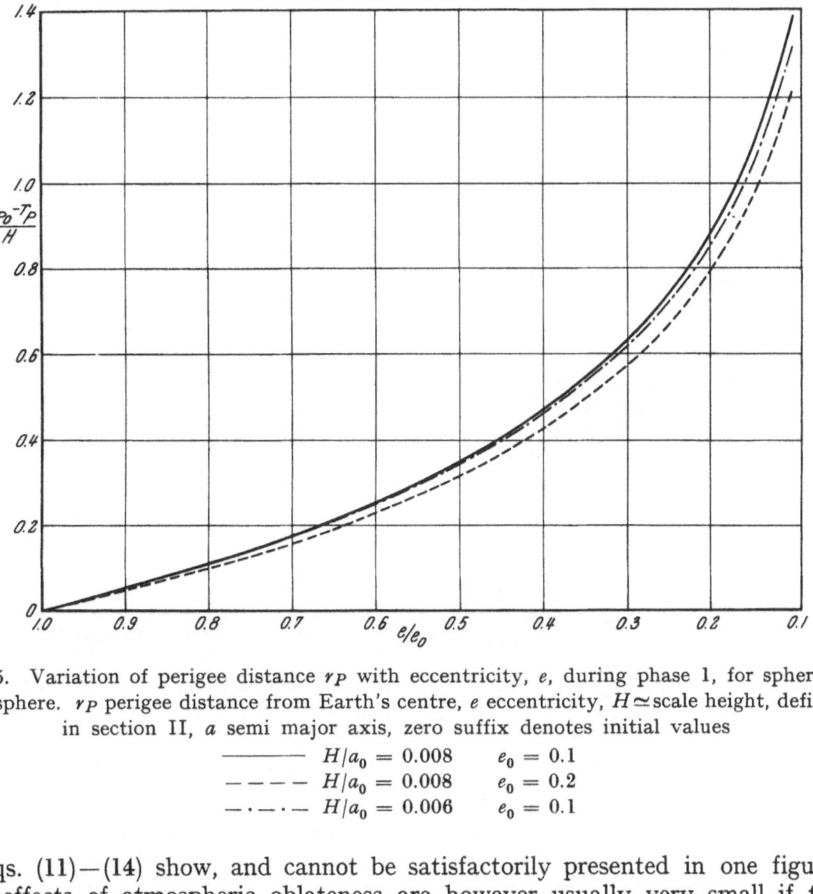

Fig. 5. Variation of perigee distance r_P with eccentricity, e, during phase 1, for spherical atmosphere. r_P perigee distance from Earth's centre, e eccentricity, $H \simeq$ scale height, defined in section II, a semi major axis, zero suffix denotes initial values

———	$H/a_0 = 0.008$	$e_0 = 0.1$
– – – –	$H/a_0 = 0.008$	$e_0 = 0.2$
—·—·—	$H/a_0 = 0.006$	$e_0 = 0.1$

as eqs. (11)—(14) show, and cannot be satisfactorily presented in one figure. The effects of atmospheric oblateness are however usually very small if the perigee makes many revolutions about the earth, i.e. if G in eq. (9) and hence u_0 in eq. (14) are large. The most interesting results are therefore those for small values of u_0, and Fig. 6 gives the variation of ζ with u/u_0 for $u_0 = 2$, which corresponds to a change in ω of about 180° while e is being reduced to 1/5 of its initial value.

The maximum value of ζ in Fig. 6 occurs as $u \to 0$. But $u/u_0 (\simeq e/e_0)$ cannot be less than 0.1, since $e > 0.02$ and $e_0 < 0.2$. So the maximum value of ζ occurs at $u/u_0 = 0.1$, and can only arise if e_0 is near 0.2. From Fig. 6 $a_0\, e_0\, \zeta / c\, H$ has a maximum value of about 15 for $u/u_0 = 0.1$, so that $a_0\, \zeta / c\, H \simeq 75$. Now, from eq. (12), $c\,H/a_0$ is equal to $(\varepsilon r_{p_0}/2\,a_0)\sin^2 i$, which is less than $(\varepsilon/2)\sin^2 i$. So the maximum value of ζ which can arise (for $u_0 = 2$) is $(75/2)\,\varepsilon \sin^2 i$, or, taking a polar orbit ($i = 90°$) and $\varepsilon = 0.00335$, $\zeta_{max} = 0.13$. The change in the spherical-atmosphere value of $(r_{p_0} - r_p)$ due to atmospheric oblateness, which, from (10), is $H\,\zeta$, is therefore never likely to exceed 13 km for $u_0 = 2$, since $H < 100$ km. And in these circumstances, with $e/e_0 \simeq 0.1$, the change in r_p/H due to a spherical atmosphere is, from Fig. 5, about 1.3, corresponding to a change in r_p of about 130 km.

The influence of atmospheric oblateness is at its maximum when ω is constant, for then ω can be kept at its "worst" value throughout the satellite's life, without any alleviation from averaging effects. So the constant-ω versions which have been given in the remainder of the equations of section IV are adequate to display the maximum effects of oblateness.

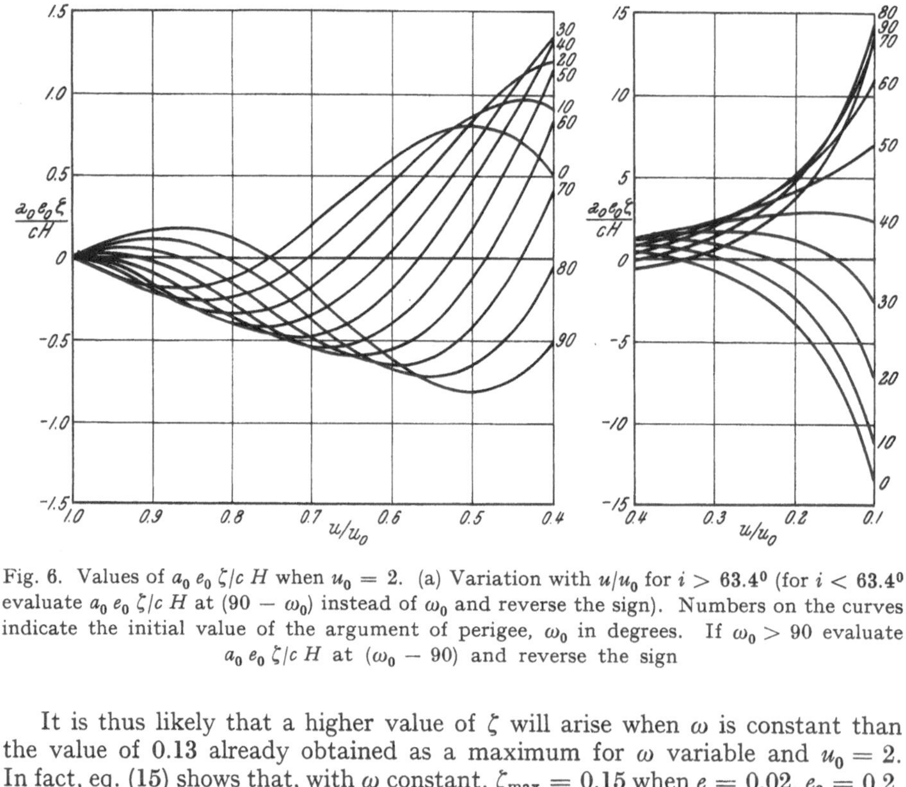

Fig. 6. Values of $a_0\,e_0\,\zeta/c\,H$ when $u_0 = 2$. (a) Variation with u/u_0 for $i > 63.4°$ (for $i < 63.4°$ evaluate $a_0\,e_0\,\zeta/c\,H$ at $(90 - \omega_0)$ instead of ω_0 and reverse the sign). Numbers on the curves indicate the initial value of the argument of perigee, ω_0 in degrees. If $\omega_0 > 90$ evaluate $a_0\,e_0\,\zeta/c\,H$ at $(\omega_0 - 90)$ and reverse the sign

It is thus likely that a higher value of ζ will arise when ω is constant than the value of 0.13 already obtained as a maximum for ω variable and $u_0 = 2$. In fact, eq. (15) shows that, with ω constant, $\zeta_{max} = 0.15$ when $e = 0.02$, $e_0 = 0.2$, $\cos 2\,\omega = -1$ and $i = 90°$, corresponding to a maximum change in perigee distance of 15 km. The main change due to spherical atmosphere is again 130 km, so there is a 12% change due to oblateness; but this percentage change can be much larger, up to 27%, when $e \simeq e_0$ and the absolute changes in perigee height are very small.

It should be emphasized that the values of ζ are usually very much less than the maxima quoted above. If each of the parameters has an average value — $i = 45°$, $\cos 2\,\omega = 2/\pi$, $e_0 = 0.1$ and $e = 0.05$ — eq. (15) gives $|\zeta| = 0.011$, less than 1/10 of its maximum value.

The variation of T with e is very little affected by atmospheric oblateness. For, since the perigee distance itself, which is always greater than 6500 km is never changed by more than 15 km as a result of atmospheric oblateness, the quantity $(r_p/r_{p_0})^{3/2}$ in eq. (16) cannot change by more than 0.35% and hence T/T_0 cannot be changed by more than 0.0035. For an "average" orbit the effect would be more than 10 times smaller. The spherical-atmosphere equation between T and e, as represented by (16) and (10) with $\zeta = 0$, provides a powerful method of determining eccentricity from period, which is frequently used in practice [17]: so it is fortunate that atmospheric oblateness has such a small effect.

Atmospheric oblateness can seriously affect the variation of eccentricity with time, as eqs. (17) and (18) show. t/t_L varies from 0 to 1 during the satellite's life, and the change in it due to atmospheric oblateness is given by ξ, which

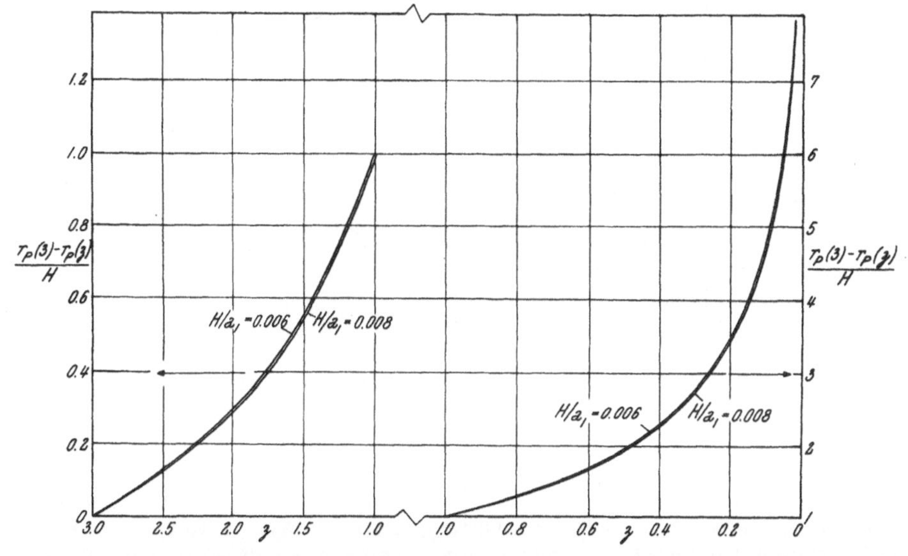

Fig. 7. Variation of perigee distance, r_P, with z, during phase 2, for spherical atmosphere

from eq. (18) can be as large as 0.1, if $e/e_0 = 0.5$, $a_0\, e_0/H = 6$, $\cos 2\,\omega = 1$, $i = 90°$, $e_0 = 0.045$. For most orbits, however, the effect is much smaller (an average value of ξ being about 0.01) and is subordinate to the changes caused by fluctuations in air density.

The lifetime formulae can also be greatly changed as a result of atmospheric oblateness, by up to 24% as eq. (19) shows, if $\cos 2\,\omega = 1$, $i = 90°$ and e_0 has its minimum value, which is taken as 0.022, since, if the initial e were any smaller than this, the orbit would be regarded as Phase 2 only. An average value for the change in t_L would be about 2%.

In the equation for air density the maximum changes due to atmospheric oblateness are half those in the lifetime formula, 12% maximum or 1% average.

3. *Results for Phase 2* $(0 < e < 0.02)$

The variation of r_p with z in a spherical atmosphere, as given by (21) with $\psi(z) = 0$, is shown in Fig. 7. In this Figure $r_p(3) - r_p(z)$ is plotted against z: to obtain $r_p(z_1) - r_p(z)$, one subtracts $r_p(3) - r_p(z_1)$ from $r_p(3) - r_p(z)$. The correction due to atmospheric oblateness to be subtracted from $(r_{p_1} - r_p)/H$, namely $\psi(z) - \psi(z_1)$, is given in similar form in Fig. 8.

The maximum value of $\psi(z) - \psi(3)$ occurs when b has its maximum value of about 0.15 and $z \to 0$. There is however a lower limit to the value of z: it can be assumed that z is never less than $z_1/6$, because if $z_1/z > 6$, the change in r_p

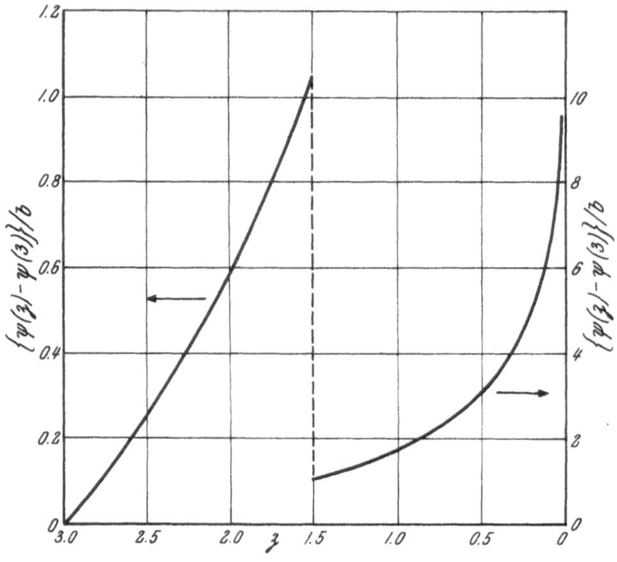

Fig. 8. Variation of ψ, the correction to $(r_{P_1} - r_P)/H$ due to atmospheric oblateness, when ω is constant

is, from Fig. 7, greater than $2\,H$. For most satellites the lifetime ends before r_p has changed by as much as this; and, if not, the value of H would have changed so much that it would usually be necessary to split Phase 2 into two parts, each with a different value of H. Thus we may assume $z > z_1/6$, so that from Fig. 8, $\psi(z_1) - \psi(z) < 4\,b < 0.6$, and the maximum change in r_p due to atmospheric oblateness is $0.6\,H$ or at most 60 km. Relative to the main spherical-atmosphere effects, the maximum change in r_p due to atmospheric oblateness occurs when $z/z_1 \to 1$; all the changes are then very small, but the inclusion of atmospheric oblateness can alter the spherical-atmosphere value by up to 30%.

The change in T due to the atmosphere can also be altered by up to 30% when atmospheric oblateness is allowed for, if z/z_1 is near 1, i.e. T/T_1 could be changed from, say, 0.997 to 0.996. The maximum absolute change in T/T_1 due to oblateness occurs when z/z_1 is small and is $(3\,H/2\,a_1)\,\{\psi(z) - \psi(z_1)\}$ or at most about 0.012, so that T/T_1 might be changed from, say, 0.90 to 0.912.

These maximum changes in r_p and T due to atmospheric oblateness are far from negligible; but it should again be emphasized that they occur only for exceptional orbits, and in general the effects are very much smaller. For an "average" orbit, with $z_1 = 2$, $z = 1$, and $b = 0.05$, the change in r_p due to atmospheric oblateness is 6 km (as compared with the maximum of 60 km) and is 9% of the spherical-atmosphere effect. The corresponding change in T/T_1 is 0.0012.

The variation of z/z_1 with τ/τ_L, as given by (24), is plotted in Fig. 9. Again it is apparent that atmospheric oblateness can cause quite large effects. For a given z, τ/τ_L canbe changed by up to 0.11 at most, i. e. the time at which a given eccentricity is attained can change by 11% of the lifetime. An average value for the change would be about 3%.

The formula for lifetime (26) can be changed as a result of atmospheric oblateness by a maximum of 14% when z_1 is near 3, or an "average" of 3%, when $z_1 = 2$, $c = 0.1$ and $\cos 2\,\omega = 2/\pi$. The formula for density at perigee can be changed by a maximum of 36%, when z_1 is near zero, or an "average" of 3%.

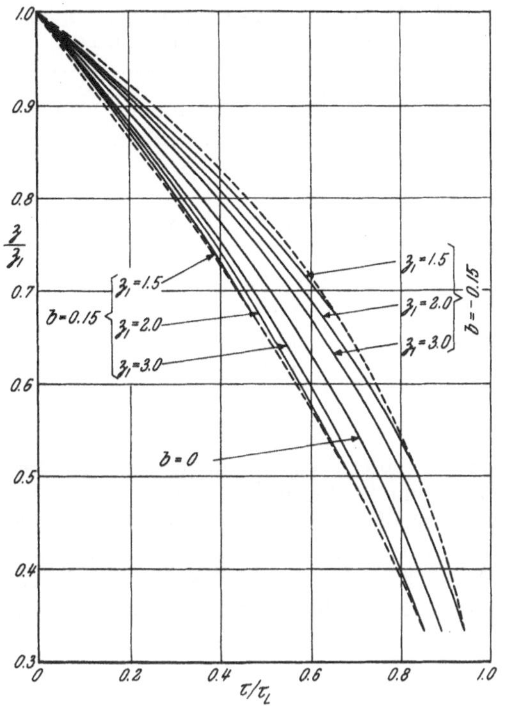

Fig. 9. Variation of z/z_1 with τ/τ_L for phase 2 and ω constant. $1 \leqslant z \leqslant 3$

VI. Conclusions

The evidence so far available indicates that at heights between 200 and 300 km, where the influence of air drag on satellite orbits is most important, and probably at greater heights too, atmospheric density is almost independent of latitude, so that the atmosphere shares the ellipticity of the earth beneath it.

The effect of this atmospheric oblateness on the contraction of a satellite orbit varies widely according to the values of the orbital parameters. If the eccentricity e exceeds 0.02, the quantity $(\varepsilon/e) \sin^2 i \cos 2\,\omega$ provides a measure of the effect, so that, if the argument of perigee ω completes several revolutions, the net effect of atmospheric oblateness is usually very small since the mean value of $\cos 2\,\omega$ is small. The effects are also small, of course, for near-equatorial orbits, with i small. If, however, e is near 0.02, i is large and the average value of $\cos 2\,\omega$ is near 1, atmospheric oblateness has an important effect on nearly all the orbital parameters. Table I summarizes the numerical values of the effects, as given in section V.

In this Table the "average" values refer to conditions when e and i have average (arithmetic mean) values and $\cos 2\,\omega$ has its average numerical value $2/\pi$, which would be appropriate for an orbit for which ω changed from say 45° to 135°. If, however, ω completes several revolutions, the mean value of $\cos 2\,\omega$ will be much lower and the effects of atmospheric oblateness correspondingly smaller. The maximum absolute values in the Table are usually attained when e is very small, whereas the maximum percentage changes usually occur when e is near its initial value; so, with the maximum values, the two halves of the Table refer to different situations.

It is again difficult to generalize about the results in this Table, but there are a few conclusions worth emphasizing. First, the variation of period T with e during Phase 1 is almost unaffected by atmospheric oblateness, and even in

Table I. *Maximum Changes in r_p, T, t, t_L and ϱ_p Due to Atmospheric Oblateness, for Any Given Eccentricity; and Average Values*

Parameter	Absolute change				Change, as % of sph. atm. effect			
	Phase 1		Phase 2		Phase 1		Phase 2	
	Max.	Av.	Max.	Av.	Max.	Av.	Max.	Av.
$r_{p_0} - r_p$	15 km	1 km	60 km	6 km	27%	3%	30%	9%
T/T_0 (or T/T_1)	0.0035	0.0003	0.012	0.0012	4%	0.3%	30%	2%
t/t_L (or τ/τ_L)	0.10	0.01	0.11	0.03	—	—	—	—
t_L (or τ_L)	0.24 t_L	0.02 t_L	0.14 τ_L	0.03 τ_L	—	—	—	—
ϱ_{p_0} (or ϱ_{p_1})	—	—	—	—	12%	1%	36%	3%

Phase 2 the effect is small. Second, the formula for the variation of time with e, and that for lifetime t_L, can be seriously affected by atmospheric oblateness, a change of up to 24% in the lifetime being possible. Third, the spherical-atmosphere formula for obtaining air density at perigee from satellite orbits [5], which has been widely used, should be corrected for the effects of atmospheric oblateness, which may alter it by up to 12% for $e > 0.02$ and up to 36% for $e < 0.02$. This correction was made when calculating the points in Fig. 2, which were obtained from the formulae given in [11] for evaluating atmospheric density, at an optimum height and in the presence of atmospheric oblateness.

References

1. D. G. King-Hele, The Effect of the Earth's Oblateness on the Orbit of a Near Satellite. Proc. Roy. Soc. A **247**, 49–72 (1958).
2. G. E. Cook, Luni-Solar Perturbations of the Orbit of an Earth Satellite. Geophys. J. Roy. Astronom. Soc. (to be published).
3. G. E. Cook, The Aerodynamic Drag of Near Earth Satellites. Ministry of Aviation A.R.C. Current Paper C.P.523. London: Her Majesty's Stationery Office, 1960.
4. D. G. King-Hele and D. C. M. Leslie, Effect of Air Drag on the Orbit of Sputnik 2: Comparison of Theory and Observation. Nature **181**, 1761–1763 (1958).
5. G. E. Cook, D. G. King-Hele, and D. M. C. Walker, The Contraction of Satellite Orbits Under the Influence of Air Drag. I. With Spherically Symmetrical Atmosphere. Proc. Roy. Soc. A **257**, 224–249 (1960).
6. T. R. Nonweiler, Effect of Solar Flares on Earth Satellite 1957 beta. Nature **182**, 468–469 (1958).
7. D. G. King-Hele and D. M. C. Walker, Irregularities in the Density of the Upper Atmosphere: Results from Satellites. Nature **183**, 527–529 (1959).
8. L. G. Jacchia, Corpuscular Radiation and the Acceleration of Artificial Satellites. Nature **183**, 1662–1663 (1959).
9. W. Priester, H. A. Martin, and K. Kramp, Earth-Satellite Observations and the Upper Atmosphere. Nature **188**, 200–204 (1960).
10. L. G. Jacchia, Satellite Drag During Events of November, 1960. Paper presented at 2nd International Space Science Symposium, Florence, 1961.

11. D. G. KING-HELE and D. M. C. WALKER, Upper-Atmosphere Density During the Years **1957** to **1961**, Determined from Satellite Orbits. Paper presented at 2nd International Space Science Symposium, Florence, **1961**.
12. D. G. KING-HELE and D. M. C. WALKER, Variation of Upper-Atmosphere Density with Latitude and Season: Further Evidence from Satellite Orbits. Nature **185**, **727**–**729** (**1960**).
13. G. V. GROVES, Latitude and Diurnal Variations of Air Densities from **190** to **280** km as Derived from the Orbits of Discoverer Satellites. Proc. Roy. Soc. A **263**, **212**–**216** (**1961**).
14. S. H. LANDBERG, Densities of the Upper Atmosphere, Derived from Discoverer Satellites. ARS Journal **31**, **155**–**157** (**1961**).
15. D. G. KING-HELE, The Earth's Gravitational Potential, Deduced from the Orbits of Artificial Satellites. Geophys. J. Roy. Astronom. Soc. **4**, **3**–**17** (**1961**).
16. D. G. KING-HELE, G. E. COOK, and D. M. C. WALKER, The Contraction of Satellite Orbits Under the Influence of Air Drag. Part II. With Oblate Atmosphere. Ministry of Aviation Report (**1960**) (to be published).
17. D. G. KING-HELE and D. M. C. WALKER, Methods for Predicting the Orbits of Near Earth Satellites. J. Brit. Interplan. Soc. **17**, **2**–**14** (**1959**).

Discussion

Mr. STERN: Do charged particles have any effect on the values of density you obtain?

Mr. KING-HELE: The effect of charged particles is believed to be small at heights between 200 and 700 km, and, at most, should not alter the results by more than 10% at 700 km and much smaller amounts at lower altitudes. At heights above 1000 km these effects are, however, more important.

Mr. KOVALEVSKY: Quelle est la valeur numérique de l'altitude de référence *H* que vous avez donné dans votre formule?

Mr. KING-HELE: *H* increases from about 40 km at a height of 200 km to about 55 km by night or about 100 km by day at a height of 600 km.

Mr. STERN: Now that you have done so many calculations can you take into account the shape of the earth?

Mr. KING-HELE: Yes, we can measure the density as a function of height above the real, oblate earth. The indications are that the contours of constant density have roughly the same ellipticity as the earth beneath.

Mr. KOOY: Does the drag coefficient vary round the orbit of the satellite?

Mr. KING-HELE: It does vary slightly with the molecular speed ratio and also with the temperature of the satellite. But both these effects are small and we think it can legitimately be taken as constant.

Mr. KOOY: How does the axis of rotation vary its direction in space?

Mr. KING-HELE: For satellites which are uncontrolled, the direction of the axis in space moves very slowly, over a period of months. The way in which it moves depends on whether it is affected more by the atmosphere or by the gravitational field. But in either case the change from one revolution to the next is very small.

Problems and Potentialities of Space Rendezvous

By

John C. Houbolt[1]

(With 14 Figures)

Abstract — Zusammenfassung — Résumé

Problems and Potentialities of Space Rendezvous. The paper gives an analysis of rendezvous operations in space, as in the soft joining of a ferry vehicle to an orbiting space station or other target, and explores the potentiality of rendezvous in accomplishing certain space missions.

The significant phases of rendezvous are examined in a sequential sense, in which problems pertinent to each phase are assessed and results of recent research studies are given. Discussed are such aspects as appropriate injection or launch trajectories for rendezvous, suitable launch times, use of midcourse guidance logic, and manned and automatic terminal guidance schemes. Attention is also given to the use of special purpose orbits, such as parking orbits, to corrective maneuvers, to fuel consumption and penalties, and to the use of different sensors (electronic, optical, etc.) in performing the various phases. The paper concludes with an examination of the mission capabilities and benefits that may be brought about through use of rendezvous.

Probleme und Möglichkeiten des Zusammentreffens im Raum. Die vorliegende Arbeit bringt eine Untersuchung derartiger Operationen; die kennzeichnenden Phasen werden behandelt und Resultate gegenwärtiger Untersuchungen angegeben. Untersucht werden unter anderem Startbahnen, geeignete Startzeit, Kurskorrekturen, bemannte und automatische Lenkung. Auch spezielle Bahnen, wie Parkbahnen für Korrekturmanöver, zum Auftanken, für Unglücksfälle usw., werden behandelt.

Problèmes et utilisations des rendez-vous dans l'espace. On donne une analyse des opérations de rendez-vous dans l'espace, comme la rencontre d'un véhicule de ravitaillement avec une station spatiale ou tout autre objectif et l'on examine les possibilités offertes par cette technique dans l'accomplissement de certaines missions spatiales.

Les phases significatives d'un rendez-vous sont examinées dans leur ordre naturel. On examine les problèmes posés à chaque phase et l'on donne les résultats des recherche récentes à leur sujet. L'auteur traite de questions telles que celle de la mise en orbite et des trajectoires de départ qui conviennent le mieux au rendez-vous, de l'heure du lancement, de l'emploi d'une correction à mi-course, des systèmes de guidage terminaux pilotés ou automatiques, ainsi que des problèmes posés par l'accostage. Sont également examinés l'emploi d'orbites spéciaux pour des buts particuliers, tels que des orbites de stationnement, la correction des maneuvres, le pourcentage de propergol consommé et ses conséquences, ainsi que l'emploi des différents dispositifs (électronique, optique, etc.) employés pour l'accomplissement de chaque phase. L'auteur termine par l'examen des bienfaits et des nouvelles possibilités que l'utilisation des rendez-vous apportera dans les missions spatiales.

---

[1] Associate Chief, Dynamic Loads Division, NASA-Langley Research Center, Langley Field, Virginia, U.S.A.

I. Introduction

This paper discusses some of the basic results that have been found in studies of the problem of rendezvous in space, involving for example the ascent of a satellite or space ferry so as to make a soft contact with another satellite or space station already in orbit. Rendezvous is of interest because it may have a vital and widespread use in many future space missions; indeed, it may make reality certain missions which are not otherwise possible. Uses already envisioned include:

Assembly of orbital units

Perform space missions with smaller launch vehicles

Personnel transfer

Rescue

Retrieval

Proper placement of special purpose satellites (24-hour orbiter, communications satellites)

Inspection

Interception

Much information on the feasibility and technical aspects of rendezvous has been generated in the past few years, as is attested by the enclosed list of references involving 59 entries which date back essentially to only 1958; and the list is by no means complete.

The main intent in this report is to highlight some of the basic advances that have been made in the understanding of rendezvous. Emphasis is given to launch timing, trajectories, guidance, and basic rendezvous schemes. Hardware items such as attitude control systems, propulsion rockets, sensors are not discussed, other than to assume they exist.

It will be noted that a large portion of the material presented is drawn from the research studies conducted at the National Aeronautics and Space Administration, primarily because of the ease of accessibility. It is remarked, however, that parallel-type work elsewhere indicates the same general findings. In this connection, and with reference to the list of references, I would like to mention that in case someone's work has been overlooked, or if proper credit has not been given or a name not recognized, no slight is intended. It simply reflects the fact that it is now rather difficult to keep up with all the literature on a subject.

Symbols

G NEWTON's universal gravitation constant
m mass of the ferry
M mass of the earth
n revolutions in chasing or parking orbit
R range between ferry and target
$\dot{R}$ range rate
$\bar{T}$ thrust vector of ferry
V_L velocity of injection at end of booster burnout
ΔV_R relative velocity between ferry and target at intercept
γ_L trajectory angle at end of booster burnout referred to local horizontal
θ_L angular position between ferry and target at instant of ferry launch
ω angular rotation rate of satellite in circular orbit

II. Rendezvous Phases

Fig. 1 depicts the commonly adopted phases of rendezvous that would be involved in launching a space ferry from the earth's surface so as to make contact with an orbiting payload or space station, hereinafter referred to as the target. In the injection phase the intent is to place the ascending vehicle as close as possible to the space station; midcourse guidance may be employed in this

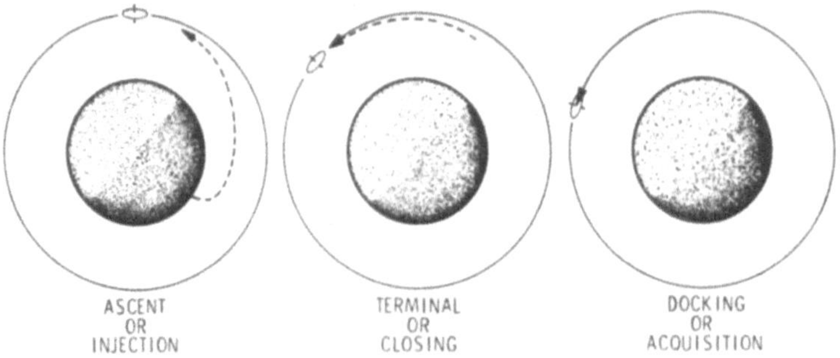

Fig. 1. Phases of rendezvous

phase. The next or terminal phase is the phase of closing on the target, involving usually only a fraction of an orbital revolution. During this phase an automatic system or the pilot of a manned vehicle may be virtually driving the craft to the station, as in the approach of a boat to a dock. The final phase, the docking or acquisition phase, involves the actual "soft" contact — the securing of lines, latches, air locks, etc.

A. Launch and Ascent

Injection Techniques

Five basic injection techniques may be employed in rendezvous as depicted in Fig. 2.

1. In-plane and Compatible Orbit. In this scheme the ferry is launched so as to ascend essentially in the orbit plane of the target as the target approaches. In general, long wait times (a month or more) may be involved before the target arrives in a position that will avoid prohibitive fuel consumption for rendezvous, see [1]. To avoid these long wait times the concept of compatible orbits has been advanced, in which the period (or altitude) of the target's orbit is chosen so that the target passes directly overhead at least once and perhaps twice each day; here the rotation of the target, precession of the orbit plane, and the earth's rotation are all considered simultaneously. Perhaps the most complete treatment that has been made on the use of compatible orbits is that of Swanson and Petersen [2].

2. Parking Orbit. Another way to avoid long wait times for in-plane launchings is to make use of parking orbits. The ferry is boosted to some sub-orbit and there gains on the target's angular position during each revolution because of the shorter orbital period. After making up the difference it is boosted out of the sub-orbit to rendezvous with the target. Some results pertaining to this technique will be given subsequently.

3. Out-of-plane Adjacency. In this scheme the launch site is in general out of target's orbital plane; the ferry is launched so that at the end of injection it is traveling adjacent to the target, at the same velocity and altitude but in an orbital plane of slightly different inclination. At the nodal point where the two planes cross (about a quarter of a revolution later) the ferry's direction is changed to coincide with that of the target. Results in [1] indicate that for launchings which are nearly East-West, rendezvous at least once a day is feasible, and further that rendezvous launchings on from **3** to **4** successive target passes are possible.

4. "Two Impulse" Scheme. Here the ferry is launched in general out of the target's orbit plane so that it follows a ballistic trajectory and meets the target just at the apogee or maximum altitude of the ballistic path. At this point another "impulse" is given to simultaneously change the direction and increase

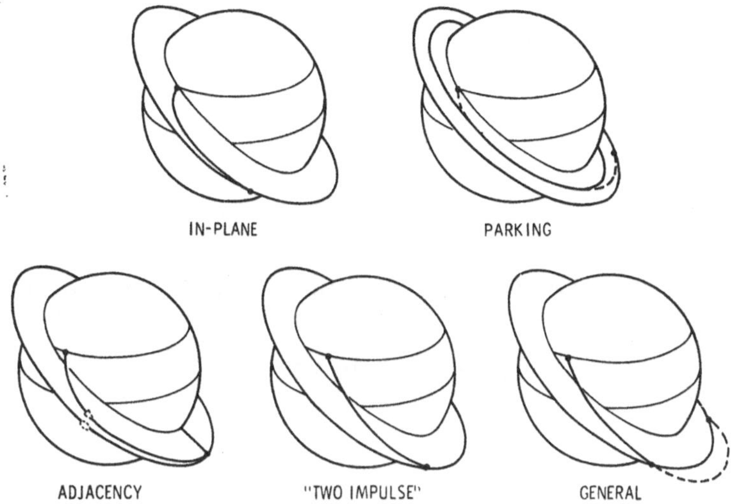

Fig. 2. Rendezvous ascent trajectories

the speed of the ferry so that its motion coincides with that of the target. Bird and Thomas [3] have advanced this scheme and show, as in the previous case, that once-a-day rendezvous is quite feasible by this method, and that three to four successive chances covering an interval of several hours are offered each day if launches of the West-East type are used. The scheme has the advantage over the previous adjacency scheme in that the plane change is made at sub-orbital speeds (the apogee point), thus realizing a substantial saving in fuel expenditure.

5. General Injection Technique. All of the previous four cases are really sub-cases of this method. In this scheme, extensively investigated by Eggleston [**4, 5, 6**] for both circular and elliptic orbits, the ferry is launched (in general, out of the target's orbit plane) with a correct combination of injection angle and velocity at booster burnout so that it follows a Keplerian type trajectory thereafter to intercept the target. At interception the necessary impulse to make direction and speed coincide is then applied. Eggleston [**5**] also studied the use of midcourse guidance during the coast trajectory phase to correct for errors which may be present at booster burnout or due to errors in the sensing equipment used. Some results are given in the next section.

Launch and Rendezvous Windows for Direct Ascent Trajectories

The left sketch on Fig. 3 shows a planar projection of the general trajectories that are possible in a direct ascent for rendezvous. At booster burnout the ferry is considered to be injected with a speed V_L and an injection angle γ_L. Depending

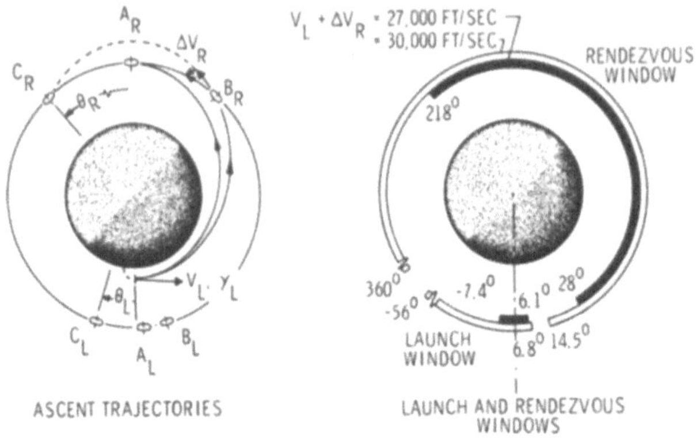

Fig. 3. Ascent paths and "windows" of rendezvous

on the combination used for these quantities three basic types of rendezvous may be made. With the target at A_L rendezvous occurs at A_R, and in this case

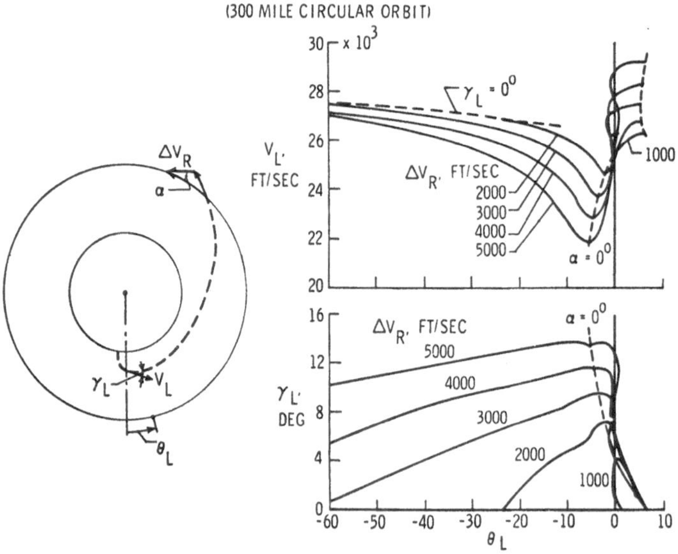

Fig. 4 a. Launch conditions for rendezvous. For ΔV_R of 1,000—5,000 ft/sec

the trajectory followed is a HOHMANN type ellipse or minimum energy path. For the target at B_L, rendezvous occurs on the outgoing leg at B_R, whereas for C_L, rendezvous occurs on the return leg at C_R. The maximum spread between B_L and C_L is determined by the performance capability of the system, and rendezvous may be achieved whenever the target is located within this spread.

It has been common therefore to call the maximum arc possible between B_L and C_L the launch window, or region of accessibility; likewise, the maximum arc possible between B_R and C_R may be termed the rendezvous window.

The right-hand sketch illustrates how these windows are governed by the performance capabilities of a system. If the sum of V_L and the impulsive velocity increment ΔV_R required at rendezvous cannot exceed 27,000 fps, then the launch window extends from $+6.1°$ to $-7.4°$, a spread of 13.5°. For a 300 mile circular orbit a good figure to keep in mind is that 4° of arc correspond to about 1 minute time of satellite travel; thus, the 13.5° launch window means that a launch for rendezvous must occur within an interval of a little more than 3 minutes. For a 30,000 fps maximum performance value, we see that the launch window has increased to 6.8 plus 56° or 62.8°, or something in the neighborhood of a 15-minute launch time available. The corresponding rendezvous windows shown show that the target may travel up to a full revolution from the time of ferry launch before rendezvous is made.

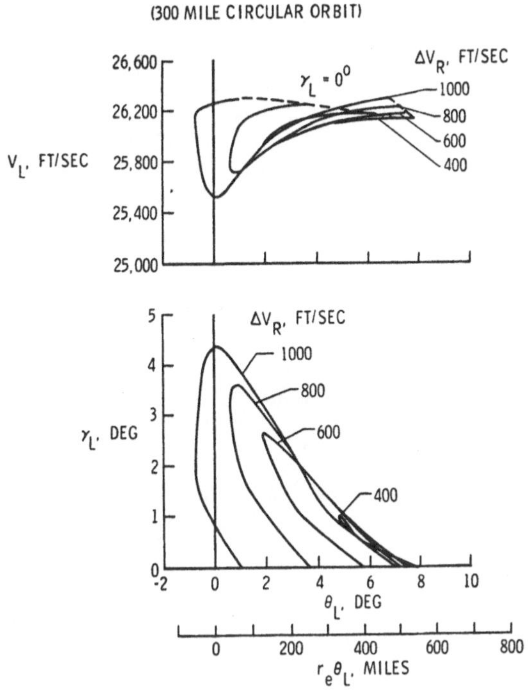

Fig. 4 b. Launch conditions for rendezvous. For ΔV_R up to 1,000 ft/sec

General results of the type determined by EGGLESTON [6, 7] are shown in Fig. 4. This figure is given in terms of the injection speed V_L and injection angle γ_L, the angular position θ_L of the target relative to the ferry at the time of ferry booster burnout, and the velocity increment ΔV_R required at rendezvous. Fig. 4 a applies for ΔV_R up to 5,000 fps, as might be involved in automatic rendezvous operations; Fig. 4 b is an enlarged portion for ΔV_R up to 1,000 fps, the range which is considered significant for manned ferry flights. The $\alpha = 0$ locus designates conditions for HOHMANN transfer ellipses, α being used to represent the angle between the ferry's path and the local horizon at rendezvous.

Circumvention of Long Holddown Intervals

Should holddown times on the launch pad be so uncertain as to preclude the use of the direct ascent trajectories of the previous section (that is, launch time uncertainties go beyond the limits of the launch windows), another recourse is available for rendezvous. Two basic ideas are involved: one is the fact mentioned previously that for nearly East-West type launchings, the target comes into overhead proximity of the launch site three to four times in succession; the second idea is that angular position deficiencies can be made up through use of chasing or parking orbits.

Fig. 5 is a typical example showing the nearness of the launching site to the orbital plane of the target (target position is here projected radially to the earth's surface) during four successive passes. Note that the total time the launch site is near the target plane is in the order of four and one-half hours. This closeness leads to the significant point that the ferry could be launched into an orbit similar to that of the target at any time during this four and one-half

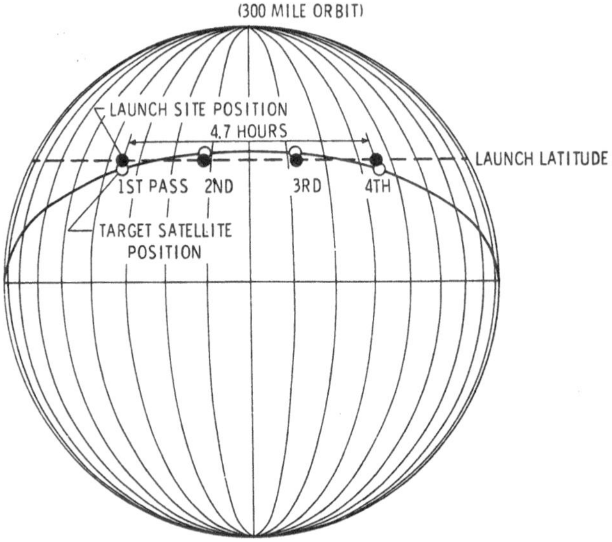

Fig. 5. Acceptable time interval that launch site is near target orbit plane

hour period (by the adjacency, two-impulse or general technique) without undue hardship or fuel expenditure. The angular position between the ferry and target may of course be large, but to get around this use is made of the chasing or parking orbit concept.

Thus, assume that the ferry is launched whenever it becomes ready during the four and one-half hour period available each day, but that the final injection velocity is held back just enough to put the ferry into a chasing orbit such that after n revolutions it has made up any angular position deficiency. Then after the position is made up, the increment in velocity not previously used is added to make ferry and target orbit coincident. This holding back of a velocity increment for subsequent addition means that no penalty in the way of additional fuel is really required over the direct ascent approach.

Fig. 6 a indicates generalized results that apply to chasing and parking orbits. In practice, one would enter at the angular position deficiency to be made up, go to the $n = 1$ curve (make up in one revolution), then to the left ordinate to determine the ΔV holdback increment. The right-hand ordinate is used to determine Δr, the maximum amount the chasing orbit will dip back toward the earth; this quantity is necessary to make sure that the ferry does not dip back into the atmosphere too deeply. If, for example, we do not want the ferry to dip back closer than 100 miles, then the maximum Δr that can be used for a 300-mile target orbit is 200 miles. If our initial calculation yields too large a Δr, then we go to a curve of higher n that yields an acceptable value of Δr, that is , we accept more revolutions to make up the positional deficiency.

The parking orbit scheme depicted on the right-hand side of Fig. 6 can be integrated into rendezvous ascent schemes as well. Here make up times are

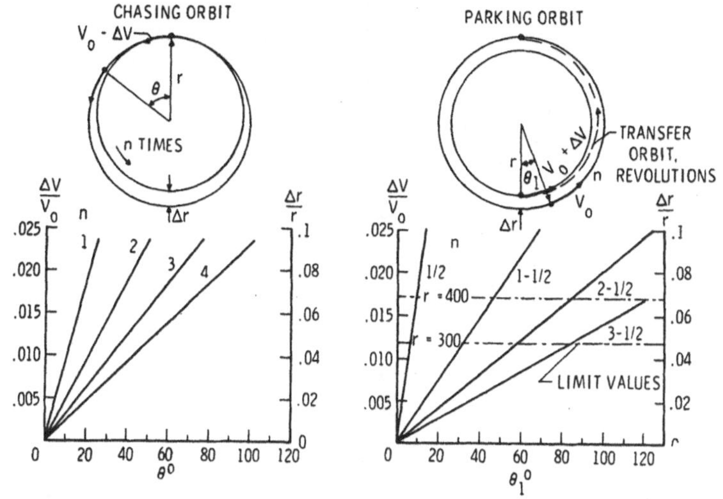

Fig. 6 a. Corrections for position mismatch. General chasing and parking orbit characteristics

better, but at the expense of one more operation — that of the impulse necessary to transfer from the parking orbit to the target orbit.

In Fig. 6 b results pertinent to chasing orbits are presented for the case of a 300-mile target orbit. The example path shown indicates that a 2.8-minute error in launch time, which corresponds to an 11° angular position deficiency, requires a holdback ΔV of 250 feet per second, and that the Δr would be about 170 miles.

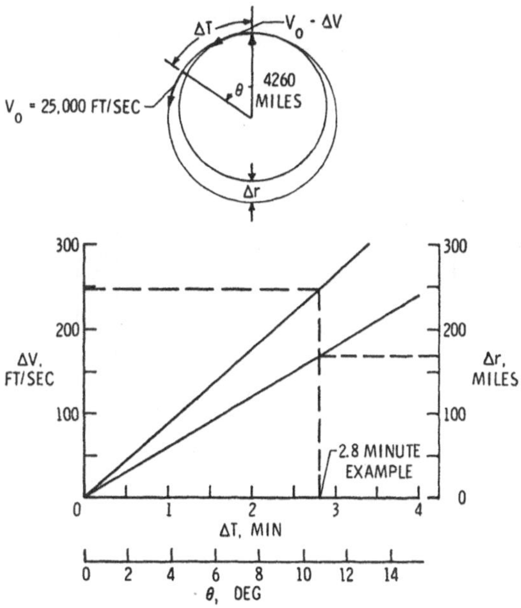

Fig. 6 b. Corrections for position mismatch. Specific chasing orbit example

B. Terminal Guidance

Guidance Equations

In the terminal phase the chief concern is the relative motion between the target and the ferry; of specific interest is the guidance of the ferry to achieve the end result of bringing both the range R and the range $\dot{R}$ to zero simultaneously at the instant of rendezvous. Table I has been prepared to show the nature of the equations that have been used in terminal phase guidance studies, and how

Table I. *Terminal Guidance Equations*

Inertially fixed axes

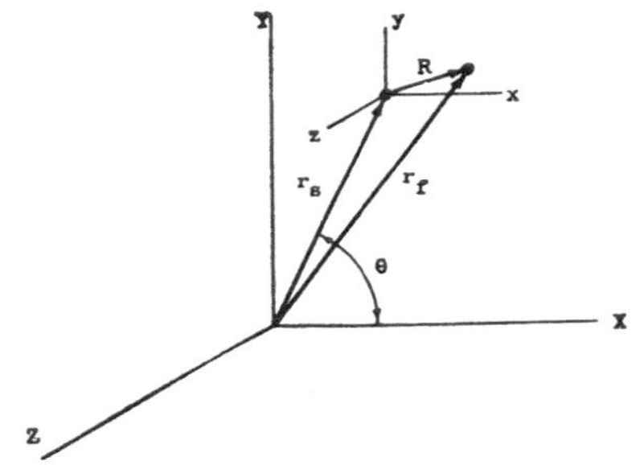

	Vector form	Rectangular coordinates
Exact	$\frac{\partial^2 \bar{r}_f}{\partial t^2} - \nabla \frac{\mu}{r_f} = \frac{\bar{T}}{m}$	Similar to form immediately below except in the expansion of $\nabla (\mu/r_f)$
Spherical earth	$\frac{d^2 \bar{r}_f}{dt^2} + \frac{GM}{r_f^3} \bar{r}_f = \frac{\bar{T}}{m}$ $\bar{r}_f = \bar{r}_s + \bar{R}$ $\bar{r}_s = (X, Y, Z)$ $= (r_s \cos\theta, r_s \sin\theta, 0)$ $\bar{R} = (x, y, z)$	$\ddot{x} + (\ddot{r}_s - r_s \dot{\theta}^2) \cos\theta - (2\, \dot{r}_s \dot{\theta} + r_s \ddot{\theta}) \sin\theta + \frac{GM}{r_f^3} (x + r_s \cos\theta) = \frac{T_x}{m}$ $\ddot{y} + (\ddot{r}_s - r_s \dot{\theta}^2) \sin\theta + (2\, \dot{r}_s \dot{\theta} + r_s \ddot{\theta}) \cos\theta + \frac{GM}{r_f^3} (y + r_s \sin\theta) = \frac{T_y}{m}$ $\ddot{z} + \frac{GM}{r_f^3} z = \frac{T_z}{m}$

	Rotating set of axes	
	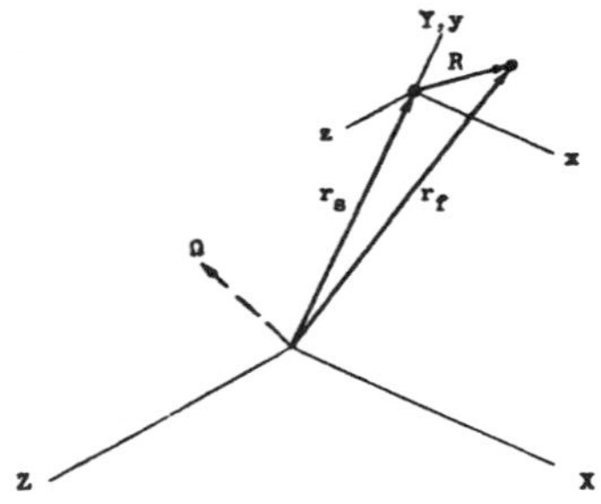	
Comments	Vector form	Rectangular coordinates
$\bar{r}_f = \bar{r}_s + \bar{R}$ μ/r_f is the gravity potential due to earth, moon, planets etc.	$\frac{\partial^2 \bar{r}_f}{\partial t^2} + 2\bar{\Omega} \times \frac{\partial \bar{r}_f}{\partial t} + \dot{\bar{\Omega}} \times \bar{r}_f + \bar{\Omega} \times \bar{\Omega} \times \bar{r}_f - \nabla \frac{\mu}{r_f} = \frac{\bar{T}}{m}$	Similar to form immediately below except in the expansion of $\nabla(\mu/r_f)$
θ is the angular velocity of station about center of earth, r_s is radial position of station	$\frac{d^2 \bar{r}_f}{dt^2} + 2\bar{\Omega} \times \frac{d\bar{r}_f}{dt} + \dot{\bar{\Omega}} \times \bar{r}_f + \bar{\Omega} \times \bar{\Omega} \times \bar{r}_f + \frac{GM}{r_f^3}\bar{r}_f = \frac{\bar{T}}{m}$ $\bar{r}_f = (x, y + r_s, z) = \bar{r}_s + \bar{R}$ $\bar{r}_s = (0, r_s, 0)$ $\bar{R} = (x, y, z)$ $\bar{\Omega} = (0, 0, \dot{\theta})$	$\ddot{x} - (y + r_s)\ddot{\theta} - 2(\dot{y} + \dot{r}_s)\dot{\theta} - x\left(\dot{\theta}^2 - \frac{GM}{r_f^3}\right) = \frac{T_x}{m}$ $\ddot{y} + x\ddot{\theta} + 2\dot{x}\dot{\theta} + \ddot{r}_s - (y + r_s)\left(\dot{\theta}^2 - \frac{GM}{r_f^3}\right) = \frac{T_y}{m}$ $\ddot{z} + \frac{GM}{r_f^3}z = \frac{T_z}{m}$ EGGLESTON [4, 5]

Table I (continued)

	Vector form	Rectangular coordinates
Spherical earth, Station in a circular orbit	$\frac{d^2\bar{R}}{dt^2} + \frac{d^2\bar{r}_s}{dt^2} + \frac{GM}{r_f^3}\bar{r}_f = \frac{\bar{T}}{m}$	$\ddot{x} - r_s\omega^2\cos\theta + \frac{GM}{r_f^3}(x + r_s\cos\theta) = \frac{T_x}{m}$ $\ddot{y} - r_s\omega^2\sin\theta + \frac{GM}{r_f^3}(y + r_s\sin\theta) = \frac{T_y}{m}$ $\ddot{z} + \frac{GM}{r_f^3}z = \frac{T_z}{m}$.
Spherical earth, Circular orbit, 1st order gravity field	$\frac{d^2\bar{R}}{dt^2} + \frac{GM}{r_s^3}\left(\bar{R} - 3\frac{\bar{r}_s\cdot\bar{R}}{r_s^2}\bar{r}_s\right) = \frac{\bar{T}}{m}$ HORD [9], KURBJUN [10], BRISSENDEN [11]	$\ddot{x} + \omega^2(x - 3x\cos^2\theta - 3y\sin\theta\cos\theta) = \frac{T_x}{m}$ $\ddot{y} + \omega^2(y - 3x\sin\theta\cos\theta - 3y\sin^2\theta) = \frac{T_y}{m}$ $\ddot{z} + \omega^2 z = \frac{T_z}{m}$
Spherical earth, Circular orbit, "Zero-order" gravity		
No gravity	$\frac{d^2\bar{R}}{dt^2} = \frac{\bar{T}}{m}$ any orbit HORD [9]	Circular orbit $\ddot{x} = \frac{T_x}{m}$ $\ddot{y} = \frac{T_y}{m}$ $\ddot{z} = \frac{T_z}{m}$

Comments	Vector form	Rectangular coordinates
$r_s = \text{Constant}$ $\dot{\theta} = \text{Constant} = \omega$ $\bar{\Omega} = \text{Constant} = (0, 0, \omega) = \bar{\omega}$ $\omega^2 = \frac{GM}{r_s^3} = \dot{\theta}^2$	$\frac{d^2\bar{R}}{dt^2} + 2\bar{\omega} \times \frac{d\bar{R}}{dt} + \bar{\omega} \times \bar{\omega} \times \bar{r}_f + \frac{GM}{r_f^3}\bar{r}_f = \frac{\bar{T}}{m}$ or $\frac{d^2\bar{R}}{dt^2} + 2\bar{\omega} \times \frac{d\bar{R}}{dt} + \omega^2 \bar{k} z + \left(\frac{GM}{r_f^3} - \omega^2\right)\bar{r}_f = \frac{\bar{T}}{m}$	$\ddot{x} - 2\omega\dot{y} - x\left(\omega^2 - \frac{GM}{r_f^3}\right) = \frac{T_x}{m}$ $\ddot{y} + 2\omega\dot{x} - (y + r_s)\left(\omega^2 - \frac{GM}{r_f^3}\right) = \frac{T_y}{m}$ $\ddot{z} + \frac{GM}{r_f^3} z = \frac{T_z}{m}$ EGGLESTON [4, 5]
$\frac{d^2\bar{r}_s}{dt^2} = -\omega^2\bar{r}_s;\ \frac{GM}{r_s^3} = \omega^2$ On the left, $\frac{GM}{r_f^3} = \frac{GM}{r_s^3}\left(1 - 3\frac{\bar{r}_s \cdot \bar{R}}{r_s^2} + \ldots\right)$ $\frac{GM}{r_f^3}\bar{r}_f \cong \omega^2\left(\bar{r}_s + \bar{R} - 3\frac{\bar{r}_s \cdot \bar{R}}{r_s^2}\bar{r}_s\right)$ On the right, $\frac{GM}{r_f^3} = \frac{GM}{r_s^3}\left(1 - 3\frac{y}{r_s} + \ldots\right)$ $\frac{GM}{r_f^3}\bar{r}_f \cong \omega^2(\bar{r}_s + \bar{R} - \bar{j}\,3y)$	$\frac{d^2\bar{R}}{dt^2} + 2\bar{\omega} \times \frac{d\bar{R}}{dt} + \bar{\omega} \times \bar{\omega} \times \bar{R} + \frac{GM}{r_s^3}(\bar{R} - \bar{j}\,3y) = \frac{\bar{T}}{m}$ or $\frac{d^2\bar{R}}{dt^2} + 2\bar{\omega} \times \frac{d\bar{R}}{dt} + \frac{GM}{r_s^3}(-\bar{j}\,3y + \bar{k}z) = \frac{\bar{T}}{m}$	$\left.\begin{array}{l} \ddot{x} - 2\omega\ddot{y} = \frac{T_x}{m} \\ \ddot{y} + 2\omega\dot{x} - 3\omega^2 y = \frac{T_y}{m} \\ \ddot{z} + \omega^2 z = \frac{T_z}{m} \end{array}\right\\}\quad (T\,1)$ WHEELON [14], CLOHESSY and WILSHIRE [15], CARNEY [17], EGGLESTON [5, 18], SPRALIN [19]
$\frac{GM}{r_f^3}\bar{r}_f \cong \frac{GM}{r_s^3}(\bar{r}_s + \bar{R})$ $\frac{GM}{r_s^3} = \omega^2$	$\frac{d^2\bar{R}}{dt^2} + 2\bar{\omega} \times \frac{d\bar{R}}{dt} + \bar{\omega} \times \bar{\omega} \times \bar{R} + \frac{GM}{r_s^3}\bar{R} = \frac{\bar{T}}{m}$ or $\frac{d^2\bar{R}}{dt^2} + 2\bar{\omega} \times \frac{d\bar{R}}{dt} + \frac{GM}{r_s^3}\bar{k}z = \frac{\bar{T}}{m}$	$\ddot{x} - 2\omega\dot{y} = \frac{T_x}{m}$ $\ddot{y} + 2\omega\dot{x} = \frac{T_y}{m}$ $\ddot{z} + \omega^2 z = \frac{T_z}{m}$ HORNBY [20]
	$\frac{d^2\bar{r}_f}{dt^2} + 2\bar{\Omega} \times \frac{d\bar{r}_f}{dt} + \dot{\bar{\Omega}} \times \bar{r}_f + \bar{\Omega} \times \bar{\Omega} \times \bar{r}_f = \frac{\bar{T}}{m}$ for any orbit $\frac{d^2\bar{R}}{dt^2} + 2\bar{\omega} \times \frac{d\bar{R}}{dt} + \bar{\omega} \times \bar{\omega} \times \bar{r}_f = \frac{\bar{T}}{m}$ for circular orbit	Circular orbit $\ddot{x} - 2\omega\dot{y} - \omega^2 x = \frac{T_y}{m}$ $\ddot{y} + 2\omega x - \omega^2(y + r_s) = \frac{T_y}{m}$ $\ddot{z} = \frac{T_z}{m}$

they evolve from the general satellite motion equations[1]. Both inertially fixed axes and rotating axes sets are indicated, with vector and rectangular coordinate forms being given for both cases. All of the equations expressed in rectangular coordinates apply to the relative motion between the ferry and the target. Rotating axes seem to be the most convenient for analytical studies, whereas inertially fixed axes lend themselves well to simulator studies. The equations which perhaps have received the most attention are those labeled (*T 1*) and which apply to a first order gravity field approximation. Undoubtedly, a reason for their popularity is the fact that these equations have a closed form solution which yields results that compare favorably with the exact results in most cases. Some of the investigators that have used the various forms are indicated on the table.

Basic Terminal Phase Schemes

A number of steering systems for terminal guidance have been proposed and investigated in the past few years. Essentially, they may be divided into two classes: those based on a proportional navigation or fire-control viewpoint,

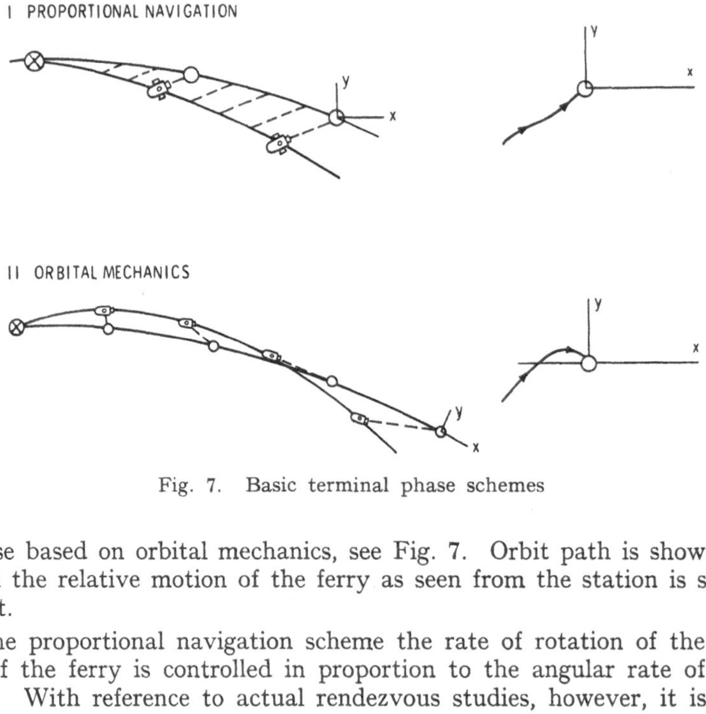

Fig. 7. Basic terminal phase schemes

and those based on orbital mechanics, see Fig. 7. Orbit path is shown on the left, and the relative motion of the ferry as seen from the station is shown on the right.

In the proportional navigation scheme the rate of rotation of the velocity vector of the ferry is controlled in proportion to the angular rate of the line of sight. With reference to actual rendezvous studies, however, it is perhaps more appropriate to consider the scheme to be based on constant bearing navigation (a limiting case of proportional navigation), since generally conditions are sought where the line of sight remains stationary in space; an inertially fixed set of axes is implied in this scheme. Wrigley [8], Hord [9], Kurbjun [10], Brissenden [11], Sears and Felleman [12], Cicolani [13] are some of the investigators of this approach.

---

[1] Appreciation is expressed here to Mr. John M. Eggleston for his assistance in setting up this table.

The orbital mechanics scheme is based essentially on the homogeneous equations of motion in a reference frame fixed to the target (the rotating axes case in Table I). These equations are solved to determine the proper course to rendezvous; impulsive corrections are given to put the ferry on a collision course, and a final impulse is given at intercept to match velocities. WHEELON [14], CLOHESSY and WILSHIRE [15], and LEVIN [16] have studied this scheme. CARNEY [17] has investigated the scheme using continuous modulated thrust

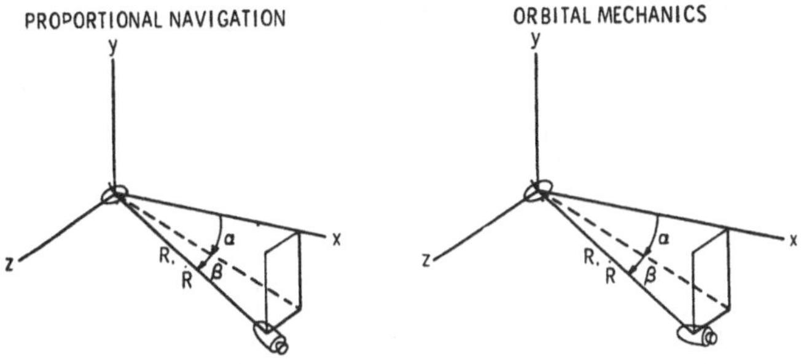

Fig. 8. Sequential operations in terminal phase

Proportional navigation	Orbital mechanics
1. Search and lock sensors to station	1. Search and lock sensors to station
2. Null $\dot{\alpha}$ and $\dot{\beta}$	2. Compute R and $\dot{R}$ errors, commence thrusting and tilting out of the horizontal plane to null errors
3. Follow braking schedule to bring $\dot{R} = 0$ at $R = 0$	3. Continue on closed loop basis to rendezvous

control, a horizon scanner in the ferry to establish a reference, and the concept of a slight tilting out of the horizontal plane to make the corrective maneuvers that lead to a collision course. EGGLESTON [18], SPRALIN [19], HORNBY [20] also make use of the orbital mechanics concept for ascent guidance.

Some schemes for terminal guidance make use of a combination of the two basic schemes, using, for example, proportional navigation for out-of-the-plane maneuvering, and orbital mechanics for in-plane flight.

Fig. 8 is given to indicate the nominal operating sequence of both schemes; the listing in the figure states the basic action required. Both systems require on-board sensors, such as radar or optical devices, to measure range, range rate, and angular rate of the line of sight. Either piloted or automatic control may be used, and thrusting (or braking) may be variable or of the on-off impulsive type.

Braking Logic

To give an idea of the thrusting or braking logic involved, some comments pertaining to the automatic braking of a system employing proportional navigation will be made. The principal feature of this system is the nominal preliminary maneuver of nulling the angular rate of the line of sight (step 2 on the left of Fig. 8); this action reduces rendezvous to essentially a one-dimensional problem. During subsequent braking action the thrust vector is tilted as needed to cancel out any angular rate that may creep in due to instrument and cutoff errors, and due to misalinements. For this case, whether the thrust control is variable or on-off, the procedure usually followed is to base braking on some convenient phase-plane law involving the range R and range rate $\dot{R}$. A law often used is the simple one-dimensional acceleration law

$$a = \frac{\dot{R}^2}{2R}.$$

Fig. 9 illustrates the use of this law together with several braking schedules that have been studied. For the variable-thrust case, path (1) shows the ferry coasting at constant $\dot{R}$ until some nominal acceleration is reached. Thrust is then applied causing the ferry to travel down the constant acceleration path to the origin; the thrust is varied in this period to account for changing mass and system errors.

For the on-off control operation depicted at the bottom, path (2), the ferry coasts until the "on" line of $a = 0.25\,g$ is reached; thrusting with an acceleration of about $0.5\,g$ then occurs up to the "off" line of $a = 0.1\,g$. The ferry steps down the band in this sequential fashion until rendezvous is achieved. The acceleration values chosen here represent cases tested, but are not otherwise significant.

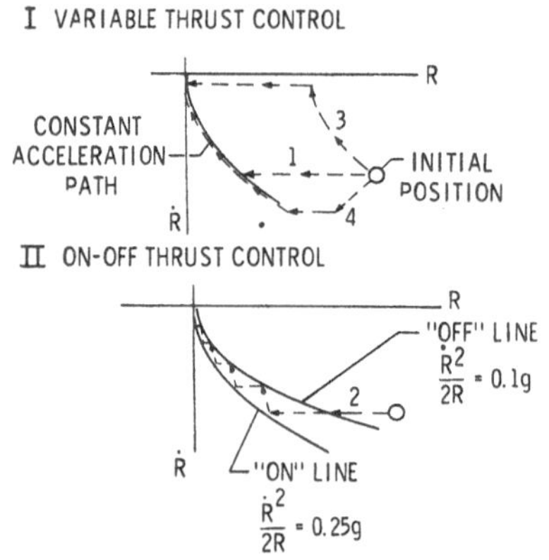

Fig. 9. Phase-plane paths in terminal guidance

Paths (1) and (2) apply to the situation wherein the transverse (angular rate) and line-of-sight velocities have been nulled separately. The most efficient approach from a fuel expenditure consideration would be to cancel most all the relative velocity (range rate as well as transverse) in the preliminary maneuver except for an infinitesimal component along the line of sight. Path (3) illustrates this maneuver for the variable-thrust case; a similar path would be followed for on-off thrusting. This scheme is efficient, as mentioned, but the time to rendezvous is lengthened considerably. In contrast to this approach, velocity may be added initially to increase range rate, giving a path similar to (4), which again applies to either variable or on-off thrusting; now time to rendezvous is decreased, but at some expense in fuel. There may be some instances where a shortened time is desirable, however, as in rescue.

For piloted thrust control it may be stated that operation similar to that shown for the on-off automatic control has been found satisfactory. The pilot simply follows a suggested program shown on a display card.

Fuel Consumption

Fig. 10 a and 10 b show, respectively, some fuel consumption results that were obtained in studies with automatic and piloted terminal phase simulators (Lineberry [17] and Kurbjun [10]); the incremental velocity ΔV that was actually required to make the rendezvous is used here as the measure of fuel consumed; however, a ΔV increment of 100 fps corresponds to a mass or fuel consumption of about 1 percent. The circled points in Fig. 10 a apply to the procedure which calls for initial nulling of angular rate of the line of sight; the 45° line represents the minimum ΔV possible by this method. The nearness of the points indicates that the system's performance was quite good. The squared points indicate the gains that can be realized by cancelling a large portion of the range rate as well as the transverse velocity during the preliminary maneuver; the lowest value possible by this approach is indicated by a dashed line. As mentioned previously, time to rendezvous is increased by this procedure, as is expected.

Fig. 10 b shows similar results as obtained in piloted studies; here, however, results are compared with the magnitude of the vector sum ΔV_R of the initial range and transverse velocities (the absolute minimum ΔV obtainable) instead of the sum of these two speeds as was used in Fig. 10 a. The main point to note is that excellent performance capabilities are demonstrated by pilots. The fact that in some cases the average value of ΔV is lower than the circled points is interesting, and indicates that there is an automatic tendency for pilots to cancel out a little range rate as well as transverse velocity in the preliminary maneuver.

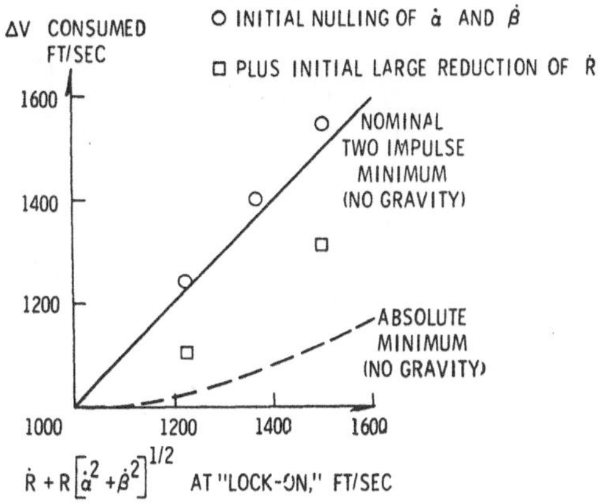

Fig. 10 a. Rendezvous fuel requirements. Automatic system

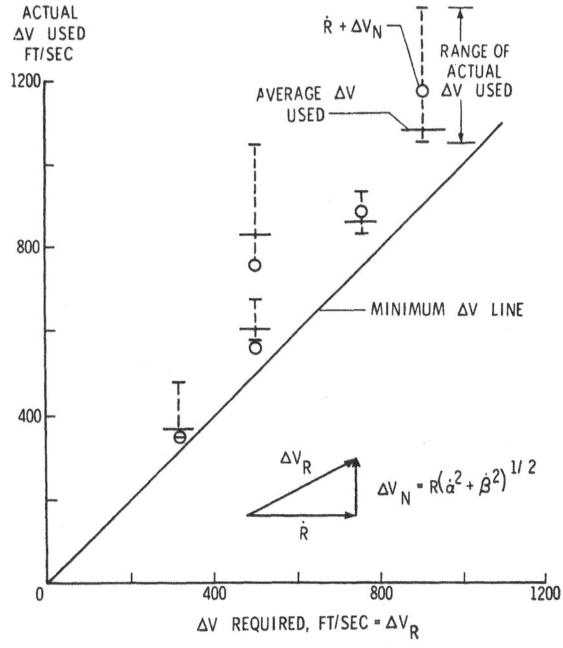

Fig. 10 b. Rendezvous fuel requirements. Piloted system

Fuel-Time Trade Off and Fuel Minimization

Fig. 11 is typical of the results obtained by EGGLESTON [21] and HORNBY [22] in assessing, via an orbital mechanics approach scheme, how fuel consumption depends on the time to rendezvous. Two situations are depicted for the example problem indicated: one pertains to intercept rendezvous the other to soft rendezvous. The curve labeled ΔV_1 is for the intercept case, and indicates the variation with time of the magnitude of a single initial impulse that is required to put the ferry on an intercept course (the magnitude of course depends on the direction of thrust application). A pronounced minimum is noted to occur at a time designated by t_1.. From the point of view of minimum fuel usage, EGGLESTON has termed this time as the optimum time to rendezvous.

The soft rendezvous case differs in that after following an intercept course a final impulse must be given to cancel all relative velocity; the magnitude of the impulse required is given by the curve labeled ΔV_2. The total impulse used in the soft rendezvous case then is the sum of the ΔV_1 and ΔV_2 curves, or the solid curve in the figure. Here again a minimum is found, at a slightly larger value of time than that for the intercept case, but in contrast to the intercept case, the curve is quite flat for times that go beyond the minimum point. The significant aspect brought out by the soft rendezvous case is that large impulse penalties are involved if rendezvous is attempted in times less than a value which leads to minimum fuel consumption, but that very little penalty is paid if greater times are used.

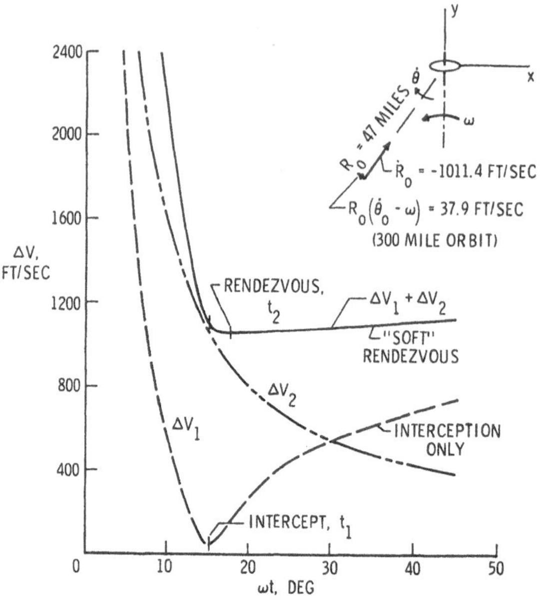

Fig. 11. Minimum fuel for rendezvous by orbital mechanics

III. Simplified Optical Schemes

To cut down on the complex hardware that may be required in terminal phase guidance, some thought has been given to the development of techniques requiring only simple instrumentation. Two schemes of interest are reviewed here.

LINEBERRY and KURBJUN [23] have explored a terminal phase sensing system which requires only a timer and a sighting scope. Fig. 12 illustrates the basic ideas of the system. In the top portion a simple scope with a grid to measure angular position is used in the ferry to sight the target vehicle and the distant star background. Thus with the scope fixed to a distant star the relative motion between the target and the ferry is immediately ascertained. Further, if the times required for the target to transverse several successive angular intervals are measured, then the range and range rate may be

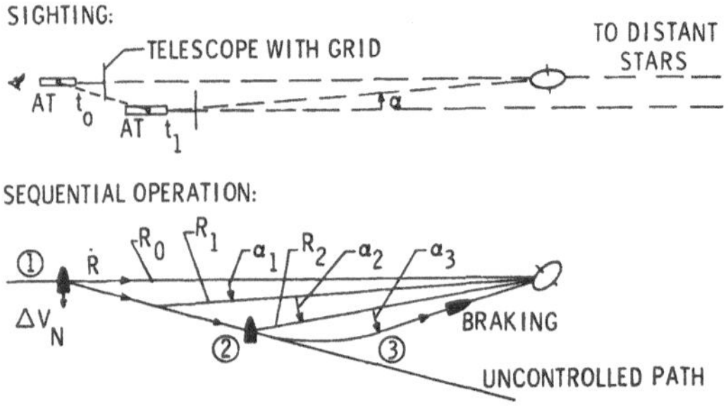

Fig. 12. Use of visual techniques in rendezvous

1. Orient vehicle with thrust axis in direction of ΔV_N
2. Measure α_1, t_1, and α_2, t_2
3. Thrust to zero angular velocity and measure α_3, t_3

then $R_2 \alpha_3 = \dfrac{a\, t_3^2}{2}$ or $R_2 = \dfrac{a\, t_3^2}{2\, \alpha_3}$

4. Using $R\,\alpha = \Delta V_N\, t$, from 2

$R_1 \alpha_1 = \Delta V_N\, t_1$ and $R_2 \alpha_2 = \Delta V_N\, t_2$

thus $\dfrac{R_1}{R_2} = \left(\dfrac{\alpha_2\, t_1}{\alpha_1\, t_2}\right)$ or $R_1 = \left(\dfrac{\alpha_2\, t_1}{\alpha_1\, t_2}\right) R_2$

5. Then $\dot{R} = \dfrac{R_2 - R_1}{t_2 - t_1}$

deduced by the sequence itemized in the bottom portion of the figure. It is noted that only elementary geometrical relations are employed. Note also that the third time interval is a period of thrusting wherein all the angular rate of the line of sight is cancelled; this angular motion and direction is of course directly evident in the scope. Thus, with the range and range rate established the pilot is ready to go into his braking schedule. Any angular rate motion that creeps back in during braking can be cancelled because it is seen immediately through the scope. Simulator studies of this scheme have shown very good results.

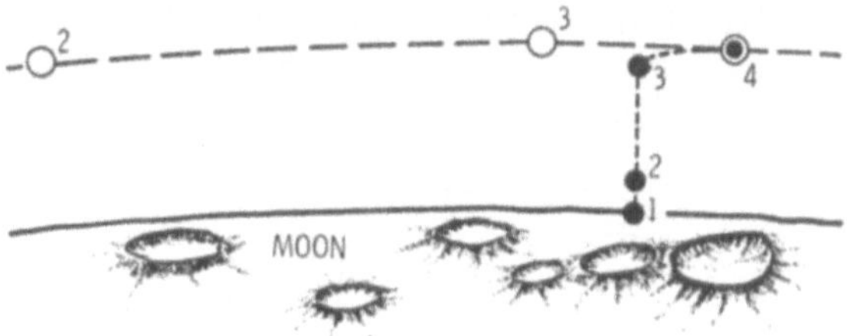

Fig. 13. Simplified lunar rendezvous (100,000 ft orbit)

1—*2* Vertical thrust (33.3 sec)
2—*3* Vertical coast (109 sec)
3—*4* Horizontal thrust (117 sec)
4 Rendezvous

As another example of the use of simplified techniques, LINA and VOGELEY [24] have advanced an elementary plan for performing lunar rendezvous. The idea embodies a plumb bob, a timer, and a sighting on distance stars, and can be described with the aid of Fig. 13. At time *1*, as indicated by a timer synchronized with the target, the ferry begins to thrust vertically; the vertical was established by a plumb bob and is held thereafter by sighting on a star vertically overhead. After thrusting and coasting, horizontal thrust of a predetermined duration begins at time *3* in a direction of a star fixed by an early sighting. At *4* rendezvous takes place. The proponents of this plan have also applied it in reserve, that is, in connection with a lunar landing from a lunar orbit. Their studies of the various errors involved indicate that the scheme is quite promising.

IV. Benefits or Potentialities of Rendezvous

The list of uses given in the Introduction is a fairly good indication in itself of the benefits that may be derived through use of rendezvous. Surely if man is to continue in his exploration of space, he will want to exploit as fully as possible all of these uses, for the advancement they offer and for their inherent achievement. It must be admitted, for example, that the successful transporting of a man to an orbiting space station, with later return, would be quite a significant and noteworthy scientific and engineering achievement. It is not too difficult to envision that such an operation may become quite routine in the future. The aspects of personnel and supply transfer, of rescue, and even of initial assembly of the space laboratory all depend on rendezvous.

But perhaps the greatest advantage offered by rendezvous is the flexibility it affords in designing space missions, that is, it relieves the growth problem which is basic to any launch vehicle. Should the payload desired exceed the capabilities of the launch vehicles, then rendezvous is a recourse. For many deep space missions, involving perhaps manned flight, rendezvous may prove invaluable.

But let us not get so far away from home in our thoughts to use rendezvous. Since man has set his sights on the moon, it is perhaps fitting to close the technical part of this paper by showing the benefits that rendezvous offers in a lunar exploration mission. Thus let us consider two schemes, each involving an exploration vehicle on the way to the moon (these exploration vehicles may be considered to be launched as complete units from the earth or to be assembled from basic units by rendezvous in earth orbit, although this consideration is immaterial to the following formulation). In one scheme the exploration vehicle is decelerated into a low-altitude circular orbit about the moon. From this orbit a lunar lander descends to the moon's surface, leaving the return vehicle in orbit. After exploration the lunar lander ascends and rendezvous with return vehicle. The return vehicle is then boosted into a return trajectory to the earth, leaving the lander behind. In the other scheme the exploration vehicle descends directly to the moon's surface with the return vehicle. Exploration is made, after which the return vehicle is surface launched back into a return trajectory. We wish to compare the size of the exploration vehicle going to the moon in both cases.

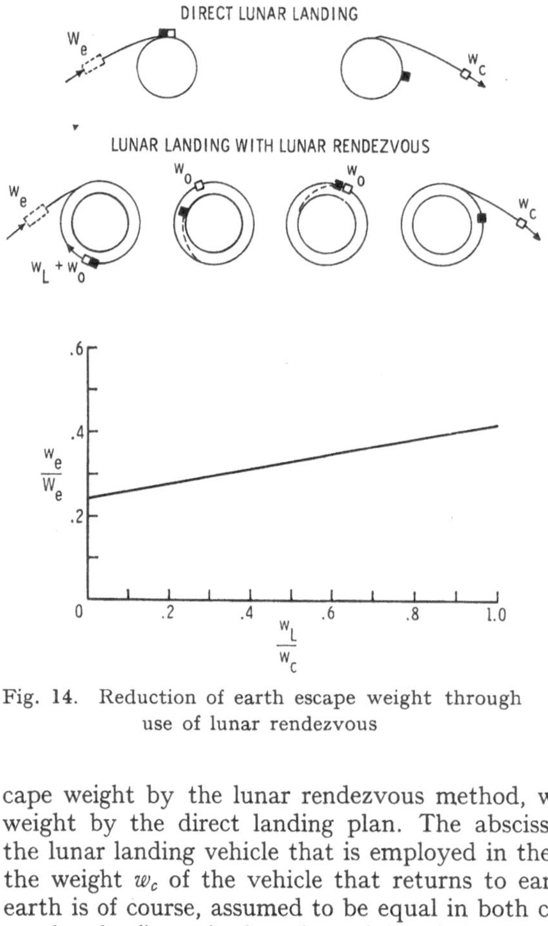

Fig. 14. Reduction of earth escape weight through use of lunar rendezvous

The two schemes are analyzed in the appendix of this paper and the basic results are given in Fig. 14. In the ordinate involving the ratio w_e/W_e, w_e is the earth escape weight by the lunar rendezvous method, whereas W_e is the earth escape weight by the direct landing plan. The abscissa is the initial weight w_L of the lunar landing vehicle that is employed in the rendezvous scheme divided by the weight w_c of the vehicle that returns to earth; the weight w_c returned to earth is of course, assumed to be equal in both cases. The prime result brought out by the figure is that the weight of the outgoing exploration vehicle can be greatly reduced by using a lunar rendezvous technique. Specifically, for a value of $w_L/w_c = 1$, the results show that the weight of the earth escape payload for the lunar rendezvous scheme is less than one-half the escape payload weight by the direct landing plan. This reduction is, of course, a direct reflection of greatly reduced energy requirements brought about by leaving a sizable mass in lunar orbit. The specific reduction depends on the particular velocity increments and rocket performance characteristic used; the values used in this example are indicated in the appendix. Similar results are indicated, however, for other choices.

V. Concluding Remarks

The intent of this report has been to single out and discuss some of the advances that have been made in the understanding of the relatively newborn and promising concept of space rendezvous, and to perhaps add some new information. Attention has been focused on the main phases of rendezvous, and basic research results pertaining to launch timing, ascent and transfer trajectories, and to terminal guidance schemes have been discussed. Benefits were indicated and brought out by examples. Perhaps conspicuous is the lack of discussion on the docking phase. This is simply because the docking phase requires experiments to find out what the problems are, and also because it appears primarily to involve hardware items, which were not intended to be discussed in this paper. It may be remarked, however, that visual aids and schemes may play a very important role in this phase of rendezvous operation, and developments along this line would be quite appropriate.

Probably the most important conclusions to be made are: (a) that rendezvous appears technically feasible, (b) that both manned and automatic control are practical, (c) that by orbital assembly of units it offers a very flexible and versatile means for performing space missions, or making possible those which could not otherwise be made, and (d) that it should play a major role in future space missions, especially for manned space flight — it must be admitted, for example, that landing on the moon is, in reality, a rendezvous problem.

Appendix

Use of Lunar Rendezvous

Basic Mass-Ratio Formula

In this section we derive a basic mass ratio formula that will be used throughout the subsequent analysis. Thus, consider the following payload-booster combination

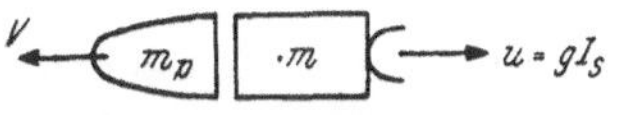

where m_p represents payload mass, m booster mass, V vehicle velocity, and u the effective jet velocity. At the end of booster burnout the fundamental rocket equation indicates that

$$\frac{m_p + m_b}{m_p + \varepsilon m_b} = e^{\Delta V/u} \tag{1}$$

where $\varepsilon\, m_b$ is the burnout mass of the booster, ε is the ratio of the booster mass at burnout to the initial booster mass, and ΔV is the ideal velocity increment due to booster burning (gravity effects, if present, are included in ΔV). The desired mass-ratio formula is found from this equation as

$$\frac{m_b}{m_p} = \frac{k-1}{1-\varepsilon\, k} = r \tag{2}$$

where $k = e^{\Delta V/u}$. In terms of m_p the total initial mass is

$$m_T = m_p + m_b = (1 + r)\, m_p = R\, m_p \tag{3}$$

Note, by eq. (2)

$$R = \frac{(1-\varepsilon)\, k}{1-\varepsilon\, k}$$

Application to Lunar Rendezvous

Consider a lunar exploration mission. Let the vehicle which approaches the moon be represented as follows

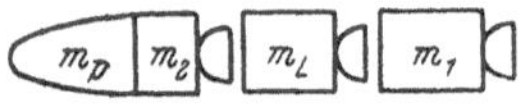

and assume the operational sequence:

1. Decelerate the vehicle into a moon orbit by m_1, ΔV_1 is the velocity change.
2. Descend to the moon by m_L, explore, ascend, and rendezvous with m_p, m_2.
3. Boost m_p back to earth return by m_2, ΔV_2 is the velocity change.

By application of eq. (2) and (3) it is found that the total earth escape mass approaching the moon is

$$m_e = R_1(R_2 m_p + m_L) \tag{4}$$

where

$$R_1 = \frac{(1-\varepsilon_1)\,k_1}{1-\varepsilon_1 k_1}\,; \qquad k_1 = e^{\Delta V_1/u}$$

$$R_2 = \frac{(1-\varepsilon_2)\,k_2}{1-\varepsilon_2 k_2}\,; \qquad k_2 = e^{\Delta V_2/u}$$

Application to Direct Lunar Landing and Take-Off

In this case, consider that the vehicle approaching the moon is represented by

and that the subsequent sequence is:

1. Land on moon by means of m_3, ΔV_3 is the ideal velocity increment.
2. From moon's surface launch m_p to earth return by m_4; ΔV_4 is the ideal velocity increment.

Eq. (2) and (3) yield for this case the following total earth escape mass approaching the moon

$$M_e = R_3 R_4 m_p \tag{5}$$

where

$$R_3 = \frac{(1-\varepsilon_3)\,k_3}{1-\varepsilon_3 k_3}\,; \qquad k_3 = e^{\Delta V_3/u}$$

$$R_4 = \frac{(1-\varepsilon_4)\,k_4}{1-\varepsilon_4 k_4}\,; \qquad k_3 = e^{\Delta V_4/u}$$

Comparison of the Two Schemes

Division of eq. (4) by eq. (5) yields the following ratio of the earth escape mass required for lunar landing when the lunar rendezvous scheme described is used to the earth escape mass required by a direct lunar landing scheme:

$$\frac{m_e}{M_e} = \frac{R_1}{R_3 R_4}\left(R_2 + \frac{m_L}{m_p}\right) \tag{6}$$

This equation was used to obtain the results shown in Fig. 14; a value $u = 12{,}880$ fps ($I_s = 400$) was used throughout and the following assumption for the ε's and the ΔV's were made:

$\eta_1 = 0.1, \quad \Delta V_1 = 8{,}800 - 5{,}300 = 3{,}500\,\text{fps}; \quad R_1 = 1.36$
$\eta_2 = 0.1, \quad \Delta V_2 = 8{,}800 - 5{,}300 = 3{,}500\,\text{fps}; \quad R_2 = 1.36$
$\eta_3 = 0.15, \quad \Delta V_3 = 10{,}500\,\text{fps}; \quad R_3 = 2.91$
$\eta_4 = 0.1, \quad \Delta V_4 = 10{,}500\,\text{fps}; \quad R_4 = 2.63$

With these values eq. (6) becomes

$$\frac{m_e}{M_e} = 0.242 + 0.178 \frac{m_L}{m_p} \tag{7}$$

References

1. J. C. Houbolt, Considerations of the Rendezvous Problems for Space Vehicles. Presented to the SAE National Aeronautical Meeting, New York, N.Y., April 5–8, 1960.
2. R. S. Swanson, and N. V. Petersen, Summary Report of Rendezvous-Compatible-Orbits. Technical Memorandum of the Astro Sciences Group ASG–TM–61–10, January 1961.
3. J. D. Bird, and D. F. Thomas, Jr., A Two-Impulse Plane for Performing Rendezvous on a Once-a-Day Basis. NASA TN D–437, 1960.
4. J. M. Eggleston, and H. D. Beck, A Study of the Positions and Velocities of a Space Station and a Ferry Vehicle During Rendezvous and Return. NASA TR R–87, 1960.
5. J. M. Eggleston, and R. S. Dunning, Analytical Evaluation of a Method of Midcourse Guidance for Rendezvous with Earth Satellites. NASA TN D–883, 1961.
6. J. M. Eggleston, The Trajectories and Some Practical Guidance Considerations for Rendezvous and Return. IAS Paper no. 61–36, presented at the IAS 29th Annual Meeting, New York, January 23–25, 1961.
7. J. M. Eggleston, Comparison of Launch Conditions and Trajectories for Manned and Unmanned Ferry Vehicles. Unpublished NASA Langley Research Center Report, February 27–28, 1961.
8. W. Wrigley, Performance of a Linear Accelerometer. Mass. Inst. of Tech., Dept. of Aero. Engr., Orbital and Satellite Vehicles, Vol. 1, Notes for a Special Summer Program, Summer Session, Aug. 6–17, 1956, Ch. 9, pp. 9–1–9–14.
9. R. A. Hord, Relative Motion in the Terminal Phase of Interception of a Satellite or a Ballistic Missile. NASA TN 4399, 1958.
10. M. C. Kurbjun, R. F. Brissenden, E. C. Foudriat, and B. B. Burton, Pilot Control of Rendezvous. IAS Paper no. 61–37, presented at the IAS 29th Annual Meeting, New York, January 23–25, 1961.
11. R. F. Brissenden, B. B. Burton, E. C. Foudriat, and J. B. Whitten, Analog Simulation of a Pilot-Controlled Rendezvous. NASA TN D–747, 1961.
12. N. E. Sears, Jr., and P. G. Felleman, Terminal Guidance for a Satellite Rendezvous. Presented at the ARS Controllable Satellite Conference, MIT, April–May 1959; ARS Preprint 778–59.
13. L. S. Cicolani, Trajectory Control in Rendezvous Problems Using Proportional Navigation. NASA TN D–772, April 1961.
14. A. D. Wheelon, An Introduction to Midcourse and Terminal Guidance. JPL Report no. 145, June 1958.
15. W. H. Clohessy, and R. S. Wiltshire, Terminal Guidance Systems for Satellite Rendezvous. IAS Paper no. 59–93, June 1959.
16. E. Levin, and J. Ward, Manned Control of Orbital Rendezvous. P–1834, the RAND Corporation, October 20, 1959.
17. T. M. Carney, and E. C. Lineberry, Automatic Terminal Guidance Logic for Rendezvous Vehicles. Unpublished NASA Langley Research Center Report, February 1961.
18. J. M. Eggleston, Optimum Time to Rendezvous. ARS Journal **30**, no. 11 (1960).

19. L. W. SPRADLIN, The Long-Time Satellite Rendezvous Trajectory. Proceedings of the National Specialists Meeting on Guidance of Aerospace Vehicles, Boston, Mass., May 25–27, 1960.
20. H. HORNBY, Least Fuel, Least Energy and Salvo Rendezvous. Presented at the 15th Annual Spring Technical Conference, Cincinnati, Ohio, April 12 and 13, 1961.
21. J. M. EGGLESTON, Extensions of Optimum Time to Rendezvous Studies. Unpublished NASA Langley Research Center Report, February 1961.
22. H. HORNBY, Problems of Rendezvousing Space Vehicles and Application in the Manned Lunar Landing Mission. NASA Prospective Report, 1961.
23. E. C. LINEBERRY, and M. C. KURBJUN, A Preliminary Study of Manned Control of the Terminal Phase of Rendezvous Using Visual Techniques. Unpublished NASA Langley Research Center Report, February 21, 1961.
24. L. J. LINA, and A. W. VOGELEY, Preliminary Study of a Piloted Rendezvous Operation from the Lunar Surface to an Orbiting Space Vehicle. Unpublished NASA Langley Research Center Report, February 21, 1961.
25. R. H. BATTIN, and J. H. LANING, JR., A Recoverable Interplanetary Space Probe. Massachusetts Institute of Technology. Instrumentation Laboratory Report R–235, Vol. IV, Appendices 1959.
26. W. E. BRUNK, and R. J. FLAHERTY, Methods and Velocity Requirements for the Rendezvous of Satellites in Circumplanetary Orbits. NASA TN D–81, 1959.
27. W. H. CLOHESSY, and R. S. WILTSHIRE, Problems Associated with the Assembly of a Multiunit Satellite in Orbit. Presented at the ASME Aviation Conference, Los Angeles, March 1959, ASME Paper 59–AV–25.
28. R. CORNOG, Economics of Satellite Supply Vehicles. Presented at the Aviation Conference, Los Angeles, Calif., March 9–12, 1959, of the American Society of Mechanical Engineers, ASME Paper 59–AV–33.
29. W. W. DUKE, E. A. GOLDBERG, and I. PFEFFER, Error Analysis Consideration for a Satellite Rendezvous. ARS Preprint no. 1198–60, May 1960.
30. E. L. DRAIN, Terminal Guidance in Satellite Rendezvous. Thesis, Astronautical Engineer, University Michigan, May 1960 (AD 239807).
31. T. N. EDELBAUM, Preliminary Comparison of Air and Ground Launching of Satellite Rendezvous Vehicles. IAS Paper 61–10, New York, January 23–25, 1961.
32. K. A. EHRICKE, Establishment of Large Satellites by Means of Small Orbital Carriers. Proc. IIIrd International Astronautical Congress, Stuttgart, 1952, pp. 111–145.
33. K. A. EHRICKE, Ascent of Orbital Vehicles. Astronaut. Acta 2, 175–190 (1956).
34. T. B. GARBER, Ascent Guidance for a Satellite Rendezvous. Proceedings of the Manned Space Stations Symposium. Sponsored by the IAS, NASA, and the Rand Corp., Los Angeles, Calif., April 20–22, 1960.
35. K. W. GATLAND, Orbital Rockets — I. Some Preliminary Considerations. J. Brit. Interplan. Soc. 1951, 97–107.
36. E. HARRISON, Some Considerations of Guidance and Control Techniques for Coplanar Rendezvous. IAS National Specialists Meeting on Guidance of Aerospace Vehicles, May 25–27, 1960.
37. A. P. HARRY, and A. L. FRIEDLANDER, Exploratory Statistical Analysis of Planet Approach-Phase Guidance Schemes Using Range, Range-Rate, and Angular-Rate Measurement. NASA TN D–268, 1960.
38. L. J. KAMM, SATRAC: Satellite Automatic Terminal Rendezvous and Coupling. ARS Paper 1497–60, Washington, December 5–8, 1960.
39. O. C. KASTE, and D. NOVAK, Study of the Rendezvous Mission. Martin Company, MLV Tech. Note 13, 1960.
40. C. L. KELLER, Satellite Ascent Paths. Sperry Engr. Rev. December 1958, 2–14.
41. T. A. MAGNESS, J. B. MCGUIRE, and O. K. SMITH, Accuracy Requirements for Inter-Planetary Ballistic Trajectories. Proceedings of the IXth International Astronautical Congress, Amsterdam, 1958, Vol. I, pp. 286–306. Wien: Springer, 1959.

42. G. W. MORGENTHALER, On Mid-Course Guidance in Satellite Interception. Astronaut. Acta **5**, **328–346** (**1959**).
43. C. J. MUNDO, JR., Trade-Off Considerations in the Design of Guidance Equipment for Space Flight. Aero/Space Engineering, June **1959**, pp. **31–34**.
44. M. L. NASON, Terminal Guidance Technique for Satellite Interception Utilizing a Constant Thrust Rocket Motor. ARS Journal **30**, no. 9 (1960).
45. M. L. NASON, Terminal Guidance and Rocket Fuel Requirements for Satellites. Presented at American Rocket Soc. Controllable Satellite Conference, MIT, April 30 – May 1, 1959, ARS paper 777–59.
46. A. W. NELSON, Manned Orbital Rendezvous. ARS Paper no. **1493–60**, December **1960**.
47. B. H. PAIEWONSKY, Transfer Between Vehicles in Circular Orbits. Jet Propulsion **1958**, **121**.
48. A. L. PASSERA, Conditional-Switching Terminal Guidance (A Terminal Guidance Technique for Satellite Rendezvous). PGANE – IRE, December **1960**.
49. R. E. ROBERSON, Gyroscopic Sensing of Satellite Yaw. Presented at the 1st Congress of the International Federation of Automatic Control, Moscow, June **1960**.
50. E. ROBERSON, Path Control for Satellite Rendezvous. Presented at the Sixth Annual Meeting, American Astronautical Society, New York, January **1960**.
51. H. G. SAFREN, Differential Correction Method of Interplanetary Navigation. Proceedings of the National Specialists Meeting on Guidance of Aerospace Vehicles, Boston, Mass., May **25–27**, **1960**.
52. A. J. SKALAFURIS, and D. H. SCHILLER, Midcourse Guidance Problem in Satellite Interception. ARS Journal **1960**, **41–46**.
53. E. SIMON, A Proposed Control System to Facilitate the Terminal Stages of Manned Rendezvous. ARS Paper no. **1480–60**, December **1960**.
54. R. A. SMITH, Establishing Contact Between Orbiting Vehicles. J. Brit. Interplan. Soc. **1951**, **295–299**.
55. P. W. SOULE, Rendezvous with Satellites in Elliptical Orbits of Low Eccentricity. AAS Preprint **60–71**, Seattle, August **8–11**, **1960**.
56. K. S. STEFFAN, A Satellite Rendezvous Terminal Guidance System. ARS Preprint no. **1494–60**, December **1960**.
57. E. A. STENIHOFF, Orbital Rendezvous and Guidance. IAS Manned Space Stations Symposium, April **20–22**, **1960**.
58. R. J. WEBER, and W. M. PAUSON, Achieving Satellite Rendezvous. ARS Journal (Tech. Notes) **29**, **592–595** (**1959**).
59. C. H. WOLOWITZ, H. M. DRAKE, and E. N. VIDEAN, Simulator Investigation of Controls and Display Required for Terminal Phase of Coplanar Orbital Rendezvous. NASA TN D–511, 1960.

Discussion

Mr. STERN: Do your different schemes and plans apply when the target satellite is not launched by the same organization? That is to say, suppose you don't know the orbit characteristics because it is not yours; can you determine the orbit of the satellite sufficiently well so that you could send up a rendezvous satellite? Then I would like to find out whether you have done any experimental tests to verify your calculations?

Mr. HOUBOLT: If the target satellite is orbital and you can observe it for one or more revolutions, there appears to be no major reason why you could not launch a rendezvous satellite, whether the target satellite is yours or someone elses. If the target satellite is in partial orbit, then we have essentially the problem of intercept and this was not intended to be treated here. The answer to your second question is no, we have not made an experiment.

Interplanetary Trajectories for Electrically-Propelled Space Vehicles

By

W. E. Moeckel[1]

(With 4 Figures)

Abstract — Zusammenfassung — Résumé

Interplanetary Trajectories for Electrically-Propelled Space Vehicles. Determination of the space-mission capabilities of electric-propulsion systems requires study of the trajectories followed by space vehicles during powered flight in gravitational fields. In the present paper, solutions to the equations of motion for such flights are discussed. Both constant- and variable-thrust acceleration are considered, and methods of matching prescribed end-point conditions are described. Particular attention is given to the problem of interplanetary transfer, for which the mass ratios needed with various thrust programs are compared.

For round-trip interplanetary missions, it is shown that large reductions in the total mission time can be achieved by following indirect heliocentric paths. For the Earth-Mars mission, it is found that electrically propelled vehicles can achieve total trip times comparable to those attainable with high-thrust rockets, with weights that may be considerably less than those possible with nuclear rockets.

Interplanetare Bahnen für elektrisch angetriebene Raumfahrzeuge. Die Ermittlung der Nutzlast von elektrisch angetriebenen Systemen erfordert zunächst die Bestimmung der Bahn des Fahrzeuges während der Antriebsperiode. In der vorliegenden Arbeit werden Lösungen der Bewegungsgleichungen für derartige Bahnen untersucht. Konstante und veränderliche Beschleunigung wird angenommen, und es werden Methoden beschrieben, um ein vorgegebenes Ziel zu erreichen.

Für interplanetare Fahrten kann die Gesamtzeit der Reise unter gewissen Umständen stark reduziert werden. Für die Erde-Mars-Fahrt zeigt sich, daß elektrisch angetriebene Fahrzeuge Reisezeiten in der Größenordnung der Reisezeit von Raketen mit hohem Schub erreichen können, wobei das Massenverhältnis wesentlich günstiger als bei Atomraketen ist.

Trajectoires interplanétaires pour des véhicules spatiaux à propulsion électrique. La détermination des missions spatiales pouvant être accomplies par des fusées à propulsion électrique nécessite l'étude des trajectoires suivies par les véhicules spatiaux durant des vols propulsés à travers des champs de gravitation. Dans cet article, sont discutées les solutions de l'équation de mouvement correspondant à ces vols. Dans les deux cas, d'une accélération constante et d'une accélération variable, l'auteur fournit des méthodes pour obtenir les conditions prescrites à l'arrivée. Une attention particulière est portée au problème des transferts interplanétaires, pour lesquels on compare les rapports de masse nécessaires, en fonction des diverses poussées envisagées.

Pour des missions interplanétaires aller et retour, il est montré que de grandes réductions de temps peuvent être obtenues au moyen de trajectoires indirectes autour

[1] Chief, Electromagnetic Propulsion Division, National Aeronautics and Space Administration, Lewis Research Center, Cleveland 35, Ohio, U.S.A.

du soleil. Pour la mission Terre-Mars, l'auteur trouve qu'un engin à propulsion électrique peut faire le voyage complet dans un temps comparable à celui obtenu par une fusée à forte poussée, alors que le poids d'une fusée à propulsion électrique peut être considérablement moindre que celui d'une fusée nucléaire.

I. Introduction

During the past few years, a large number of studies have been published on the trajectories followed by space vehicles during powered flight. Most of these studies have been motivated by the need for determining the mission capabilities of space vehicles propelled by low-thrust propulsion systems such as electric rockets. The results of these studies are of use, however, for all propulsion systems that operate with ratios of thrust to weight less than unity, as well as for evaluating the effect of other external forces on the motion of the vehicle. Some of the methods of solution of the equations of motion may be of interest in other fields of science in which the primary force is inversely proportional to the square of the distance.

The present paper, however, like most of the others, is concerned primarily with the trajectories followed by space vehicles propelled by electric rockets. It is further restricted to a discussion of those methods which are particularly useful for preliminary studies of one-way and round-trip interplanetary missions. For such studies, a succession of central gravitational fields is assumed (planet-sun-planet), with thirdbody effects and other perturbations ignored. It is usually assumed that the planetary orbits are circular. These assumptions are justified on the grounds that the major energy changes involved in the actual mission are included in this simplified model.

As is well-known, the ascent from a low orbit about a planet to escape velocity, and the descent from escape velocity to a low planetary orbit are achieved in a nearly-optimum manner by application of constant thrust, directed parallel to the velocity (tangential thrust). Charts are available [1] for calculating the required trajectory parameters for these phases of the mission. For the Earth-planet orbit transfer phases, however, constant tangential thrust is not the most efficient or desirable thrust programming, and many possible alternatives have been investigated. For the one-way trip, an optimization procedure was developed in [2], whereby the time history of the magnitude and direction of the thrust vector could be calculated along the path that produced least mass consumption for a given transfer time. This solution, while providing an excellent reference point for comparison with other programs, has the disadvantage that very large variations in thrust magnitude and specific impulses are required. Such large variations may be difficult to achieve in practice, particularly at the constant jet power level assumed, since efficiency of an electric rocket is a strong function of specific impulse. Consequently, it is pertinent to examine other more realistic thrust programs for interplanetary orbit transfer to determine whether the propellant required is significantly greater than the best possible. Such studies have recently been carried out in [3], where a variational procedure is evolved for optimizing the interorbital transfer using trajectory segments for which the thrust magnitude is constant, but the thrust angle is variable. These segments are joined by zero-thrust matching paths.

For round-trip missions, no optimum programming has as yet been developed. The problem is much more complex than the one-way mission, in that a double planetary rendezvous must be considered. If mission time were not a significant factor, this consideration would not add appreciably to the complexity, since optimum one-way transfers could be used to and from the destination, with waiting time at the destination as an arbitrary parameter to effect the return rendezvous. However, to reduce the total mission time, it is necessary, with low acceleration as well as high acceleration, to follow indirect Earth-planet paths [**4**]. These indirect paths introduce a number of additional variables that make absolute optimization of the entire round-trip trajectory a very tedious undertaking. In lieu of complete optimization, however, it is possible to choose a particular simple family of trajectories, for which it is easy to vary trip times, transit angle, and perihelion or aphelion distance, and to use combinations of this family that minimize the initial weight for a given total trip time. This procedure, which was introduced and evaluated in [**4**], enables one to undertake extensive preliminary analyses of round-trip interplanetary missions with a facility almost comparable to that possible with high-thrust trajectories. Since such an analysis yields results that are not necessarily near optimum, a proponent of electric propulsion has the satisfaction of stating that it should be possible to do even better than this with further studies.

II. Equations of Motion and Mass Ratio

The equations of motion of a vehicle propelled by thrust force F in a central gravitational field are:

$$\ddot{r} - r\,\dot{\theta}^2 + \frac{\mu}{r^2} = \frac{F}{m}\sin\delta \tag{1}$$

$$2\,\dot{r}\,\dot{\theta} + r\,\ddot{\theta} \equiv \frac{1}{r}\,\frac{d(r^2\,\dot{\theta})}{dt} = \frac{F}{m}\cos\delta \tag{2}$$

where

- r radial coordinate
- θ angular coordinate (true anomaly)
- μ gravitational constant of field-producing body (GM)
- F thrust
- m mass
- δ angle between thrust vector and normal to radius vector
- t time

If these equations are nondimensionalized in terms of the radius r_0, gravitational acceleration g_0, and circular velocity v_{co}, at some reference orbit, they become:

$$\varrho'' - \varrho\,\theta'^2 + \varrho^{-2} = a\sin\delta \tag{3}$$

$$\varrho\,\theta'' + 2\,\varrho'\,\theta' \equiv \frac{1}{\varrho}\,\frac{d(\varrho^2\,\theta')}{d\tau} = a\cos\delta \tag{4}$$

where

$$\varrho = \frac{r}{r_0}, \qquad \tau = \frac{v_{c0}\,t}{r_0}, \qquad a = \frac{F}{m\,g_0}$$

and where primes indicate derivatives with respect to τ. The nondimensional velocity is $V = v/v_{co}$, and the radial and circumferential velocity components are

$$V_r \equiv \frac{v_r}{v_{c0}} = \varrho' \qquad \text{and} \qquad V_\theta \equiv \frac{v_\theta}{v_{c0}} = \varrho\,\theta' \tag{5}$$

Other parameters of importance for powered trajectory studies are the dimensionless angular momentum H, total energy E, eccentricity ε, and semimajor axis A. These are given by

$$H \equiv \frac{r\, v_\theta}{r_0 v_{c0}} = \varrho\, V_\theta = \varrho^2\, \theta' \tag{6}$$

$$E = \frac{1}{v_{c0}^2}\left(\frac{v^2}{2} - \frac{\mu}{r}\right) = \frac{V^2}{2} - \frac{1}{\varrho} = \frac{1}{2}\,[\varrho'^2 + \varrho^2\,\theta'^2] - \frac{1}{\varrho} \tag{7}$$

$$\varepsilon = (1 + 2\,E\,H^2)^{1/2} \tag{8}$$

$$A = -\frac{1}{2\,E} \tag{9}$$

From eqs. (3), (4), (6), and (7), the time derivatives of H and E can be written [5]

$$H' = a\,\varrho\cos\delta \tag{10}$$

$$E' = a\,[\varrho'\sin\delta + \varrho\,\theta'\cos\delta] \tag{11}$$

From eqs. (10) and (11), the following statements can be made regarding the direction of the thrust vector δ:

(1) Circumferential thrust ($\delta = 0$) maximizes the rate of change of angular momentum.

(2) Radial thrust ($\delta = \pi/2$) maintains constant angular momentum.

(3) Tangential thrust (thrust parallel to the velocity vector, $\tan\delta = \varrho'/\varrho\,\theta'$) maximizes the rate of change of E.

(4) Thrust normal to the velocity vector ($\tan\delta = -\,\varrho\,\theta'/\varrho'$) maintains constant energy.

Mathematically, the simplest assumption regarding the magnitude of the thrust vector (as distinguished from the direction) is that it varies in such a manner that the thrust acceleration a is constant. This, in turn, implies that the thrust decreases as the mass m is consumed.

Slightly more complicated equations of motion are obtained if the thrust magnitude (rather the thrust acceleration) is assumed to be constant. In this case, the mass m, in eqs. (3) and (4), is given by [1]:

$$m = m_0 - \frac{F\,t}{v_j} = m_0\left(1 - \frac{a_0\,\tau}{V_j}\right) \tag{12}$$

where $a_0 = F_0/m_0\,g_0$, v_j is jet velocity, and $V_j = v\,/v_{c0}$. The thrust acceleration, a, is then

$$a = \frac{a_0}{1 - \dfrac{a_0\,\tau}{V_j}} \tag{13}$$

Obviously, solutions for the case $a =$ constant are obtained as a special case ($V_j = \infty$) of the solutions for constant thrust magnitude. (A slightly more general case, in which mass other than propellant is utilized and continuously ejected during the powered flight, is considered in [1] and [4].)

An excellent survey of the methods used by various authors to solve the equations of motion for the preceding special cases and the results of these methods, is contained in [5]. For interplanetary missions, the constant tangential thrust (or thrust acceleration) is the most useful of these special cases, since some phases of such a mission require simply that escape velocity, or a specified hyperbolic excess velocity, be attained, as economically as possible, relative to the central body. With constant thrust or acceleration, the least propellant mass is consumed when these energies are attained in the least possible time.

For the Earth-planet transfer phase of an interplanetary mission, however, it is necessary to achieve a specified end-point velocity vector. For this program, none of the above special cases, or combinations thereof, are likely to produce the most economical transfer path. It is shown in [2], in fact, that the most economical path for such a transfer involves large variations in thrust magnitude and direction. The mass-ratio equation for such variable-thrust programs is somewhat more complex than eq. (12). For arbitrary thrust programs (but constant jet power), this equation is obtained as follows [2]). The jet power, P_j, is given by

$$P_j = \frac{1}{2}\dot{m} v_j^2 = \frac{1}{2} F v_j = \text{constant} \tag{14}$$

and the mass decrement is

$$dm = -\frac{F\,dt}{v_j} = -\frac{F^2}{F v_j}\,dt = -\frac{m^2 a^2 g_0^2}{2 P_j}\,dt \tag{15}$$

Integration of (15) yields:

$$\frac{1}{m_0} - \frac{1}{m_1} = -\frac{g_0^2}{2 P_j}\int_0^{t_1} a^2\,dt$$

or

$$\frac{m_1}{m_0} = \left[1 + \frac{m_0 g_0 v_{c0}}{2 P_j}\gamma\right]^{-1} = \left[1 + \frac{\gamma}{a_0 V_{j0}}\right]^{-1} \tag{16}$$

where

$$\gamma = \int_0^{\tau_1} a^2\,d\tau \tag{17}$$

This integral, γ, of the square of the nondimensional acceleration determines the mass ratio, and should be minimized as much as possible for each mission.

For the case of constant thrust acceleration ($a = a_0$), the mass-ratio eq. (16) becomes:

$$\frac{m_1}{m_0} = \left(1 + \frac{a_0 \tau_1}{V_j}\right)^{-1} \tag{18}$$

which also is somewhat different from the equation for constant thrust magnitude [eq. (12)]. For mission phases employing constant thrust acceleration or thrust magnitude, the quantity $a_0 \tau$, which is the dimensionless total impulse, must be determined. For more general thrust programming, the quantity γ must be obtained.

A particular family of variable-thrust trajectories which yield a simple expression for the mass-ratio parameter, γ, required for transfer between circular orbits has recently been studied by the author [4]. The assumption was made that $\theta' = \varrho^{-3/2}$ at all points on the trajectory. This assumption implies that the circumferential component of the velocity vector is equal to the local circular velocity at all times. The equations of motion (3) and (4) then assume the particularly simple form

$$\varrho'' = a \sin \delta \tag{19}$$

$$\frac{1}{2}\varrho^{-3/2}\varrho' = a \cos \delta \tag{20}$$

and the expression for γ becomes

$$\gamma = \int_0^{\tau_1} \left[(\varrho'')^2 + \frac{1}{4} \varrho^{-3} (\varrho')^2 \right] d\tau \tag{21}$$

Although the assumption $\theta' = \varrho^{-3/2}$ defines the relation between the two independent variables θ and ϱ, it still permits arbitrary variations of ϱ as a function of time. For transfer between two circular orbits, an expression for ϱ must be used that satisfies the following boundary conditions:

$$\left.\begin{array}{llll} \text{At} & \tau = 0: & \varrho = 1, & \varrho' = 0 \\ \text{At} & \tau = \tau_1: & \varrho = \varrho_1, & \varrho' = 0 \end{array}\right\} \tag{22}$$

The simplest expression that satisfies these conditions is

$$\varrho = 1 + (\varrho_1 - 1) \left(\frac{\tau}{\tau_1}\right)^2 \left(3 - 2\frac{\tau}{\tau_1}\right) \tag{23}$$

Denoting τ/τ_1 by ξ, the expressions for ϱ and its derivatives are

$$\varrho = 1 + (\varrho_1 - 1)\, \xi^2 (3 - 2\,\xi) \tag{24}$$

$$\varrho' = \frac{6(\varrho_1 - 1)}{\tau_1} \xi (1 - \xi) \tag{25}$$

$$\varrho'' = \frac{6(\varrho_1 - 1)}{\tau_1^2} (1 - 2\,\xi) \tag{26}$$

Substitution of eqs. (24) to (26) into (21) yields:

$$\gamma = \frac{3(\varrho_1 - 1)^2}{\tau_1} \left[\frac{4}{\tau_1^2} + 3\,F(\varrho_1) \right] \tag{27}$$

where

$$F(\varrho_1) = \int_0^1 \frac{\xi^2 (1 - \xi)^2\, d\xi}{[1 + (\varrho_1 - 1)\, \xi^2 (3 - 2\,\xi)]^3} \tag{28}$$

Other expressions for ϱ as function of time may be found to satisfy boundary conditions for problems other than transfer between circular orbits.

In undertaking mission studies, it is also necessary to know the angular distance, θ_1, travelled during the transfer. For the $\theta' = \varrho^{-3/2}$ family of trajectories,

$$\theta_1 = \int_0^{\tau_1} \varrho^{-3/2}\, d\tau$$

$$= \tau_1 \int_0^1 \varrho^{-3/2}\, d\xi \tag{29}$$

Evaluation of the integrals in eqs. (28) and (29) showed that, for considerable ranges of ϱ_1, excellent approximate expressions for γ and θ are

$$\gamma = \frac{3(\varrho_1 - 1)^2}{\tau_1} \left[\frac{4}{\tau_1^2} + 0.10\, \varrho_1^{-3/2} \right] \tag{30}$$

$$\theta_1 = \tau_1\, \varrho_1^{-3/4} \text{ radians} \tag{31}$$

In [4], eq. (30) is shown to be valid to very good approximation for $0.1 < \varrho_1 < 20$, while eq. (31) is very good for $0.3 < \varrho_1 < 3.0$. These ranges of validity are more than adequate for use in evaluating Earth-Mars or Earth-Venus transfers. This family of trajectories, although not necessarily optimum, is particularly useful for analyzing round-trip missions for which it is necessary to follow indirect interplanetary paths if moderate mission times are desired. Such indirect paths are made up of two transfers between circular orbits, the first to an arbitrary radius, ϱ_a, inside or outside of the destination orbit, and the second from ϱ_a to the destination radius, ϱ_c. If a portion, K, of the total transfer time τ_c is allotted to the first transfer segment, the value of γ for the entire indirect transfer becomes:

$$\gamma = \frac{3(1-\varrho_a)^2}{K\tau_c}\left[\left(\frac{2}{K\tau_c}\right)^2 + 0.10\,\varrho_a^{-3/2}\right] + \frac{3(\varrho_c-\varrho_a)^2}{(1-K)\tau_c}$$
$$\left\{\left[\frac{2}{(1-K)\tau_c}\right]^2 + 0.10\,(\varrho_a\varrho_c)^{-3/2}\right\} \tag{32}$$

As shown in [4], the value of K that minimizes γ for gives values of ϱ_a, ϱ_c and τ_c is given to good approximation by:

$$\frac{1-K}{K} = A^{1/2} \tag{33}$$

where

$$A = \left|\frac{\varrho_c - \varrho_a}{1-\varrho_a}\right| \tag{34}$$

The resulting minimum value of γ for the indirect transfer is then

$$\gamma = \tau_c^{-3} f_1(\varrho_c, \varrho_a) + \tau_c^{-1} f_2(\varrho_c, \varrho_a) \tag{35}$$

where

$$f_1(\varrho_c, \varrho_a) = 12(1-\varrho_a)^2(1+\sqrt{A})^4 \tag{36}$$

$$f_2(\varrho_c, \varrho_a) = 0.30\,(1-\varrho_a)^2(\varrho_a)^{-3/2}(1+\sqrt{A})\left[1+\left(\frac{A}{\varrho_c}\right)^{3/2}\right] \tag{37}$$

The angular distance, θ, for the indirect trip becomes

$$\frac{\theta}{\tau_c} = \frac{1+\varrho_c^{-3/4}\sqrt{A}}{\varrho_a^{3/4}(1+\sqrt{A})} \tag{38}$$

Even more useful for interplanetary round-trip mission studies is the lead angle, φ, acquired during the transfer by the vehicle relative to the initial planet (Earth). Any lead angle acquired during the transfer to the destination planet (and during the time spent in the vicinity of that planet), must be reduced to zero or augmented to 2π on the return trip if the rendezvous with the initial planet is to be achieved. The lead angle for the transit is

$$\varphi = \theta - \frac{v_{c,E}}{r_E}\,t_c \tag{39}$$

where $v_{c,E}$ and r_E are the orbital speed and radius of the Earth. With curves of φ/t_c as functions of ϱ_a for a given Earth-planet radius ratio ϱ_c, the intermediate radius ϱ_a required for the return trip can be determined when the outward trip and waiting times are specified. The required relation is

$$\varphi_{\text{return}} = 2\,n\pi - \varphi_{\text{out}} - \varphi_{\text{wait}} \tag{40}$$

For fast round trips, the case $n = 0$ is of most interest.

Several methods have been used to serve as a guide in determining the thrust programming required to satisfy the end-point condition for interplanetary transfer. One of these, due to RODRIGUIZ [6], is a plot of energy as functions of angular momentum. Shown in Fig. 1 are the paths in an $E - H$ coordinate system of several of the special thrust programs discussed in this section. This figure shows that, in addition to the variable-thrust programs, the interplanetary transfer phase of a mission can be accomplished by combinations of segments

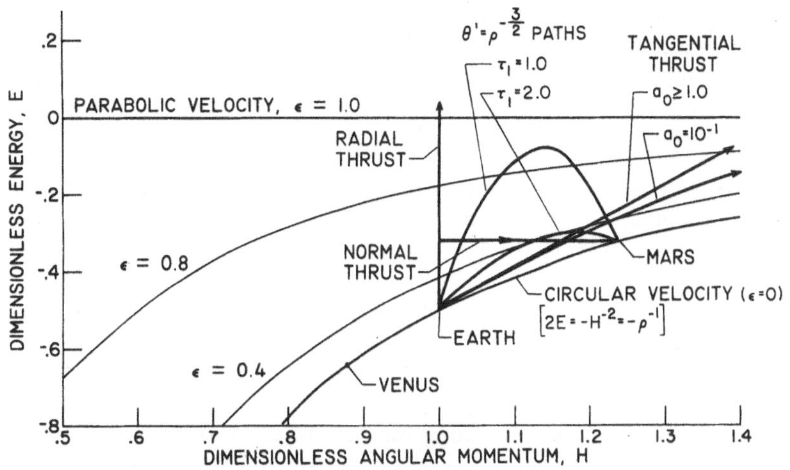

Fig. 1. $E - H$ diagram for several thrust programs

of constant thrust (tangential, radial, or normal). Since any point on the $E - H$ plot corresponds to a particular zero-thrust coasting trajectory, it is possible to match the powered segments with an appropriate unpowered segment. Such plots, of course, yield no information on the efficiency of the particular thrust programs.

III. Applications and Results

Among the many missions for which electric-propulsion systems appear very promising, for reducing initial weight requirements, are interplanetary space probes and, eventually, manned expeditions to the near planets. Such missions can generally be divided into seven phases: (1) Earth escape, (2) Earth-planet transfer, (3) descent to planetary orbit (or surface), (4) waiting or exploration period, (5) escape from planet, (6) planet-Earth transfer, and (7) descent to Earth orbit or surface. For one-way interplanetary probes, only the first two or three of these phases need be considered, while for round-trip missions, all must be included. With high-thrust propulsion systems, only phases (1), (3), (5), and (7) involve propulsion periods, while with low-thrust systems, all phases except phase (4) may involve continuous propulsion. Phases (1), (3), (5), and (7) are best accomplished with constant tangential thrust, and can be analyzed with the aid of charts such as those of [1]. For phases (2) and (6), any of the available methods of calculating transfers between circular orbits can be considered, although only the method of [4] is so far available for indirect paths such as those needed for fast round-trip missions.

To illustrate the procedures and results of powered-flight trajectory studies, two rather specific missions will be considered: (1) a one-way Mars probe, and (2) a round-trip manned Mars expedition.

1. One-Way Mars Probe Mission

During the present decade, it may become feasible to perform this mission with electric-propulsion systems now entering the development stage. Some preliminary weight and performance estimates for such systems are available. It is contemplated that the so-called SNAP—8 nuclear turboelectric system will produce about 60 kilowatts of electric power with a total weight that may be in the vicinity of 3000 pounds. A space-probe vehicle weighing about 9000 pounds could be launched into a low orbit (about 400 miles) around the Earth by an Atlas-Centaur rocket. Using these figures, we can estimate the weight that can be delivered into a low orbit around Mars as function of the total trip time.

The procedure consists of first calculating the mass ratio as a function of escape time for the Earth-escape phase (phase 1). For this purpose it is necessary to calculate a_0 and V_j as a function of escape time so that $(a_0 \tau)$ can be obtained from the charts of [1]. If the ratio of thrust to initial weight F/W_i is denoted by a_{00}, eq. (14) yields:

$$a_{00} = \frac{F}{W_i} = \frac{45.9}{I} \frac{P_j}{W_i} = \frac{45.9\,\eta}{I} \frac{P_E}{W_i} \tag{41}$$

where power is expressed in kilowatts, specific impulse I in seconds, and weight in pounds. The factor η is the efficiency of conversion of electric power P_E to jet power. This efficiency is a strong function of the type of electric thrust unit used and of the specific impulse I. For an ion rocket of the electron-bombardment type, [7] indicates that the following variation of efficiency with specific impulse should be attainable with little additional development:

I	1000	1500	2000	2500	3000	4000	5000	6000
η	0.165	0.31	0.45	0.55	0.64	0.76	0.83	0.86

These values are somewhat higher than have presently been attained, but are probably lower than can eventually be achieved, particularly for the higher values of specific impulse.

The initial thrust acceleration for phase (1) is

$$a_{01} = \frac{F}{m_i g_1} = a_{00} \frac{g_{00}}{g_1} \tag{42}$$

where g_{00} and g_1 are the gravitational accelerations on the Earth's surface and in the initial Earth orbit, respectively. The dimensionless jet velocity is

$$V_{j1} = \frac{g_{00} I}{v_{c1}} \tag{43}$$

With a_{01} and V_{j1} known, the value of $(a_0 \tau)$ required to reach escape conditions is found from [1]. For the range of specific impulses shown in the above table, values of $(a_0 \tau)$ between 0.66 and 0.87 are obtained. The escape time is then found from

$$t_1 = \frac{(a_0 \tau)_1}{a_{01}} \frac{r_1}{v_{c1}} \tag{44}$$

and the mass ratio for phase (1) is

$$\frac{m_1}{m_0} = 1 - \frac{(a_0 \tau)_1}{V_{j1}} \tag{45}$$

For an initial orbit at $r_1 = 4360$ miles, $v_{c1} = 4.69$ miles/second, and $g_1 = 26.6$ feet/second², these equations become

$$a_{01} = 45.9\left(\frac{60}{9000}\right)\left(\frac{\eta}{I}\right) \tag{46}$$

$$V_{j1} = 1.3 \times 10^{-3}\, I \tag{47}$$

$$t_1 = 0.0107\,\frac{(a_0\,\tau)_1}{a_{01}}, \quad \text{days} \tag{48}$$

Variations of the parameter I then permit calculation of mass at the end of phase (1) as a function of escape time t_1.

Using several selected values of t_1, the mass at the end of phase (2) is now calculated as functions of t_2, the Earth-Mars transfer time. The calculation varies, depending on the thrust programming assumed for this phase. If the variable-thrust transfers of [2] or [4] are used, eq. (16) yields the mass ratio. However, this equation is obtained under the assumption that the jet power (and hence the efficiency) is constant, even though large variations in specific impulse are required. To be conservative, the constant jet power can be taken as the smallest value attained during the transfer. Eq. (16) then becomes

$$\frac{m_2}{m_1} = \left[1 + \frac{m_1\, g_2\, v_{c2}\, \gamma_2}{0.28\,\eta P_E}\right]^{-1} \tag{49}$$

where g_2 and v_{c2} are the heliocentric gravitational acceleration and orbital velocity at the Earth's orbit (0.0196 ft/sec², and 18.5 miles/sec, respectively). Curves of m_2 as functions of $(t_1 + t_2)$ can now be plotted for each of the selected values of t_1. The envelope of these curves yields the largest values of m_2 attainable as a function of $(t_1 + t_2)$. Several points on the envelope can now be selected as initial conditions for phase (3), the descent to a low orbit around Mars. The procedure is similar if other phase (2) thrust programs are used, such as the combination of constant thrust and coast segments of [3], except that the appropriate form of the mass-ratio equation may differ.

For phase (3), the expressions for a_{03}, V_{j3}, and t_3 are:

$$a_{03} = \frac{F}{m_3\, g_3} = a_{00}\,\frac{m_i\, g_{00}}{m_3\, g_3} \tag{50}$$

$$V_{j3} = \frac{g_{00}\, I}{v_{c3}} \tag{51}$$

$$t_3 = \frac{(a_0\,\tau)_3}{a_{03}}\,\frac{r_3}{v_{c3}} \tag{52}$$

Since the appropriate value of a_0 for descent phases is the value attained when the final orbit is reached, the final mass m_3 appears in eq. (50). Although this mass is not initially known, little difficulty is encountered in practice, because $(a_0\,\tau)$ is quite insensitive to moderate variations in a_0. Consequently, a_{03} can be estimated sufficiently closely to determine $(a_0\,\tau)$, which, together with V_{j3}, yields the final mass m_3.

The problem of finding the final mass as a function of total trip time is essentially completed when curves of m_3 are plotted as functions of $(t_1 + t_2 + t_3)$ for several initial values of $(t_1 + t_2)$. The envelope of such curves then yields the maximum value of m_3 attainable as function of trip time, as well as the optimum distribution between t_1, t_2, and t_3. Results of such an analysis are shown in Fig. 2 for three types of phase (2) trajectory programming. The highest mass ratio was obtained with the variable thrust optimum programming of [2]. Net payloads, in addition to electric power supply, are obtained for trip times greater than 240 days. Considerably longer trip times are required, for the same mass ratio, using the optimized constant-thrust programming of [3]. In this programming, the direction of the thrust vector, but not its magnitude, varies

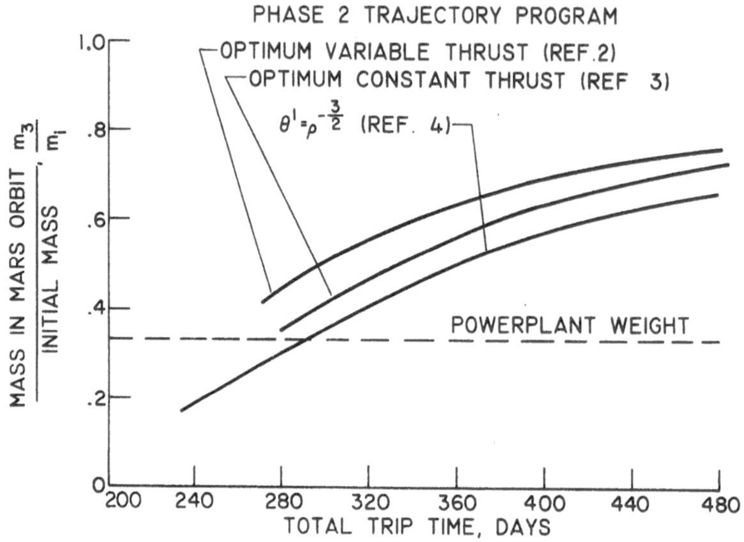

Fig. 2. Weight ratio for Mars probe mission. Initial weight, 9000 lbs; electric power, 60 KW

continuously during powered portions of the flight, and a coasting period is inserted between the powered segments. The duration of the coasting period was also optimized. The authors of [3] found, in addition, that the mass ratio obtained with the variable-direction programming could be very closely duplicated by using powered-flight segments in which the direction of the thrust vector, as well as its magnitude, was fixed. For this programming, the best thrust angles, relative to the velocity vector, were in the range $+ 20°$ to $60°$ during departure from the Earth's orbit, and $- 20°$ to $- 60°$ on approach to the Mars orbit.

The lowest of the three mass-ratio curves is obtained with the $\theta' = \varrho^{-3/2}$ trajectories. This family is therefore of little use for one-way interplanetary missions. This result indicates also that the weight ratios obtained for round-trip missions in the following section using $\theta' = \varrho^{-3/2}$ trajectories may be pessimistic, and that considerable future improvement is to be expected in the weight ratios as functions of trip time using electric-propulsion systems.

2. Round-Trip Mars Mission

Round-trip Mars missions using electric-propulsion systems have been analyzed as functions of total trip time only for the $\theta' = \varrho^{-3/2}$ family of transfer trajectories. The procedure, which is described in detail in [4], is similar to that used for one-way missions, except that more optimization processes are required because of the additional phases of the mission. In addition, since round-trip missions are primarily of interest for manned expeditions, the reduction in supplies due to consumption en route and the abandonment of exploration equipment at Mars should be considered. These features of manned missions are included in [4], but will be omitted in the present paper.

Table I shows the near-optimum breakdown of times required for the phases of a round-trip Mars mission for various total round-trip times. These results were obtained for a value of $t_w = 50$ days, where t_w is the time spent in the

Table I. *Near-Optimum Parameters for Earth-Mars Round Trips for $t_w = 50$ Days Using $\theta' = \varrho^{-3/2}$ Trajectories*

Total trip time, days	t_1	t'	t_2	t_3	t_4	t_5	t_6	t_7	Γ_2, kw/lb	Γ_6, kw/lb	$\varrho_{a,6}$
300	10	275	64	15	25	10	161	15	0.100	0.170	0.58
350	15	320	81				189	15	0.050	0.100	0.60
400	15	370	98				222	15	0.0315	0.063	0.61
450	20	410	110				250	20	0.023	0.044	0.62
500	25	455	122				283	20	0.0165	0.031	0.62
550	30	495	130				315	25	0.0132	0.0235	0.625
600	35	540	148				342	25	0.0095	0.0185	0.63
700	50	620	160				410	30	0.0060	0.0124	0.63
800	70	700	190	↓	↓	↓	460	30	0.0043	0.0090	0.63

vicinity of Mars, and includes phases (3), (4), and (5) of the mission. The time t' in Table I is the round-trip time excluding escape from and descent to the Earth [phases (1) and (7)]. The Γ's of Table I are given by

$$\Gamma = \frac{v_{c0}}{0.28} \frac{g_0}{g_{00}} \gamma, \qquad \frac{\text{kw}}{\text{lb}} \tag{53}$$

which, for the Earth-Mars transfer, is

$$\Gamma = 0.04\,\gamma \qquad \text{kw/lb} \tag{54}$$

The mass ratio equation, in terms of Γ, is [from eqs. (53) and (16)]:

$$\frac{m_2}{m_1} = \left(1 + \frac{m_1}{m_i} \frac{\alpha}{\beta} \Gamma\right)^{-1} \tag{55}$$

where α is the specific powerplant weight (powerplant weight per kw of jet power), and β is the ratio of powerplant weight to initial total vehicle weight.

A typical round-trip trajectory is shown in Fig. 3. It was found in [4] that best results were obtained if one of the Earth-planet transfers was direct and the other indirect. For the 500-day trip shown in Fig. 3, the outward trip is the direct one. On the indirect return trip, the trajectory passes inside the Venus orbit. The minimum radius on the return trip $\varrho_{a,6}$ was found to be rather insensitive to total mission time, as shown in Table I. Also shown in Fig. 3 is the optimum breakdown of the 500-day trip time with 25-day waiting period for a vehicle propelled by high-thrust rockets [4].

Shown in Fig. 4 are the ratios of initial mass of the space vehicle (the mass launched into orbit around the Earth) to the payload mass (not including electric powerplant) returned to Earth as functions of total mission time for several values of specific powerplant weight α. A value of β of 0.2 was assumed for these calculations. This value is near optimum for this mission.

The range of α from 5 to 10 pounds per kilowatt is generally considered to be feasible for nuclear turboelectric systems in the megawatt power range of interest for these missions [8]. Values as low as 2.5 pounds per kilowatt might be possible if the weight of the waste heat radiator can be reduced by schemes such as the moving belt [9].

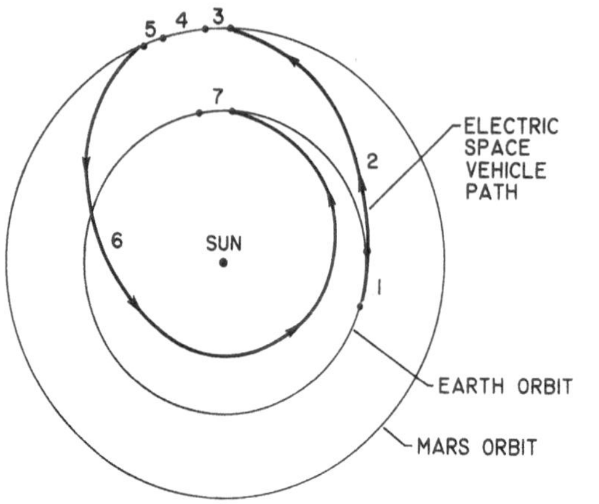

Fig. 3. 500-day Mars round trip path using $\theta' = \varrho^{-3/2}$ trajectories

Also shown in Fig. 4 for comparison are weight ratios obtained for vehicles propelled with nuclear heat-transfer rockets having specific impulses of 800 and 1000 seconds. These curves are obtained using the optimum high-thrust round-trip trajectories found in [10]. The two branches of these curves result from the fact that different families of high-thrust paths turn out to be optimum in the two ranges of total mission time shown.

Time in days

Propulsion System	Escape from Earth	Earth to Mars	Descent to Mars	Time at Mars	Escape from Mars	Mars to Earth	Descent to Earth	Total
Electric	25	122	15	25	10	283	20	500
High-Thrust	—	226	—	25	—	249	—	500

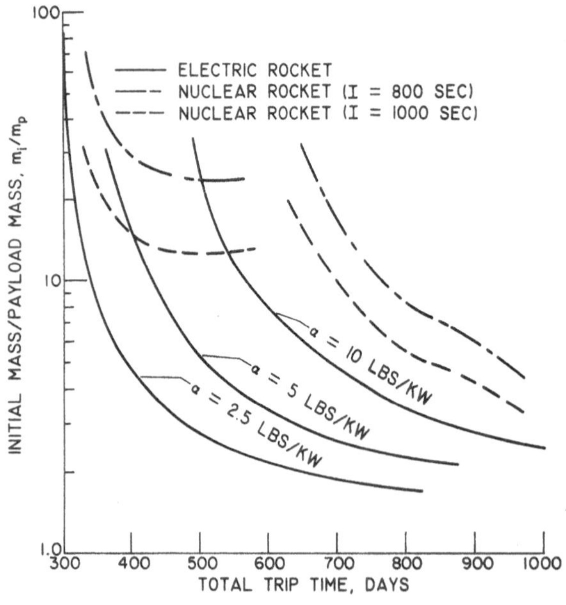

Fig. 4. Weight-ratio comparison for Mars round trips

Fig. 4 shows that, for the range of mass ratios below 10, electrically propelled space vehicles can accomplish a round-trip Mars mission faster than nuclear-rocket-propelled vehicles, provided that specific powerplant weights of the order of 10 pounds per kilowatt or less can be developed. This result is somewhat contrary to common belief that electrically propelled vehicles, because of their low thrust, are not competitive timewise with high-thrust systems for trips to the near planets. This result is perhaps even more impressive when it is recognized that no attempt has yet been made to optimize the interplanetary transfer phases of the electric-rocket trajectories.

IV. Concluding Remarks

Studies of interplanetary missions using electrically propelled vehicles are basically more tedious than similar studies with high-thrust systems. Closed-form solutions for the trajectory equations are available for only a few special types of thrust programming, and these are rarely the best possible for the missions of interest. The tangential-thrust programming, which is nearly optimum for transfer between circular and parabolic or hyperbolic orbits with no end-point requirements, does not have a closed-form solution, but can fortunately be generalized sufficiently to permit construction of charts from which all necessary variables can be evaluated. Such charts greatly simplify analyses of some phases of interplanetary missions. For the remaining phases, and for optimization of the relation between phases, variational methods and iterative numerical integration must be relied upon to select, out of the infinity of possible programs, those most likely to produce the best results. Such numerical studies are now proceding in several laboratories. The results of these studies are likely to lead to many simple approximate methods that can be used for preliminary mission analysis.

Beyond this, there is the need for complete, precision trajectory and mission studies in three dimensions, taking into account all significant perturbations and three-body effects. Computer programs for such studies are already available, and will be used initially to determine which of the factors neglected in the coplanar, two-body studies is significant, and the magnitude of the resulting errors.

References

1. W. E. Moeckel, Trajectories with Constant Tangential Thrust in Central Gravitational Fields. NASA TR R-53, **1960**.
2. J. H. Irving and E. K. Blum, Comparative Performance of Ballistic and Low-Thrust Vehicles for Flight to Mars. In: Vistas in Astronautics. London: Pergamon Press, Inc., **1960**.
3. J. A. MacKay, L. G. Rossa, and A. V. Zimmerman, Optimum Low-Acceleration Trajectories for Earth-Mars Transfer. Presented at IAS Meeting on Vehicle Systems Optimization, Garden City, Nov. 28—29, **1961**.
4. W. E. Moeckel, Fast Interplanetary Missions with Low-Thrust Propulsion Systems. NASA TR R-79, **1961**.
5. E. Levin, Low Acceleration Transfer Orbits. Rand Corporation Report P-2038, July 15, **1960**. (To become a chapter in: Handbook of Astronautical Engineering. New York-Toronto-London: McGraw-Hill.)
6. E. Rodriguiz, Method of Determining Steering Programs for Low-Thrust Interplanetary Vehicles. ARS Journal **29**, 783 (1959).
7. H. R. Kaufman and P. D. Reader, Experimental Performance of Ion Rockets Employing Electron Bombardment Ion Sources. ARS Paper No. 1374—60, Presented at ARS Electrostatic Propulsion Conference, Monterey, Calif., Nov. 3—4, **1960**.
8. R. E. English, H. O. Slone, D. T. Bernatowicz, E. H. Davison, and S. Lieblein, A 20,000-Kilowatt Nuclear Turboelectric Power Supply for Manned Space Vehicles. NASA MEMO 2—20—59E, **1959**.
9. R. C. Weatherston and W. E. Smith, A Method of Heat Rejection from Space Powerplants. ARS Journal **30**, 268 (1960).
10. J. F. Dugan, Jr., Analysis of Trajectory Parameters for Probe and Round-Trip Missions to Mars. NASA TN D-281, **1960**.

Discussion

Mr. MOECKEL: I may mention that Professor LAWDEN has published several studies of low-thrust trajectories. He found, I believe, that optimal thrust direction for escape is not tangential.

Mr. LAWDEN: Yes, for a given thrust, optimal conditions are obtained by directing the thrust between the circumferential and the tangential direction, but actually the superiority of that optimal program is very marginal indeed and certainly would not be of any practical utility.

Mr. MOECKEL: There is very little difference up to escape conditions, but if excess hyperbolic velocities are required, these differences become increasingly large.

Mr. FRAEIJS DE VEUBEKE: Have you compared your results with trajectories using simple navigation device like keeping the same line of sight between the space vehicle, the target planet, and the earth?

Mr. MOECKEL: A group at my laboratory is trying to find systems of programmed thrust which are simpler than the optimal programming, and which yield propellant weight requirements not too much greater than the optimal values. They have found, as I mentioned in the paper, that programs using constant thrust magnitude and direction interspersed with coasting periods, can be used, and that the weight penalties are rather small, at least for one-way planetary missions. Similar studies for round-trip mission are now being carried out.

Mr. STERN: I would like to know with what kind of rocket you can realize these trajectories. Of which kind of practical rocket do you think when you do your calculations?

Mr. MOECKEL: There are numerous research and development programs under way which should produce the powerplants and thrust units needed for these trajectories. At present, for example, we have an ion thrust unit at my laboratory which produces efficiencies in the range 60–80 percent of the specific impulses needed for one-way Mars or Venus trips. The so-called SNAP-8 nuclear-electric power system, which is now under development, is intended to supply the powerplant needed for these early interplanetary trips. As time goes on, we hope to develop larger systems with weight and performance needed for the manned round-trip missions.

The Stability of the 24-Hour Satellite

By

Ladislav Sehnal[1]

(With 7 Figures)

Abstract – Zusammenfassung – Résumé

The Stability of the 24-Hour Satellite. The perturbations of the 24-hour satellite, produced by the Moon, Sun and the aspherical shape of the Earth, are investigated. The most important influence of the lunar gravitational attraction is studied with the aid of numerical integration of the equations of motion. The shift of the satellite from its initial position is studied for an ideal 24-orbit. The principal shift will have a periodical character with an amplitude of about 10°.

Die Stabilität des 24-Stunden-Satelliten. Die Bahnstörungen des 24-Stunden-Satelliten, bewirkt durch Mond, Sonne und durch Abweichungen des irdischen Gravitationsfeldes von der Kugelsymmetrie, werden untersucht. Der wichtigste Fall, der Einfluß des Gravitationsfeldes des Mondes, wird numerisch behandelt. Die Abweichungen von der 24-Stunden-Bahn werden untersucht. Der Hauptanteil der Drift ist periodisch mit einer Amplitude von etwa 10°.

La stabilité du satellite de 24 heures. L'auteur étudie le mouvement du satellite de 24 heures dans le champ de gravité de la Terre et l'influence de la Lune sur son orbite. Le calcul est fait par intégration numérique et l'on calcule la trajectoire du satellite durant une révolution de la Lune autour de la Terre.

I. Introduction

One of the important tasks, for which the artificial satellites are launched, is to establish a communication system. The utilizing of the "24-hour" satellite for this purpose was suggested in many projects of such a system. This satellite would move in the plane of the Earth's equator with the period of 24 sidereal hours, so that it would remain stable with regard to certain place on the Earth's surface.

However, the orbit of such a satellite is also of interest in astrodynamics since, owing to the commensurability of the period of the satellite with the rotation of the Earth, there arise longperiodic perturbations as a result of the asphericity of the Earth's equator. On the other hand, the orbit of the satellite is at such a great distance from the Earth, that the influence of the Moon's and the Sun's gravity will be considerable — the main perturbing force being the lunar attraction. Therefore, in this study of the stability of the 24-hour satellite, the influence of the lunar gravitational attraction is studied in detail, by means of the numerical integration of the equations of motion. The elements of the integrated orbit have been then taken as a basis for further investigations.

[1] C. Sc., Astronomical Institute of the Czechoslovak Academy of Sciences, Ondřejov, Č.S.S.R.

II. The Orbit in the Earth-Moon System

The influence of the Moon's gravity was studied by numerical integration of the equations, describing the motion of a body in the restricted three-body system — the Earth, the Moon and the satellite. The orbit of the satellite was chosen so that the period of the first revolution be exactly 24 sidereal hours and the shape of the orbit be as nearly circular as possible. Therefore the selected orbit is an idealized one. The initial coordinates of the satellite's position and velocity components were computed from the equations describing the motion of two bodies, the satellite around the Earth. The numerical constants, such as the mass of the Earth and the Moon, the gravitational constant, the distance Earth-Moon were taken from the paper by Herrick, Baker and Hilton [1]. The radius of the satellite's circular orbit is $a = 42\,161.7$ km and the initial velocity is 2.98 km/sec. The inclination of the orbit was obtained as the difference between the inclinations of the Earth's equator and the Moon's orbit to the ecliptic.

The reference frame used is plotted in Fig. 1. The origin of the coordinate system is in the centre of gravity of the Earth and the Moon and the x, y-plane lies in the plane of the lunar orbit. The reference frame rotates uniformly with the motion of the two bodies around the centre of gravity, the z-axis being the axis of rotation. Then the initial coordinates of the bodies are as follows: the Earth $(-a_1, 0, 0)$, the Moon $(a_2, 0, 0)$, the satellite (x_0, y_0, z_0), where

$a_1 = 4666.54$ km,
$a_2 = 379\,740.2$ km,
$x_0 = 35\,363.03$ km,
$y_0 = 0$, $z_0 = 13\,237.84$ km.

The inclination of the orbit is $i = 18°17'57''$.

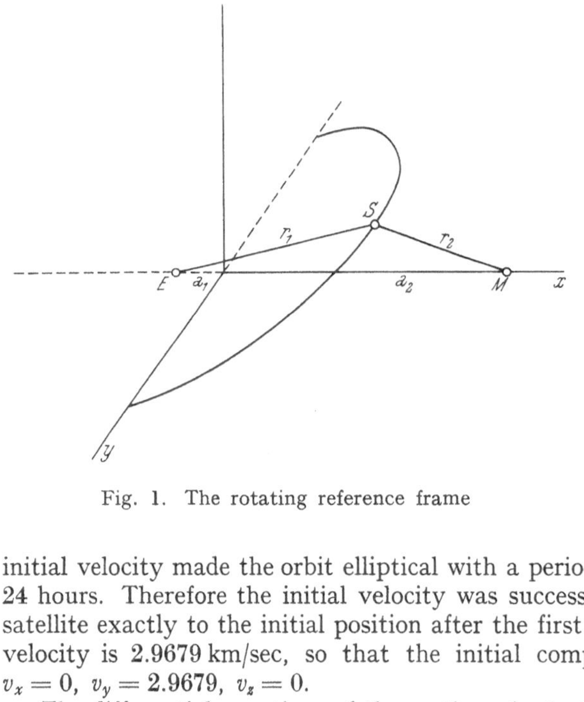

Fig. 1. The rotating reference frame

It turned out after the first revolution, that the initial velocity made the orbit elliptical with a period substantially greater than 24 hours. Therefore the initial velocity was successively diminished to get the satellite exactly to the initial position after the first revolution. This new initial velocity is 2.9679 km/sec, so that the initial components of the velocity are $v_x = 0$, $v_y = 2.9679$, $v_z = 0$.

The differential equations of the motion of a body in a restricted problem of three bodies were given in a convenient manner by Message [2]. They are

$$\begin{aligned} \frac{d^2x}{dt^2} - 2n'\frac{dy}{dt} - n'^2 x &= -\frac{\mu}{r_1^3}(x + a_1) - \frac{\mu k}{r_2^3}(x - a_2) \\ \frac{d^2y}{dt^2} + 2n'\frac{dx}{dt} - n'^2 y &= -\frac{\mu}{r_1^3} y - \frac{\mu k}{r_2^3} y \\ \frac{d^2z}{dt^2} &= -\frac{\mu}{r_1^3} z - \frac{\mu k}{r_2^3} z \end{aligned} \tag{1}$$

where x, y, z are the coordinates of the satellite, n' is the velocity of the rotation of the system Earth-Moon, $\mu = G \cdot M_e$, $k = M_m/M_e$, G being the constant of gravitation, M_m and M_e the masses of the Moon and the Earth, respectively. The radii-vectors r_1 and r_2 are given by

$$\begin{aligned} r_1^2 &= (x + a_1)^2 + y^2 + z^2 \\ r_2^2 &= (x - a_2)^2 + y^2 + z^2 \end{aligned} \tag{2}$$

The equations of motion were solved by numerical integration, using the RUNGE-KUTTA method with the aid of the "Ural 1" electronic digital computer of the Institute of the Theory of Informations and Automatization of the Czechoslovak Academy of Sciences. There were computed 28 revolutions of the Moon around the Earth. The JACOBIAN constant

$$J = \left(\frac{dx}{dt}\right)^2 + \left(\frac{dy}{dt}\right)^2 + \left(\frac{dz}{dt}\right)^2 - n'^2 (x^2 + y^2) - \frac{2\mu}{r_1} - \frac{2\mu k}{r_2} \tag{3}$$

was computed after every step as a check of the accuracy of the results. The step of integration was chosen so as to maintain J constant to six significant figures.

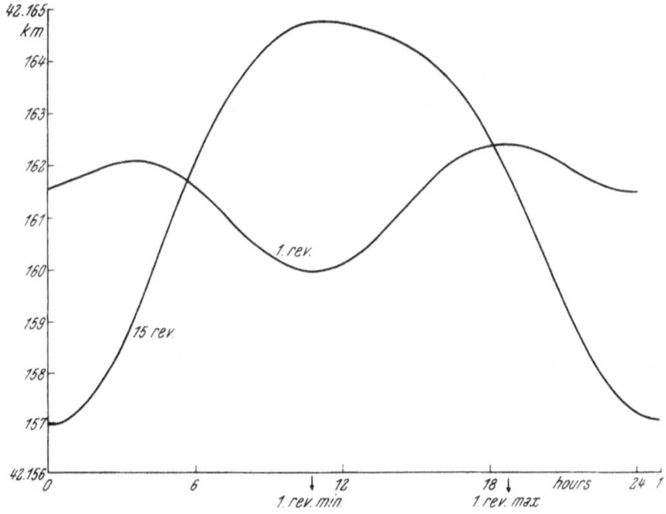

Fig. 2. The course of the radii-vectors of the satellite in the first and 15th revolution

During the first revolution the satellite revolves in a very nearly circular orbit, which is distorted by the lunar gravitational attraction. The distance of the satellite from the Earth (the radius-vector r_1) has two peaks and two minima. The eccentricity of the orbit becomes elliptical. If we take the principal maximum of r_1 in the first revolution as the apogee and the principal minimum as the perigee, the effective eccentricity of the orbit is $e_1 = 0.275 \cdot 10^{-4}$. In the 15th revolution the eccentricity becomes $e_{15} = 0.907 \cdot 10^{-4}$ and then it diminishes again till the value $e_{28} = 0.284 \cdot 10^{-4}$ in the 28th revolution. The course of r_1 in the first and 28th revolution is plotted in Fig. 2, where also the principal maximum and minimum in the first revolution is indicated.

The eccentricity has a periodical course with a period of 1 sidereal month and the amplitude of $0.6 \cdot 10^{-4}$, which is shown in Fig. 3. The variation of the heights of the perigee and apogee is shown in Fig. 4. Both these quantities show the same principal period, 1 sidereal month, which is, in the case of the apogee, interrupted by a small terminal decrease in the peak of the basic curve.

However, we must notice that all these changes of the distance of the satellite from the Earth, the course of the perigee and apogee and also the daily course

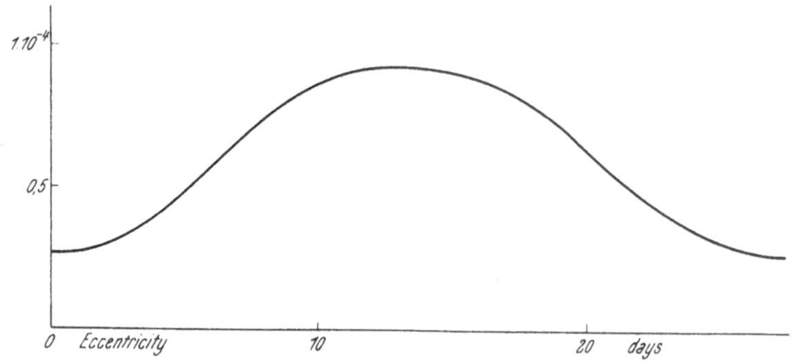

Fig. 3. The course of the eccentricity

of the r.-v. r_1, occur above nearly the same place on the surface of the Earth, so that the height of the satellite above the subsatellite point changes relatively rapidly within the range of about 8 km.

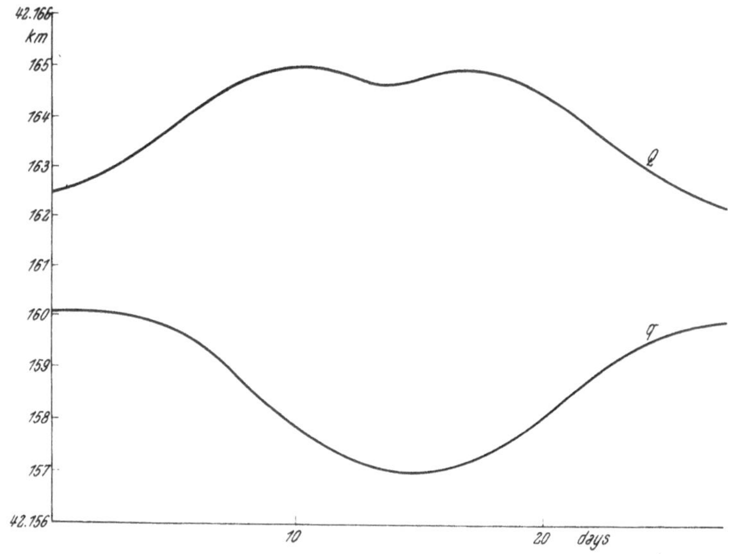

Fig. 4. The variations of the heights of the perigee (q) and apogee (Q)

An interesting feature shows the course of the distance of perigee ω, which is always on the other side of the Earth than the Moon. So the distance of perigee has a secular change and on Fig. 5 we see that it has also secondary periodical variation with a period of 1 sidereal month.

The course of the inclination shows only very small periodical changes with a period 1/2 month and with the amplitude of only 22″, as is plotted on Fig. 6.

The data about the changes of these elements enable us to construct the projection of the satellite on a sphere from the centre of the Earth as the centre

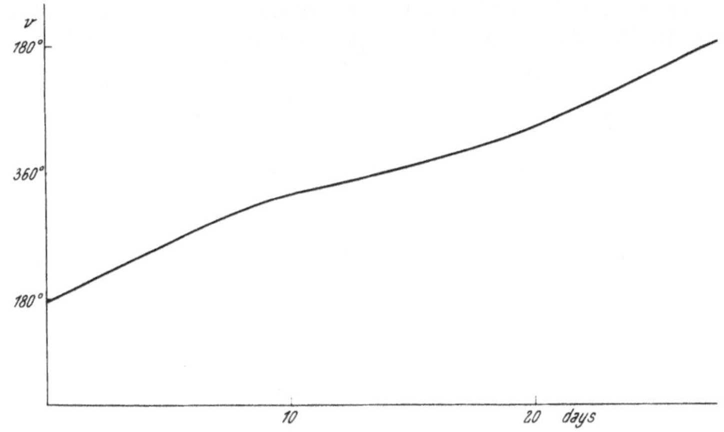

Fig. 5. The course of the distance of the perigee (ω)

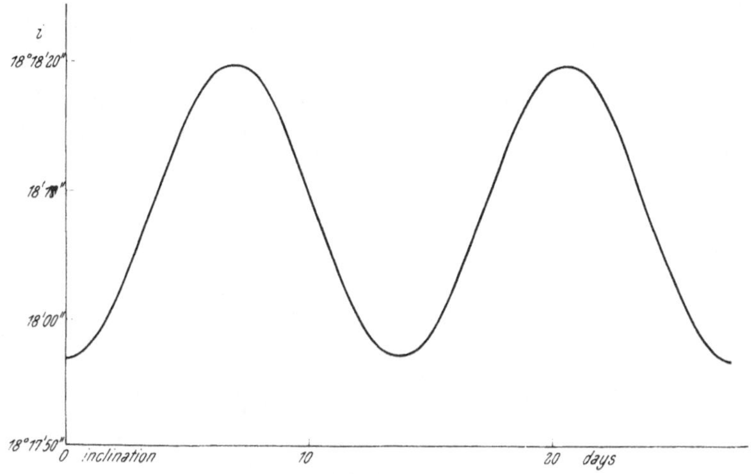

Fig. 6. The variation of the inclination (i)

of projection. This is plotted in Fig. 7. The scale of the y-axis is 72 times greater than the scale of the x-axis; the projection of the equator would be a line lying at the altitude of 0° in the direction of the x-axis. The satellite makes a pendulum motion around the central position with a semi-amplitude of 6.8° and a period of 1 sidereal month. As seen from the subsatellite point, the amplitude would be a little greater. This motion arises as a result of the changes of the satellite's period.

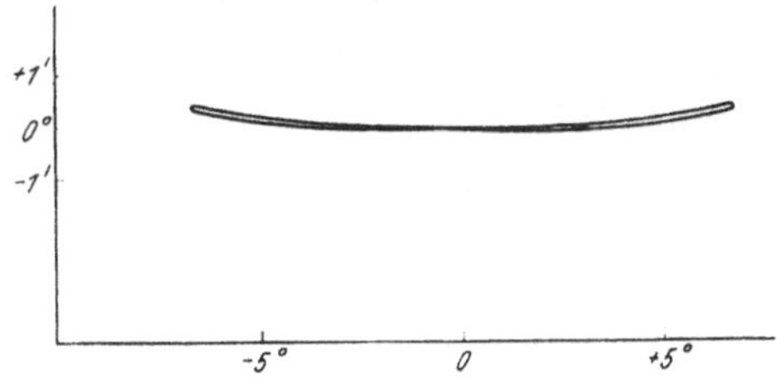

Fig. 7. The projection of the orbit of the satellite on the celestial sphere

The lunar gravitational attraction produces the periodical changes in the elements of the satellite's orbit and it is evident that similar changes will be produced by the attraction of the Sun. Since the disturbing force of the Sun is in our case less then half the disturbing force of the Moon, we may expect changes with about the half-amplitude of the previous changes and with a period of 1 year. The disturbing forces can sometimes act in the same sense and then we may expect a shift of the satellite from its initial position by about 10 degrees.

III. The Secular Lunisolar Perturbation

A rough picture of the secular changes of the orbital elements can be obtained by the comparing of the values of the elements at the beginning and at the end of each period. But in our case it is better to get the secular changes analytically as we have only one or two periods of the periodical changes at our disposal. So we try to derive the formulae for the secular changes under some simplifying assumptions.

The disturbing function is given in the form

$$R = G M_M \left(\frac{1}{\Delta} - \frac{r r' \cos \psi}{r'^3}\right) \tag{4}$$

where Δ is the distance of the satellite from the disturbing body, r and r' are the r.-v. of the satellite and the Moon respectively, measured from the centre of the Earth, ψ is the angle subtended by the radii-vectors r and r'. We assume circular orbit of the Moon and a nearly circular orbit of the satellite, so that we can retain only the terms till e^2 in the development of the disturbing function. We also limit the expansion to the terms up to $(r/r')^3$. For the computation of the changes of the elements we use the LAGRANGIAN equations, where we can drop the derivations in the mean anomaly, as we restrict ourselves to the secular changes only. The equations are then as follows:

$$\frac{de}{dt} = -\frac{\sqrt{1-e^2}}{n a^2 e}\frac{\partial R}{\partial \omega} \tag{5}$$

$$\frac{di}{dt} = \frac{\operatorname{cotg} i}{n a^2 \sqrt{1-e^2}}\frac{\partial R}{\partial \omega} - \frac{1}{\sin i}\frac{1}{n a^2 \sqrt{1-e^2}}\frac{\partial R}{\partial \Omega} \tag{6}$$

$$\frac{d\Omega}{dt} = \frac{1}{\sin i}\frac{1}{n a^2 \sqrt{1-e^2}}\frac{\partial R}{\partial i} \tag{7}$$

$$\frac{d\omega}{dt} = \frac{\sqrt{1-e^2}}{n a^2 e}\frac{\partial R}{\partial e} - \frac{\operatorname{cotg} i}{n a^2 \sqrt{1-e^2}}\frac{\partial R}{\partial i} \tag{8}$$

The direct part of the disturbing function will be developed in the series of LEGENDRE polynomials:

$$\frac{1}{\Delta} = \frac{1}{a'}\left[1 - 2\frac{r}{a'}\cos\psi + \left(\frac{r}{a}\right)^2\right]^{1/2} = \frac{1}{a'}\sum_{n=0}^{n=3}\left(\frac{r}{a'}\right)^n P_n(\cos\psi) \tag{9}$$

After substituting into R we get the disturbing function in the form

$$R = \frac{G M_M}{a'}\left[-\frac{1}{2}\left(\frac{r}{a'}\right)^2 + \frac{3}{2}\left(\frac{r}{a'}\right)^2\cos^2\psi - \frac{3}{2}\left(\frac{r}{a'}\right)^3\cos\psi + \frac{5}{2}\left(\frac{r}{a'}\right)^3\cos^3\psi\right] \tag{10}$$

The $\cos\psi$ is given by the expression

$$\cos\psi = A\cos u + B\sin u \tag{11}$$

where $u = (v + \omega)$, v being the true anomaly in the satellite's orbit and ω is the distance of the perigee. The coefficients A, B are the functions of the elements of the orbits of the disturbing and disturbed body; $A, B = F(\Omega, \Omega', i, i', u')$. After the substitution of $\cos\psi$ into R we can develop the expressions depending on the true anomaly v according to the CAYLEY's tables [3], and retain only the terms which are not depending on the mean anomaly of the satellite's orbit. So we get the expression for the disturbing function R in the form

$$R = \frac{G M_M}{a'} \left\{ -\frac{1}{2}\left(\frac{a}{a'}\right)^2 \left(1 + \frac{3}{2} e^2\right) + \frac{3}{4}\left(\frac{a}{a'}\right)^2 \left(1 + \frac{3}{2} e^2\right)(A^2 + B^2) + \frac{15}{4}\left(\frac{a}{a'}\right)^2 e^2 \left[\frac{1}{2}(A^2 - B^2)\cos 2\omega + A B \sin 2\omega\right] + \frac{15}{4}\left(\frac{a}{a'}\right)^3 e\,[A\cos\omega + B\sin\omega] - \frac{75}{16}\left(\frac{a}{a'}\right)^3 e\,[A(A^2 + B^2)\cos\omega + B(A^2 + B^2)\sin\omega] \right\}. \tag{12}$$

However, we must also remove the terms depending on the mean anomaly of the orbit of the disturbing body. The coefficients A and B and their sums and multiples can be expressed as functions of the mean anomaly of the disturbing body:

$$\begin{aligned} A &= K\cos u' + l\sin u' \\ B &= m\cos u' + n\sin u' \end{aligned} \tag{13}$$

The secular terms of these expressions can be easily computed. They are

$$\begin{aligned} A^2 + B^2 &= \tfrac{1}{2}(k^2 + l^2 + m^2 + n^2) = K \\ A^2 - B^2 &= \tfrac{1}{2}(k^2 + l^2 - m^2 - n^2) = L \\ A B &= \tfrac{1}{2}(k m + l n) = M \end{aligned} \tag{14}$$

These secular coefficients K, L, M can be computed from the formulae

$$\begin{aligned} k &= \cos(\Omega - \Omega') \\ l &= \sin(\Omega - \Omega')\cos i' \\ m &= -\sin(\Omega - \Omega')\cos i' \\ n &= \cos(\Omega - \Omega)\cos i\cos i' + \sin i\sin i' \end{aligned} \tag{15}$$

Now we can substitute into the LAGRANGIAN equations and the expressions for the changes of elements become

$$\frac{de}{dt} = \frac{\sqrt{1 - e^2}\, G M_M}{n a^2 e \quad a'} \left\{\frac{15}{8}\left(\frac{a}{a'}\right)^2 e^2 [-L\sin 2\omega + M\cos 2\omega]\right\}$$

$$\begin{aligned} \frac{di}{dt} = {} & \frac{\operatorname{cotg} i}{n a^2\sqrt{1 - e^2}} \frac{G M_M}{a'} \left\{\frac{15}{8}\left(\frac{a}{a'}\right)^2 e^2 [-L\sin 2\omega + M\cos 2\omega]\right\} - \\ & - \frac{1}{\sin i} \frac{1}{n a\sqrt{1 - e^2}} \frac{G M_M}{a'} \left\{\frac{3}{8}\left(\frac{a}{a'}\right)^2 \left(1 + \frac{3}{2} e^2\right) \frac{\partial}{\partial\Omega}(K) + \right. \\ & \left. + \frac{15}{16}\left(\frac{a}{a'}\right)^2 e^2 \left[\cos 2\omega \frac{\partial}{\partial\Omega}(L) + \sin 2\omega \frac{\partial}{\partial\Omega}(M)\right]\right\} \end{aligned} \tag{16}$$

$$\frac{d\omega}{dt} = \frac{\sqrt{1-e^2}\,G\,M_M}{n\,a^2\,e\qquad a'}\left\{-\frac{3}{2}e\left(\frac{a}{a'}\right)^2 + \frac{3}{8}e\left(\frac{a}{a'}\right)^2 K + \right.$$
$$\left. + \frac{15}{8}\left(\frac{a}{a'}\right)^2 e\,[\cos 2\,\omega\,L + \sin 2\,\omega\,M]\right\} -$$
$$- \frac{\operatorname{cotg} i}{n\,a^2\sqrt{1-e^2}}\frac{G\,M_M}{a'}\left\{\frac{3}{8}\left(\frac{a}{a'}\right)^2\left(1+\frac{3}{2}e^2\right)\cos 2\,\omega\,\frac{\partial}{\partial i}(K) + \right. \tag{17}$$
$$\left. + \frac{15}{16}\left(\frac{a}{a'}\right)^2 e^2\left[\cos 2\,\omega\,\frac{\partial}{\partial i}(L) + \sin 2\,\omega\,\frac{\partial}{\partial i}(M)\right]\right\}$$

$$\frac{d\Omega}{dt} = \frac{1}{\sin i}\frac{1}{n\,a^2\sqrt{1-e^2}}\frac{G\,M_M}{a'}\left\{\frac{3}{8}\left(\frac{a}{a'}\right)^2\left(1+\frac{3}{2}e^2\right)\frac{\partial}{\partial i}(K) + \right.$$
$$\left. + \frac{15}{8}\left(\frac{a}{a'}\right)^2 e^2\left[\cos 2\,\omega\,\frac{\partial}{\partial i}(L) + \sin 2\,\omega\,\frac{\partial}{\partial i}(M)\right]\right\} \tag{18}$$

These formulae can be used for the computation of the secular perturbation of the orbits of artificial satellites under the conditions stated above, i. e. the circular orbit of the disturbing body and the nearly-circular orbit of the satellite.

From the obtained expressions we can now compute the change of elements of our satellite's orbit, e. g. the change of the perigee height and the change of the inclination. We can neglect the terms with e^2, with respect to the small excentricity of our orbit. Some terms also vanish owing to the vanishing inclination of the orbit:

$$\frac{\partial}{\partial\Omega}(K) = 0 \tag{19}$$

The semi-major axis of the satellite's orbit is not subject to secular perturbations, so we can obtain the change of the height of the perigee (q) from the expression for the change of the eccentricity. The results of the substitution of numerical values are given in Table I. There are the changes per one month. The changes,

Table I

	Moon (n. i.)	Moon	Sun
$[\Delta q]$	− 190 m	− 20 m	− 9 m
$[\Delta i]$	< 1″	< 1″	< 1″

found from the numerical integration are given in the second column. The values of the changes, computed from the analytical equations (16) are given in the third (Moon) and the fourth (Sun) column. We can see that the changes of the elements are very small, which is certainly partly due to the very small eccentricity of the satellite's orbit.

IV. The Influence of the Non-Spherical Shape of the Earth

The influence of the non-spherical shape of the Earth could be also traced in the motion of the satellite, in spite of its great distance from the Earth. The equatorial ellipticity of the Earth influences the motion of the satellite owing to the commensurability of the period of the satellite with the rotation of the Earth.

The secular change of the inclination will be also influenced by the third harmonics of the Earth's gravitational potential.

The gravitational field of the Earth is described by the equation

$$F = \frac{G M_E}{r}\left[1 + \left(\frac{r_0}{r}\right)^2 c_{20} P_{20} (\sin \delta) + \right.$$

$$+ \left(\frac{r_0}{r}\right)^2 (c_{22} \cos 2\lambda - d_{22} \sin 2\lambda) P_{22} (\sin \delta) + \tag{20}$$

$$\left. + \left(\frac{r_0}{r}\right)^3 c_{30} P_{30} (\sin \delta)\right]$$

where P_{20} and P_{30} are the zonal harmonics and P_{22} is the tesseral harmonics, λ and δ are the geocentric coordinates of the satellite. c_{30}, c_{22} and d_{22} are the numerical coefficients which determine the values of the asphericity of the Earth. The coordinates must be expressed as the functions of the elements of the orbit of the satellite. Then we can develop the expressions $r^n \cdot \cos m \cdot v$ (v is the true anomaly of the satellite's motion) in the terms of the mean anomaly. If we retain the long-period terms only in the expression for the equatorial ellipticity of the Earth, we have for the disturbing function

$$F_{22} = \frac{3}{4} G M_E \frac{r_0^2}{a^3}\left\{\frac{9}{2} e^2 \sin^2 i \,[c_{22} \cos 2 (M + \Omega) - d_{22} \sin 2 (M + \Omega)] + \right.$$

$$+ \left(1 - \frac{5}{2} e^2\right) (1 + 2 \cos i + \cos^2 i) \,[c_{22} \cos 2 (M + \Omega + \omega) +$$

$$+ d_{22} \sin 2 (M + \Omega + \omega)] \tag{21}$$

Only the secular term in the third harmonics is of interest here, namely

$$F_{30} = \frac{3}{2} G M_E \frac{r_0^3}{a^4} c_{30} \cdot e \cdot \sin i \cdot \sin \omega \left(\frac{5}{4} \sin^2 i - 1\right) \tag{22}$$

In the long-period disturbing function F_{22} there are partly the terms with argument $(M + \Omega)$ and partly the terms with argument $(M + \Omega + \omega)$, where M is the mean anomaly of the satellite's motion. The former terms are not influenced in their long-periodicity by the motion of the distance of the perigee and so they preserve their long-periodical character even if there is a motion of the distance of the perigee. But this term depends on e^2 so that it will be very small for nearly-circular orbits. From these equations we can derive the changes of the elements that are of importance for the stability of the satellite above the fixed subsatellite point.

$$\Delta P = \frac{9}{2} \pi \left(\frac{r_0}{a}\right)^2 \frac{1}{n + \Omega}\left\{\frac{9}{2} e^2 \sin^2 i \,[c_{22} \cos 2 (M + \Omega) - d_{22} \sin 2 (M + \Omega)] + \right.$$

$$+ \left(1 - \frac{5}{2} e^2\right) (1 + 2 \cos i + \cos^2 i) \,[c_{22} \cos 2 (M + \Omega + \omega) + \tag{23}$$

$$\left. + d_{22} \sin (M + \Omega + \omega)]\right\}$$

$$\Delta i = \frac{3}{2} \frac{e}{\sqrt{1 - e^2}} \frac{n}{a^3} r_0^3 c_{30} \left(\frac{5}{4} \sin^2 i - 1\right) \sin \omega \tag{24}$$

The equations are written for $\omega = \text{const}$. After the substitution of the numerical values of the elements we obtain the changes of elements of our orbit. The coefficient c_{30} has been obtained from the measurements made on artificial satellites; according to KOZAÏ [4] it is $c_{30} = 2.29 \cdot 10^{-6}$. The coefficients c_{22} and d_{22} are given from gravimetrical measurements, e. g. in the paper by JONGOLOVICH [5], although the first attempt to obtain these coefficients has already been performed (IZSAK [6]). But these new values by IZSAK do not introduce serious discrepancies in the results. If we also substitute the numerical value of the secular motion of the distance of perigee, which was found by means of numerical integration, we obtain for the shift of the satellite from its initial position a periodical change of a semi-amplitude of 46 seconds of arc with the period of one sidereal month. All these perturbations are of course very small, but the small value of the eccentricity of the orbit plays here an important role.

Concluding this study, we can say that the main perturbations, which cause the shift of the 24-hour satellite from its initial position are the luni-solar perturbations, of which the perturbations with the period of one sidereal month and the semi-amplitude of 6.8° are the most important ones.

I am greatly indebted to the staff of the Institute of Theory of Information and Automatisation of the Czechoslovak Academy of Sciences and especially to ing. I. MATĚJOVSKÝ for the help with the computations on the "Ural 1" computer.

References

1. S. HERRICK, R. M. L. BAKER, JR., and C. G. HILTON, Gravitational and Related Constants for Accurate Space Navigation. VIIIth International Astronautical Congress, Barcelona **1957**, p. **197**. Wien: Springer, **1958**.
2. P. J. MESSAGE, Some Periodic Orbits in the Restricted Problem of Three Bodies and Their Stabilities. Astronom. J. **64**, **226** (**1957**).
3. A. CAYLEY, Tables for Development of Functions in the Theory of Elliptic Motion. Mem. RAS **29**, **191** (**1861**).
4. Y. KOZAÏ, The Gravitational Field of the Earth Derived from the Motion of Three Satellites. Astronom. J. **66**, **8** (**1961**).
5. I. D. JONGOLOVICH, Potentiel de l'attraction terrestre. Bull. Inst. Theor. Astronomy **6**, **505** (**1957**).
6. I. G. IZSAK, A Determination of the Ellipticity of the Earth's Equator from the Motion of Two Satellites. Smiths. Inst. Spec. Report, No. **56**, **11** (**1961**).

Discussion

Mr. KOVALEVSKY: I notice that you have found a small change in the height of the satellite and therefore there must be also only a small change in the semi-major axis. Then it is a little surprising that the shift in the longitude of the satellite is greater than one might expect.

Mr. SEHNAL: The nature of the oscillations in the longitude of the satellite cannot be exactly explained from the results obtained by numerical integration. In this computation we obtain only the rectangular coordinates of the satellite and the origin of the changes of the elements are very difficult to determine.

Mr. KING-HELE: I should like to ask what is your opinion about the principal gravitational perturbative force, acting on the resonance satellites; is that the lunisolar influence or the irregularities in the gravitational field of the Earth?

Mr. SEHNAL: I can say that in the case of the 24-hour satellite the principal perturbations will be caused by the lunisolar disturbing force. But there might be also the resonance satellites which are in orbits nearer to the Earth and then the influence of the irregularities of the Earth's gravitational field will be increasing.

Mr. Kovalevsky: I think that there are long-period terms which might be of some importance, related to the motion of lunar node and apogee, especially in the inclination.

Mr. Sehnal: The next step in the study of the lunisolar perturbations of the 24-hour satellite will be the determination of periodic perturbations from the development of the disturbing function. Then it will be possible to distinguish between individual parts of the perturbations.

Mr. Kovalevsky: And I think, it would be very interesting to find out the reason of the shift in longitude that you find.

Practical Aspects of Re-Entry Problems[1]

By

Antonio Ferri[2] and **Lu Ting**[3]

(With 14 Figures)

Abstract — Zusammenfassung — Résumé

Practical Aspects of Re-Entry Problems. Entry trajectories for vehicles without and with lift are investigated. The existence and the significance of the second peak deceleration in re-entry trajectories are presented. The second peak may be greater than the first one. For a given entry velocity, the first peak deceleration can be a discontinuous function of entry angle. With proper lift modulation of the body-type vehicle, i.e. by changing the angle of attack of the vehicle, the range of the entry angle for non-skip trajectories can be increased significantly, while the first peak deceleration remains less than 5 times the gravitational acceleration.

A constant elevation maneuver is presented and its effectiveness in reducing the second peak deceleration is explained. This maneuver, followed by a constant lift modulation, permits the control of the point of landing over a wide range.

The practical application of the rate of change of deceleration in deciding the proper lift modulation is demonstrated.

Praktische Aspekte des Rückkehrproblems. Es werden Rückkehrbahnen für Fahrzeuge mit und ohne Auftrieb untersucht; hierbei wird auf das Auftreten einer zweiten Verzögerungsspitze hingewiesen. Diese zweite Verzögerungsspitze kann höher sein als die erste. Weiters kann die erste Verzögerungsspitze bei gegebener Eintauchgeschwindigkeit eine diskontinuierliche Funktion des Eintrittswinkels sein.

Ferner wird gezeigt, daß durch ein Aufsteigemanöver die zweite Verzögerungsspitze verringert werden kann, wobei auch noch eine Variation des Landungspunktes möglich ist.

Aspects pratiques de problèmes de rentrée. Les investigations de trajectoires de rentrée avec ou sans portance montrent l'importance d'un second pic de décélération. Il peut être supérieur au premier. A vitesses de rentrée égales, le premier pic peut être une fonction discontinue de l'angle de rentrée. Une modulation de portance permet d'accroître sensiblement la plage de variation de l'angle de rentrée et de réduire le premier pic en dessous de 5 *g*.

Une manoeuvre d'élévation constante est effective pour réduire l'importance du second pic et, suivie d'une portance constante, permet de contrôler l'impact dans un domaine étendu.

L'utilisation pratique de la variation d'accélération subie pour le contrôle de la modulation de portance à appliquer est démontrée.

---

[1] This research was supported by the United States Air Force through the Air Force Office of Scientific Research, Air Research and Development Command, under Contract No. AF 49(638)-445, Project No. 9781.

[2] Professor of Aerodynamics, Director of Aerodynamics Laboratory, Head of Department of Aerospace Engineering and Applied Mechanics, Polytechnic Institute of Brooklyn, 333 Jay Street, Brooklyn, N.Y., U.S.A.

[3] Research Professor of Aeronautical Engineering, Polytechnic Institute of Brooklyn, 333 Jay Street, Brooklyn, N.Y., U.S.A.

I. Introduction

In many investigations [1 — 5] on re-entry trajectories with or without lift modulations, the limit on the peak deceleration is usually considered to be ten or fifteen times the gravitational acceleration at sea level, i.e. 10 g or 15 g. The astronauts can endure such a high deceleration only for a few seconds [6, 7], although they have been specially selected and trained for years. In the future it must be expected that some space vehicles will carry scientists and other personnel who usually will not be able to take even 10 g. Therefore, it can be of interest to investigate a method of entry while the peak deceleration is well below 10 g. In principle, the peak deceleration can always be reduced if the vehicle experiencing re-entry has either a retrorocket, a variable drag device, a large lifting surface or any combination of them. However, any one of these devices implies a reduction in the payload of the vehicle. In the present paper we will show that without paying this penalty of the weight of any special device, a typical body-type vehicle can re-enter along a non-skip trajectory without exceeding 5 g by lift modulation, that is, by changing the angle of attack of the body. With entry velocities ranging from near circular speed to twice this speed the range of the entry angles corresponding to such trajectories will be greater than 2 degrees; hence we will not encounter a severe guidance problem.

The aerodynamic characteristics of a cylindrical body-type vehicle with a nose cone and proper fairings are expressed in the following equations based on test data:

$$C_D = C_{D_0} + k\, C_L^2 = 0.4 + 0.625\, C_L^2 \tag{1}$$

$$\frac{W}{A} = 25 \text{ lbs/sq. ft.} \tag{2}$$

$$C_L = 0.02\,\alpha \tag{3}$$

where C_L and C_D are the lift and drag coefficients, W is the weight of the re-entry vehicle, α is the angle of attack in degrees and A is the reference area. The last equation is not required in the determination of the trajectories if C_L is specified instead of α. The existence and the significance of a second peak deceleration in non-lifting trajectories will first be demonstrated. Then a proper lift modulation will be selected in order to increase the entry angle, while keeping the first peak deceleration below 5 g, and to reduce the second peak deceleration. The point of landing will be controlled after all the peak decelerations have been reduced. Finally, the practical application of the rate of deceleration in deciding the required lift modulation will be discussed.

II. Non-lifting Trajectories

Fig. 1 shows the trajectories for various entry angles of the body-type vehicle re-entering at hyperbolic speed, nearly $\sqrt{3}$ times the circular orbiting velocity. Also shown in the figure are the lines of constant G, the lines of constant dG/dt and the three lines of maximum G meeting at a triple point where G is the non-dimensional deceleration with respect to the gravitational acceleration g. Both G and dG/dt can be measured accurately inside the vehicle. Their practical applications will be discussed in the last section of this paper.

At small entry angle, say 9°, the trajectory passes through a relatively low peak G and leaves the limit of atmosphere. It will re-enter in a multiple pass trajectory with several peak G's. For larger entry angle, say 9.8°, the trajectories will skip and pass through two peak G's. The second peak G can be larger than the first peak G. Without lift modulation the first peak G of a nonskip trajectory will exceed 5. For larger entry angles, the first peak G becomes continuously larger and the second peak G disappears as the trajectory passes through the left side of the triple point.

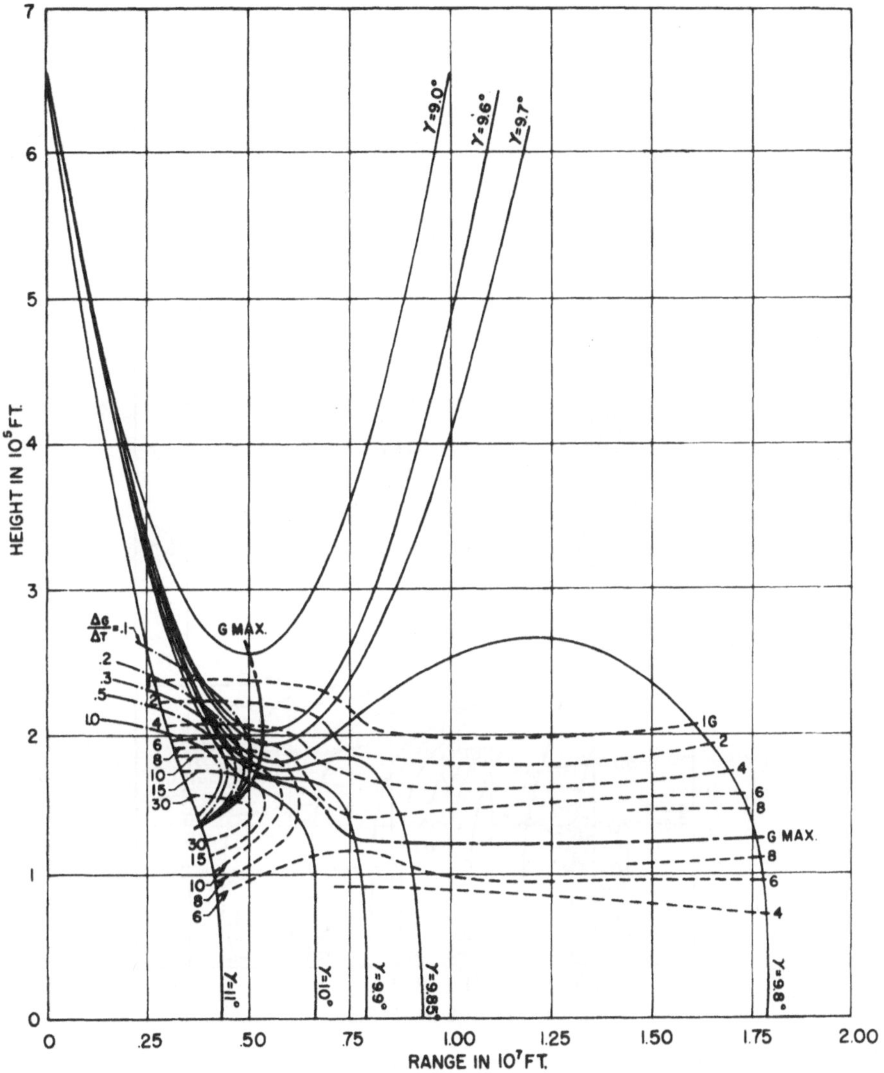

Fig. 1. Entry angle and trajectories with $V = 44,250$ ft/sec ($\bar{V} = 1.705$) at 200 km. $W/(C_{D_0}A) = 62.5$ lbs/ft^2, $C_L = 0$

Fig. 2 shows the trajectory of the same vehicle re-entering at near parabolic speed. The lines of constant G and the lines of constant dG/dt which are not shown are similar to those in Fig. 1. The three lines of maximum G also meet at a triple point. For smaller entry angles, the trajectories pass through the right side of the triple point and possess two peak decelerations. The second one is larger than the first. As the entry angle increases, the first peak G increases continuously. The second peak G will not be present for trajectories on the left side of the triple point.

Part of the results of Figs. 1 and 2 has been reported by CRISP [4]. The limit of atmosphere in [4] has been assumed to be at 200 km as shown in Figs. 1 and 2.

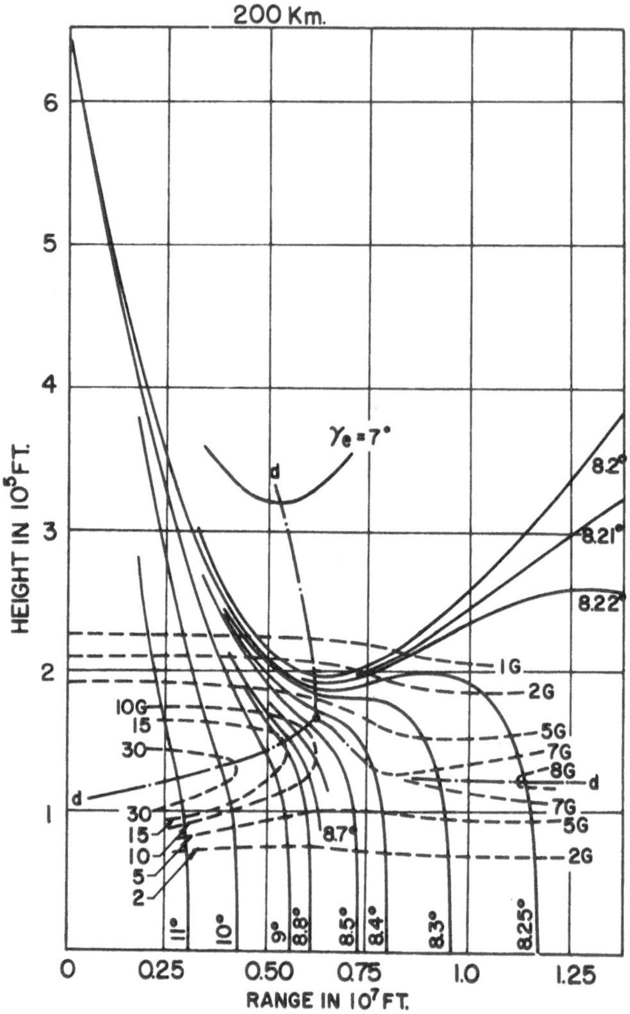

Fig. 2. Entry angle and trajectories with $V = 35{,}540$ ft/sec ($\bar{V} = 1.370$) at 200 km. $W/(C_{D_0} A) = 62.5$ lbs/ft², $C_L = 0$

Since the density of air is still extremely low at 200 km the limit of atmosphere for all the subsequent calculations in this paper has been assumed to be at 400,000 ft.

Fig. 3 shows the trajectories with velocity 1.2 times the circular orbiting velocity. Although the trajectories and the lines of constant G are similar in shape as to those in Figs. 1 and 2, one of the lines of maximum G stops in the middle instead of meeting the other lines of maximum G which form a triple point.

For the trajectory with entry angle 0.076 rad., there are two peak decelerations. The first is 2.2 g and the second is 7.4 g. For the trajectory with entry angle 0.077 the first peak G becomes 7.2 g, which actually belongs to the group of

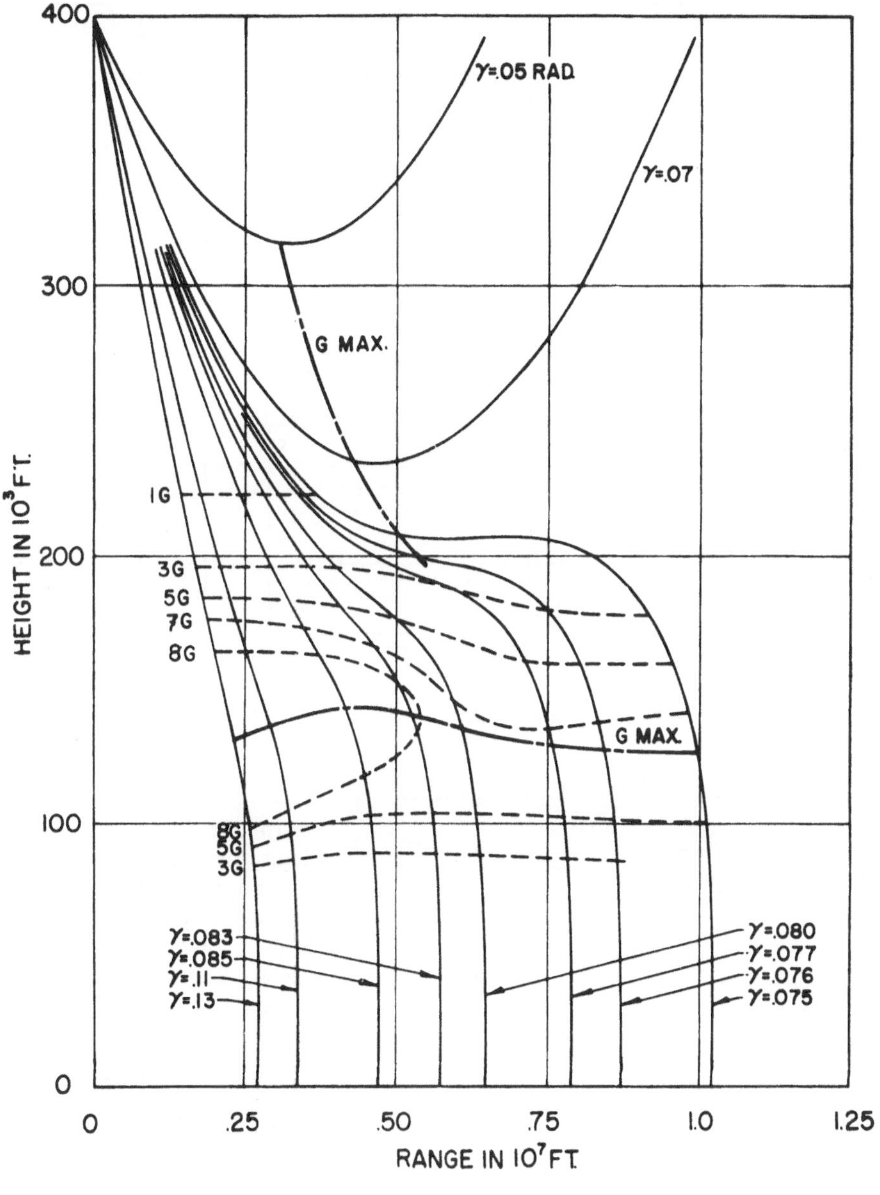

Fig. 3. Entry angles and trajectories with $V = 31{,}120$ ft/sec ($\bar{V} = 1.2$) at 400,000 ft. $W/(C_{D_0} A) = 62.5$ lbs/ft², $C_L = 0$

second G's. This strange behavior can be clearly explained by Fig. 4 in which the dimensionless deceleration G is plotted against elevation for each trajectory in Fig. 3. For small entry angles there are two peak decelerations. As the entry angle increases the first peak becomes a point of inflection and disappears. The first peak along these trajectories actually belongs to the group of second peak deceleration.

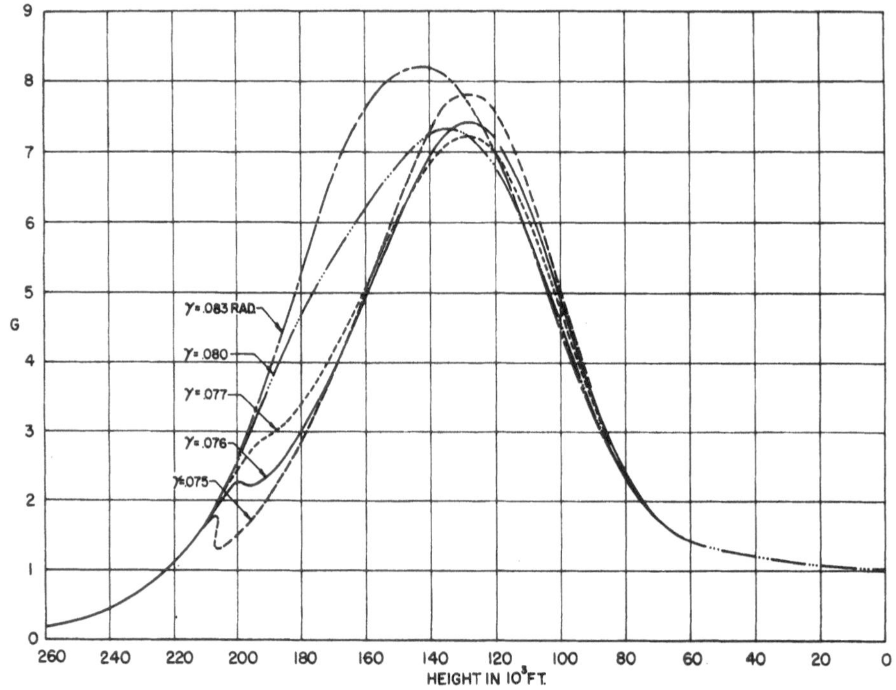

Fig. 4. Deceleration along the trajectories with $V = 31{,}150$ ft/sec ($\bar{V} = 1.2$) at 400,000 ft. $W/(C_{D_0} A) = 62.5$ lbs/ft², $C_L = 0$

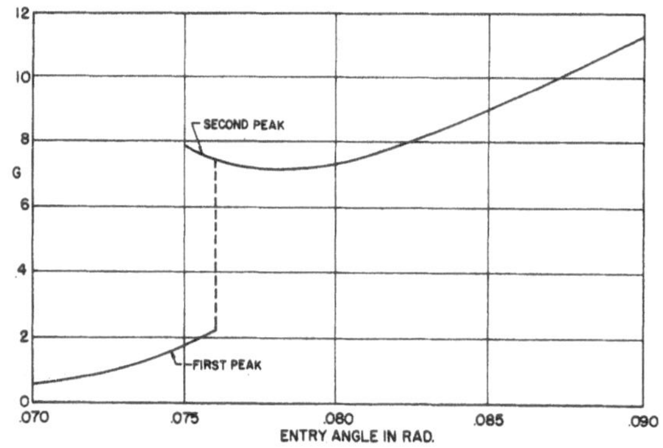

Fig. 5. First and second peak G's versus the entry angle

In Fig. 5, the value of G along the lines of maximum G in Fig. 3 is plotted against the entry angle. It is clear that for trajectories with entry velocity 1.2 times the circular orbiting velocity, the first peak G is a discontinuous function of the entry angle. Without lift modulation the first peak G will be either greater than 7 or less than 2 for all entry angles.

III. Maximum Entry Angle with Lift Modulation

To obtain the maximum entry angle for a given limit on the first peak deceleration we must solve an indirect problem. In the direct problem the entry velocity and angle are given, while the first peak G and the corresponding elevation or time are obtained either by numerical integration of the differential equations or by one of the approximate analytical solutions [3, 8, 9, 10], whichever is applicable. In the indirect problem, the first peak G and its elevation are prescribed and we look for the corresponding entry velocity and angle.

The governing equations are

$$\bar{\varrho} = \text{nondimensional density} = \varrho/\varrho_s = e^{-\beta h} \quad (4)$$

$$\frac{d\bar{V}}{dt} = -K C_D \bar{\varrho} \bar{V}^2 + \frac{g}{V_s \bar{r}^2} \sin\gamma \quad (5)$$

$$\bar{V}\frac{d\gamma}{dt} = -K C_L \bar{\varrho} \bar{V}^2 - \left(\frac{g}{V_s}\right)\left(\frac{\bar{V}^2}{\bar{r}} - \frac{1}{\bar{r}^2}\right)\cos\gamma \quad (6)$$

$$\frac{dh}{dt} = -V_s \bar{V} \sin\gamma \quad (7)$$

where

$$K = \frac{\varrho_s V_s A}{2W} \; (= 56.7 \text{ sec}^{-1} \text{ for } W/A = 25 \text{ lb/ft}^2)$$

In Eq. (4), the density is expressed as an exponential function of height h. Eqs. (5) and (6) are the equations of motion in the directions tangential and normal to the velocity, where $\bar{V}$ is the nondimensional velocity with respect to V_s, which is the circular orbiting velocity at sea level, and $\bar{r}$ is the nondimensional radial distance with respect to r_s, which is the radius of earth. Eq. (7) is the kinematic equation.

With density $\bar{\varrho}$ as the independent variable, approximate analytic solutions for constant C_L and C_D have been obtained [10] in the following forms:

$$\cos\gamma = f_1(\bar{\varrho}, \bar{V}_e, \gamma_e), \qquad \bar{V} = f_2(\bar{\varrho}, \bar{V}_e, \gamma_e) \quad (8)$$

where the subscript e refers to the entry condition and f_1 and f_2 are simple functions of density. If C_L is a third order polynomial of density, f_1 and f_2 will involve elliptical integrals.

The approximate analytic solutions, together with the justification of the approximations and their accuracy as compared with solutions using numerical integration in several direct problems, have been presented in detail in [10]. For the present indirect problem, the maximum entry angle computed from the analytic solutions of Eq. (8) for constant C_L will differ from the results of numerical integration in Fig. 6 by less than one-eighth of a degree.

The value of the angle γ and the velocity $\bar{V}$ corresponding to the given peak deceleration at a given elevation are determined by the equation for the acceleration, Eq. (9), and the necessary condition for a maximum

$$G = B\sqrt{1 + \Lambda^2}\, C_D \bar{\varrho} \bar{V}^2 \quad (9)$$

where

$$B = \frac{\varrho_s V_s^2 A}{2 W g} \qquad \text{and} \qquad \Lambda = C_L/C_D$$

The necessary condition, $dG/dt = 0$, for constant C_L or $dC_L^2/dt = 0$ at maximum G, i.e., G_m, is

$$\sin\gamma = \frac{2G_m}{\sqrt{1+\Lambda^2}\left[\beta r_s \bar{V}^2 + \dfrac{2}{\bar{r}^2}\right]} \sim \frac{2G_m}{\sqrt{1+\Lambda^2}\,\beta r_s \bar{V}^2} \quad (10)$$

Eq. (9) expresses velocity $\bar{V}$ in terms of density $\bar{\varrho}$ for a given value of G, which is the nondimensional deceleration.

The condition of extreme, $dG/dt = 0$, expresses the angle γ in terms of the velocity $\bar{V}$. In the approximation we note that $\beta\, r_s$ is of the order of one thousand, 10^3, and $\bar{V}$ is of the order of unity.

Eqs. (9) and (10) have been introduced by Broglio [11]. The velocity $\bar{V}$ and the angle $\bar{\gamma}$ obtained from Eqs. (9) and (10) give the solution for a local maximum deceleration if $d^2G/dt^2 < 0$. With the same approximation of Eq. (10) and that $\cos\gamma \sim 1$ for small γ we obtain the inequality (11) when $d^2(C_L{}^2)/dt^2 = 0$ at G_m:

$$\bar{V}^4 - b\,\bar{V}^2 + c > 0 \, ^1 \tag{11}$$

where

$$b = 1 - \frac{\Lambda\, G_m}{\sqrt{1+\Lambda^2}} \qquad \text{and} \qquad c = \frac{4\, G_m{}^2}{\beta\, r_s\,(1+\Lambda^2)}\,.$$

The left side of the inequality (11) is a quadratic expression of $\bar{V}^2$. The larger root x_1 is

$$x_1 = \frac{b + \sqrt{b^2 - 4\,c}}{2} \tag{12}$$

The inequality holds if $\bar{V}^2$ is greater than the larger root x_1 [2]. We note that b, c, and x_1 depend only on the ratio of lift and drag coefficients.

As the density $\bar{\varrho}$ at the instant of maximum deceleration increases velocity $\bar{V}$ decreases as the entry velocity and angle decrease. The curve of entry velocity and entry angle will terminate at the point where $d^2G/dt^2 = 0$, or $\bar{V}^2 = x_1$. In other words, for a given G_m and $C_{L_{\max}}$ there exists a particular entry velocity if x_1 is real and positive. When the entry velocity is less than that particular velocity, the first peak G will be either less or greater than the given G_m for all entry angles.

This phenomenon has already occurred in the preceding section for nonlifting trajectories. When the nondimensional entry velocity is equal to 1.2, the first peak G will be either greater or less than 5 for all entry angles as shown in Figs. 3—5. Eqs. (11) and (12) show that when C_L is more negative, b and also x_1 increase. Therefore the point of termination will move towards the larger entry velocity for larger negative C_L.

Solution of Eqs. (9) and (10) will always be a local maximum if x_1 is negative or imaginary, that is, if $b < 2\sqrt{c}$.

For $C_L = 0$, or $\Lambda = 0$, x_1 is imaginary when $G > \sqrt{\beta\, r_s/16} = 7.71$. This also explains why curves of entry angle and velocity for zero C_L do not terminate in the middle when the first maximum deceleration is 10 or 15.

For a given value of G_m, x_1 is imaginary or negative for

$$\frac{C_L}{C_D} = \Lambda > \sqrt{1 - \frac{1}{G_m{}^4} + \frac{16}{\beta\, r_s\, G_m{}^2} - \frac{4}{\sqrt{\beta\, r_s}}}$$

---

[1] Detail steps in deriving Eq. (11) can be found in [12].

[2] When $\bar{V}^2$ is less than the smaller root x_2 of the quadratic expression for $\bar{V}^2$, the inequality (11) holds. It will be again a local maximum deceleration corresponding to the second peak. Furthermore when $\bar{V}^2 < x_2$ which may be much less than unity, the corresponding angle γ or $\sin\gamma$ is no longer small. Hence the approximation which leads $(d^2G/dt^2) < 0$ to the inequality (11) is no longer accurate. In this case we have to resort to the completed expression for $(d^2G/dt^2) < 0$ and also that for $(dG/dt) = 0$ since the approximation $\beta\, r_s\, \bar{V}^2 \gg 2$ is not very good either. These expressions and additional results can be found in [12].

and it is a function of G only. ($\Lambda > 0.068$ for $G = 5$.) This shows that with a very small C_L or C_L/C_D, the solutions of Eq. (9) and (10) will always be a local maximum.

The simplest lift modulation is with constant C_L or constant angle of attack. Fig. 6 shows the curves of maximum entry angle vs. entry velocity for the first peak G equal to 5 and for various values of constant C_L.

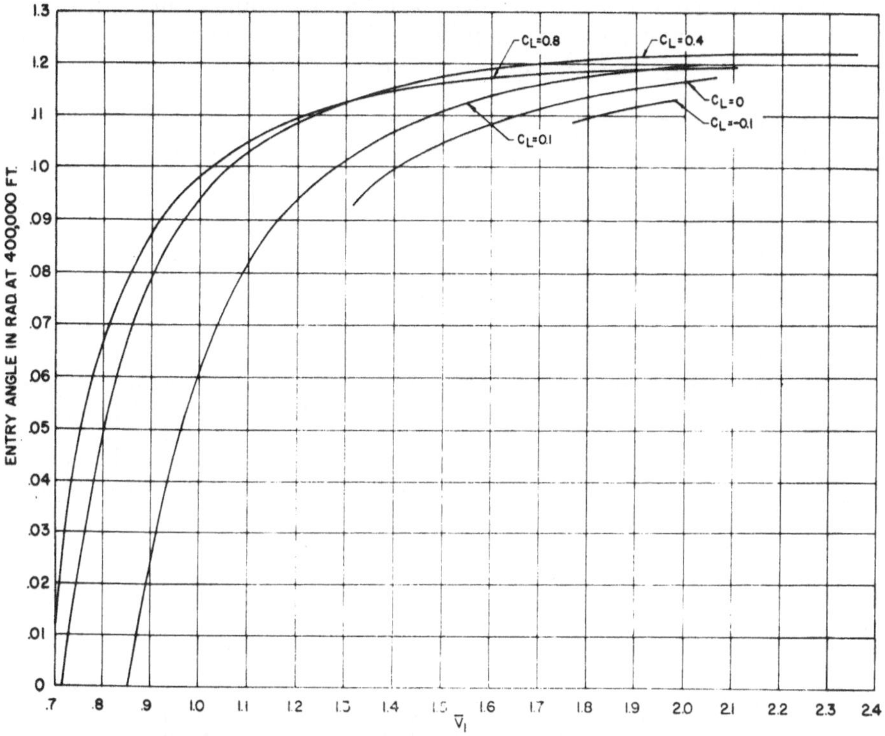

Fig. 6. Entry angle for the first peak $G = 5$ with constant C_L

For $C_L = 0$ and -0.1 the curves stop at a finite value of γ. The corresponding velocity is larger for more negative C_L. For $C_L = 0.1$, 0.4 and 0.8, they fulfill the requirement of $\Lambda > 0.068$, therefore the corresponding curves end at $\gamma = 0$ instead of at finite values of γ. These phenomena are in agreement with the preceding analysis.

From Fig. 6 it can be seen that the maximum entry angle is increased when C_L changes from negative to positive. However, the curves for $C_L = 0.8$, 0.6 and 0.4 intersect each other. Therefore $C_L = 0.8$, when C_L/C_D is maximum, is not always better than the smaller values of C_L for increasing the entry angle or reducing the first peak G.

The problem of finding the optimum lift modulation for maximum entry angle with given entry velocity and first peak G belongs to the calculus of variations. In the present paper a variation of C_L based on physical reasoning will be selected in order to demonstrate that the variable C_L is superior to constant C_L in raising the entry angle for a given first peak G.

It is desirable that at high elevation or low density we use a larger C_L to reduce the angle, while at higher densities we use smaller C_L to reduce $\sqrt{C_L^2 + C_D^2}$

and thereby decrease the deceleration given by Eq. (9). The following expression for C_L has the required characteristic:

$$C_L = a + b\left(\frac{\varrho_* - \varrho}{\varrho_* - \varrho_e}\right)^n \tag{13}$$

where ϱ_e is the density at the limit of atmosphere (400,000 ft) and ϱ_* is the density at the first peak G.

Symbols a and $a + b$ will be values of C_L at maximum G and entry, respectively; n is the number which decides the rate of decreases of C_L as density increases. The rate is larger for larger values of n.

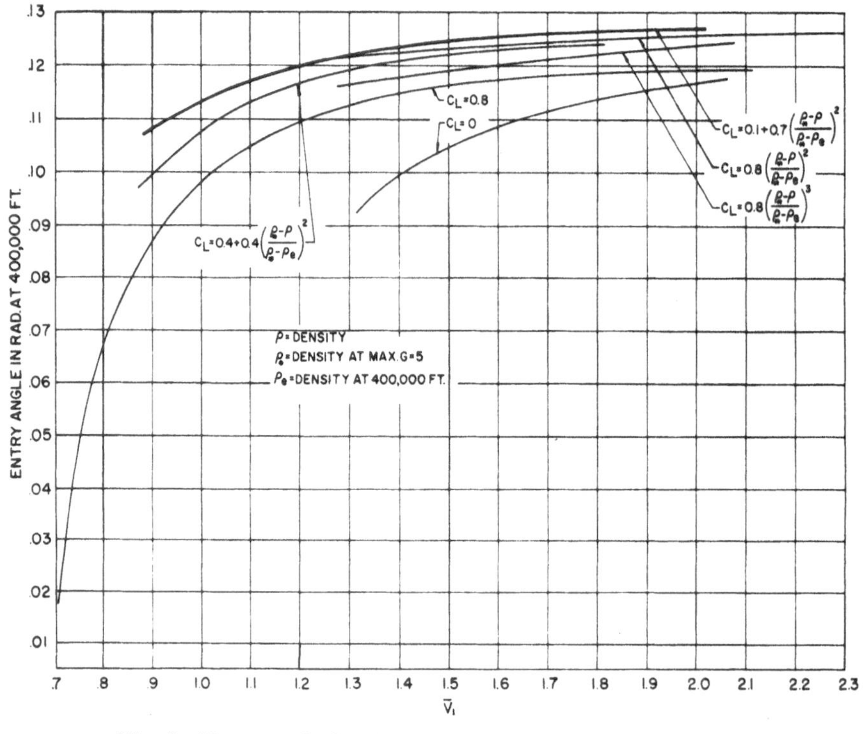

Fig. 7. Entry angle for the first peak $G = 5$ with variable C_L

Fig. 7 shows the curves of maximum entry angle vs. entry velocity with C_L given by Eq. (13) for various combinations of values of a, b and n.

First, the cases with $a = 0$, i.e., $C_L = 0$ and $C_D = C_{D_0}$ at $G_{\max}$ will be considered.

When the lift coefficient varies as the square power of the density ratio ($n = 2$) from 0.8 to 0 as ρ increases from ρ_e to ρ_*, the entry angle vs. velocity curve is above the curves for constant C_L. It is also above the corresponding curve with C_L decreasing from 0.4 to 0 as square power of density ratio. Also, the curve is lower when C_L decreases from 0.8 to 0 as cube power of density ratio ($n = 3$) instead of square power, due to the fact that C_L decrease from 0.8 to 0 faster for $n = 3$ than for $n = 2$.

The curve for C_L, varying as linear power of density ratio, is not shown since the deceleration at ρ_* is a local minimum because of the contribution of the variation of C_L in the second derivative of G.

Since the deceleration attains a maximum at ρ_* with $C_L = 0$, $dC_L^2/dt = 0$ and $d^2C_L^2/dt^2 = 0$, these three curves terminate at points corresponding to the point for the non-lifting cases when the point of maximum becomes a point of inflection.

To avoid such a point of termination, a very small C_L is needed at G_{max} as shown by Eqs. (11) and (12). When the C_L decreases from 0.8 to 0.1 as a square power of density ratio ($a = 0.1$, $b = 0.7$, $n = 2$), i.e., $C_L = 0.1$ at density ρ_* and $C_D \sim C_{D_0}$, the curve extends beyond the point of termination into the region where the entry velocity is less than the circular orbiting velocity.

The curve is above the one for C_L decreasing from 0.8 to 0.4 which in turn is above the curve for the limiting case of constant $C_L = 0.8$. The figure shows that the entry angle can be increased by lift modulation, especially when the velocity is near to the circular orbiting velocity ($\bar{V} = 1$).

It is worthwhile to point out that in the range of velocity $\bar{V} = 1$ to 2 the entry angle is nearly constant; therefore, it is possible to select the optional transfer from a space orbit to an entry orbit such that the entry orbit has the proper entry angle without specifying accurately the entry velocity.

IV. Constant Elevation Maneuver

As it has been mentioned before, when the first peak deceleration is small, the trajectory will either exit or will have a second peak deceleration which may be larger than the limit. There are, of course, many ways of lift modulations to reduce the second peak deceleration and also to change some skip or exit trajectories to non-skip trajectories. Here only the constant elevation maneuver will be discussed. In this maneuver the velocity will be reduced gradually as lift increases to C_L maximum in order to maintain a constant elevation. After such a maneuver, the vehicle will descend with lower velocity, hence smaller peak deceleration.

This special maneuver is of interest not only because of the mathematical simplicity of the analysis, but also on the basis of the following physical reasoning. If the C_L is kept below the C_L necessary for the horizontal flight, the vehicle will descend earlier at higher velocity, hence will encounter higher G. On the other hand, if C_L is increased too rapidly the vehicle will ascend in a skip trajectory and finally return to original elevation at a finite angle. Along the skip trajectory the vehicle will pass through a region of lower density which is less effective in reducing velocity in a given time or distance. Therefore we would assume that such a skip trajectory will be used as a means of changing the impact point with respect to the point given by a non-skip trajectory with part of it horizontal. This argument will be substantiated by the numerical results in Fig. 11 which will be discussed in detail in the next section.

In the constant elevation maneuver, h, $\bar{r}$, and ρ remain unchanged while $\gamma = 0$ and $d\gamma/dt = 0$.

The normal component of the equation of motion, Eq. (6), yields the simple relationship between C_L and $\bar{V}$ in which

$$C_L = \frac{1 - \bar{V}^2 \bar{r}}{X^2 \bar{V}^2 \bar{r}^2} \tag{14}$$

where $X = \rho V_s^2 A/(2 W g)$.

The tangential component of the equation of motion, Eq. (5), yields

$$\left(\frac{V_s}{X g}\right) \frac{d\bar{V}}{dt} = - C_D \bar{V}^2 \tag{15}$$

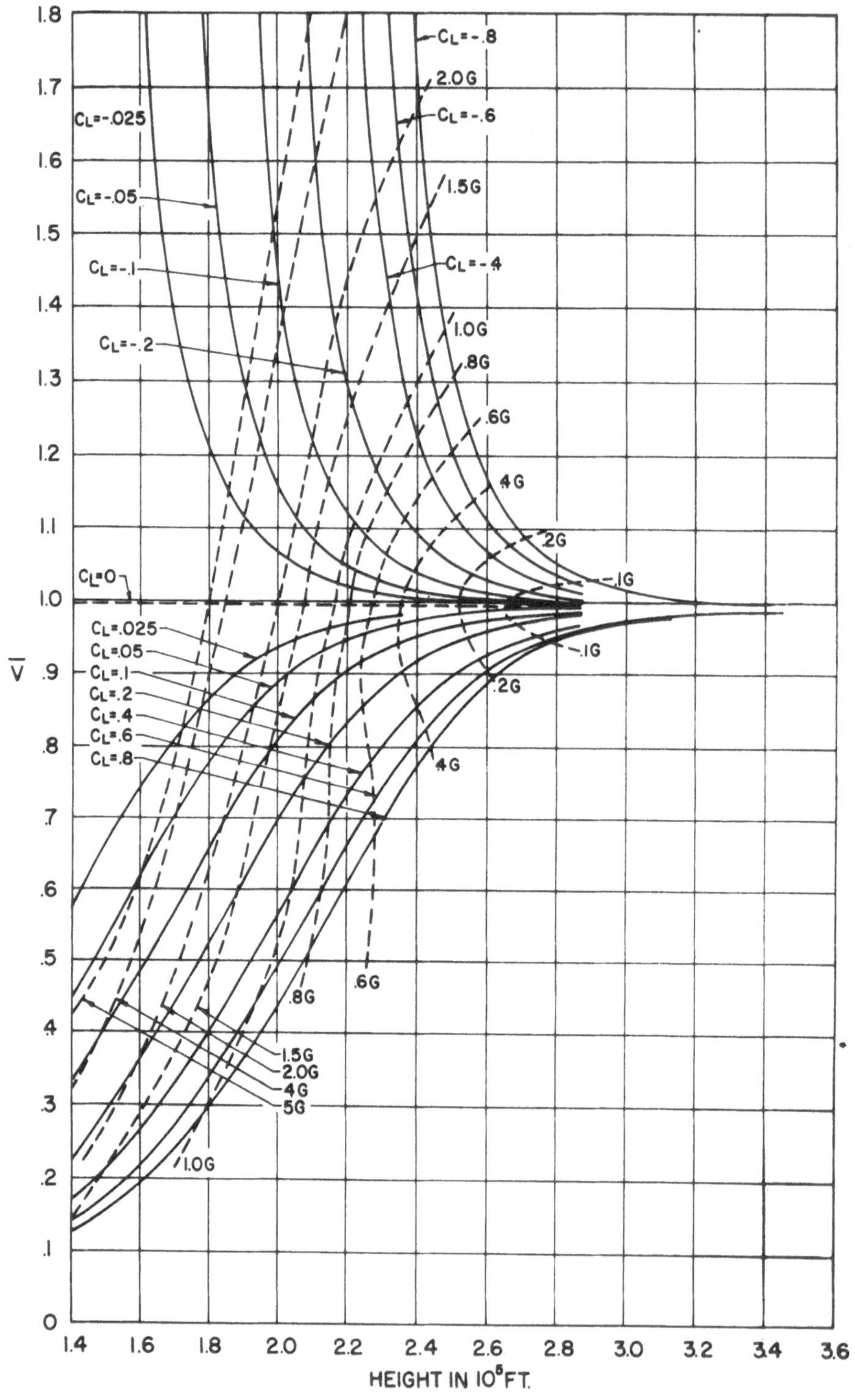

Fig. 8. Range of velocities in constant elevation maneuvers

Here C_D is related to C_L by Eq. (3) while C_L is related to $\bar{V}$ by Eq. (14). Eq. (15) can be integrated and gives the relationship between t and $\bar{V}$:

$$-\left(\frac{g}{X\,V_s}\right)(t-t_0) = \frac{1}{4\,f\,d}\ln\left(\frac{d\bar{V}^2 - f\,\bar{V} + e}{d\bar{V}^2 + f\,\bar{V} + e}\right)\Bigg|_{\bar{V}_0}^{\bar{V}} + \frac{1}{4\,d^2}\arctan\left(\frac{2\,\bar{V}\,j}{e - \bar{V}^2\,d}\right)\Bigg|_{\bar{V}_0}^{\bar{V}} \tag{16}$$

where

$$e = \sqrt{k/\bar{r}^2} \qquad d = \sqrt{C_{D_0}X^2 + k/\bar{r}^2}$$
$$f = \sqrt{2(e^2\,\bar{r} + de)} \qquad j = \sqrt{de - f^2/4}$$

and $\bar{V}_0$ is the nondimensional velocity at the instant t_0 when the maneuver begins.

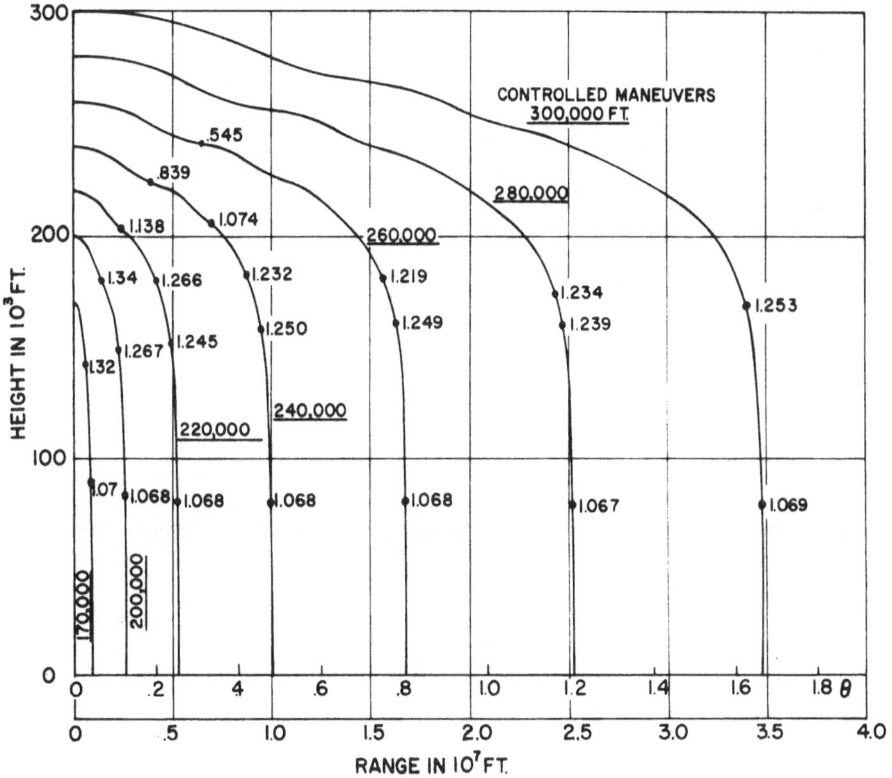

Fig. 9. Descending trajectories with $C_L = 0.8$ after constant elevation maneuvers

This equation together with Eq. (14) relates t to C_L, i.e., they yield the required lift program. Detailed steps in the derivation of Eq. (16) can be found in [12].

The range or the angle φ swept by the radial vector is obtained by the integration of the kinematic equation

$$\left(\frac{r_s}{V_s}\right)\bar{r}\,\frac{d\theta}{dt} = \bar{V} \tag{17}$$

The result is an analytic relation between θ and C_L:

$$\theta = \frac{1}{2\,[C_{D_0} X^2 \bar{r}^2 + k]} \left\{ \sqrt{\frac{k}{C_{D_0}}} \arctan\left(\sqrt{\frac{k}{C_{D_0}}}\, C_L \right) + \ln\left[(1 + X\, C_L\, \bar{r}) \sqrt{C_{D_0}/C_D} \right] \right\} \quad (18)$$

with $\theta = 0$ at $C_L = 0$. Detail steps can be found in [12] and also in [5].

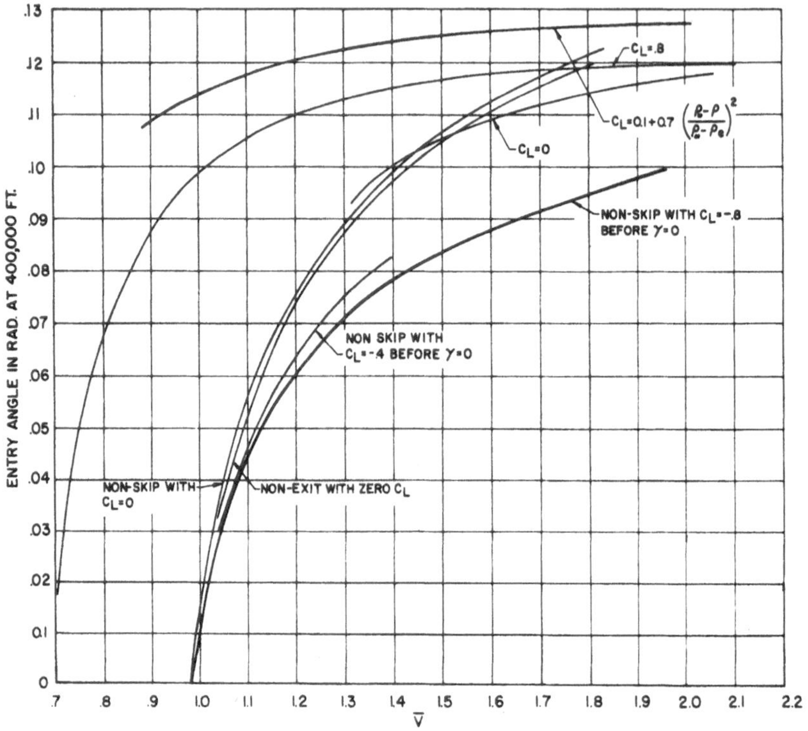

Fig. 10. Entry angle for trajectories with the first peak $G = 5$ and for nonskip trajectories (with lift modulation)

The effectiveness of this maneuver is indicated by the results shown in Fig. 8. It shows the changes in velocity in constant elevation maneuvers due to different ranges of lift coefficients. For example, at an elevation of 200,000 ft $\bar{V}$ decreases from 1.4 to 0.8 as C_L changes from -0.1 to $+0.1$.

If the velocity is reduced all the way to the lowest curve $C_L = 0.8$ and then the vehicle descends with const. $C_L = 0.8$ the peak deceleration is always less than 1.35, as shown in Fig. 9. Therefore, in order to keep the peak deceleration less than 5, it is possible to start the descent earlier, i.e., before C_L reaches 0.8.

When the entry angle is at the upper limit corresponding to the curve C_L varying from 0.8 to 0.1 in square power of density ratio in Fig. 7, the deceleration reaches the maximum 5 g at $C_L = 0.1$. The deceleration will then decrease with C_L kept constant at 0.1, and the angle γ will decrease to zero at an elevation and velocity which is well inside the range of the constant elevation maneuver.

The upper range of the maneuver is given by the curve in Fig. 8 with $C_L = -0.8$, if this is the limit of negative C_L. With a point on this curve as the initial data we can go backward to find the entry velocity and angle for the non-skip curve as shown in Fig. 10.

With velocity and entry angle corresponding to the point on this curve, the trajectory will descend with $C_L = -0.8$ until $\gamma = 0$, then will continue with constant elevation maneuver to the final descent.

For entry conditions above this curve, skip can be avoided with smaller negative C_L. The data in the same figure shows that the non-skip curve, with $C_L = -0.4$ until $\gamma = 0$, is higher.

On the figure the non-skip curve and nonexit curve for zero C_L are also shown. These curves are very close to the curve of $G_{max} = 5$. With zero C_L, the initial velocity and angle for a non-skip or nonexit trajectory with first peak deceleration less than 5 are restricted to a very small area. This area is now substantially enlarged to the region between the two heavy curves when lift modulation is employed.

V. Control of Point of Landing

It has been pointed out that the subsequent peak deceleration can be reduced to below 5 g with partial constant elevation maneuvers, that is, it is possible to start the descent with a smaller C_L or pull out with a larger C_L. In doing so it

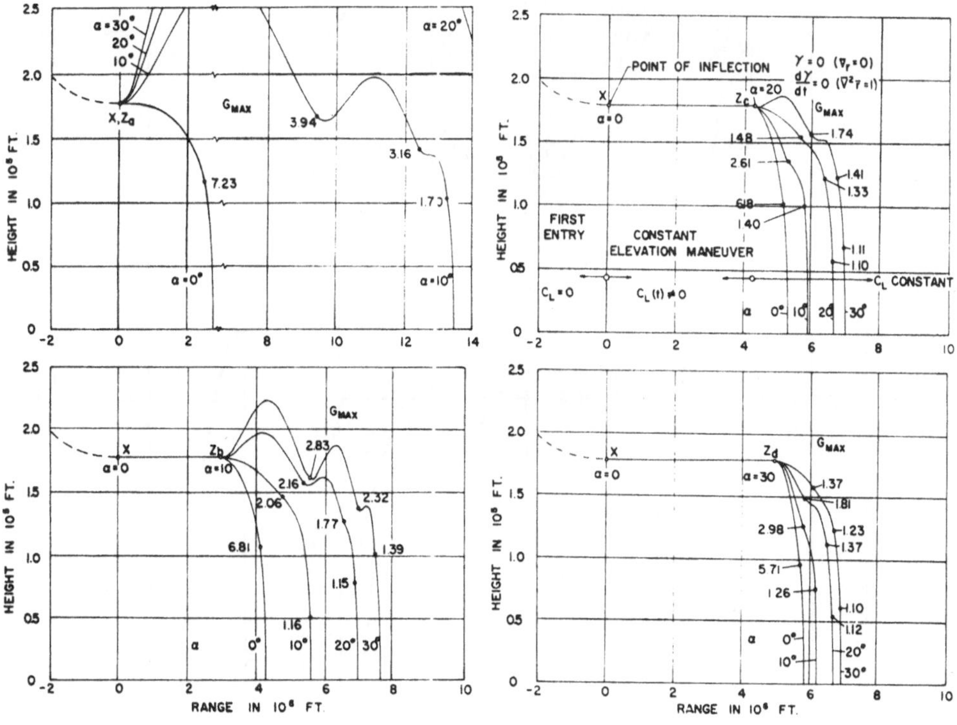

Fig. 11. Variation of range for trajectories with constant elevation maneuver and constant C_L

is possible to control the point of landing. In Fig. 11, the variation of the ranges of trajectories which all begin with the local circular velocity at elevation 187,000 ft. is shown. On the upper left figure, skip trajectories with $\alpha = 10°$, 20°, 30°, corresponding to $C_L = 0.2$, 0.4, 0.6 respectively are shown. For all trajectories the range is greater than 12×10^6 ft. In the lower left figure, a constant elevation maneuver is used until $\alpha = 10°$ or $C_L = 0.2$ and is followed by the subsequent descent or skip trajectory, the lower range is nearly 5×10^6. In the two figures on the right side, the constant elevation maneuvers are carried

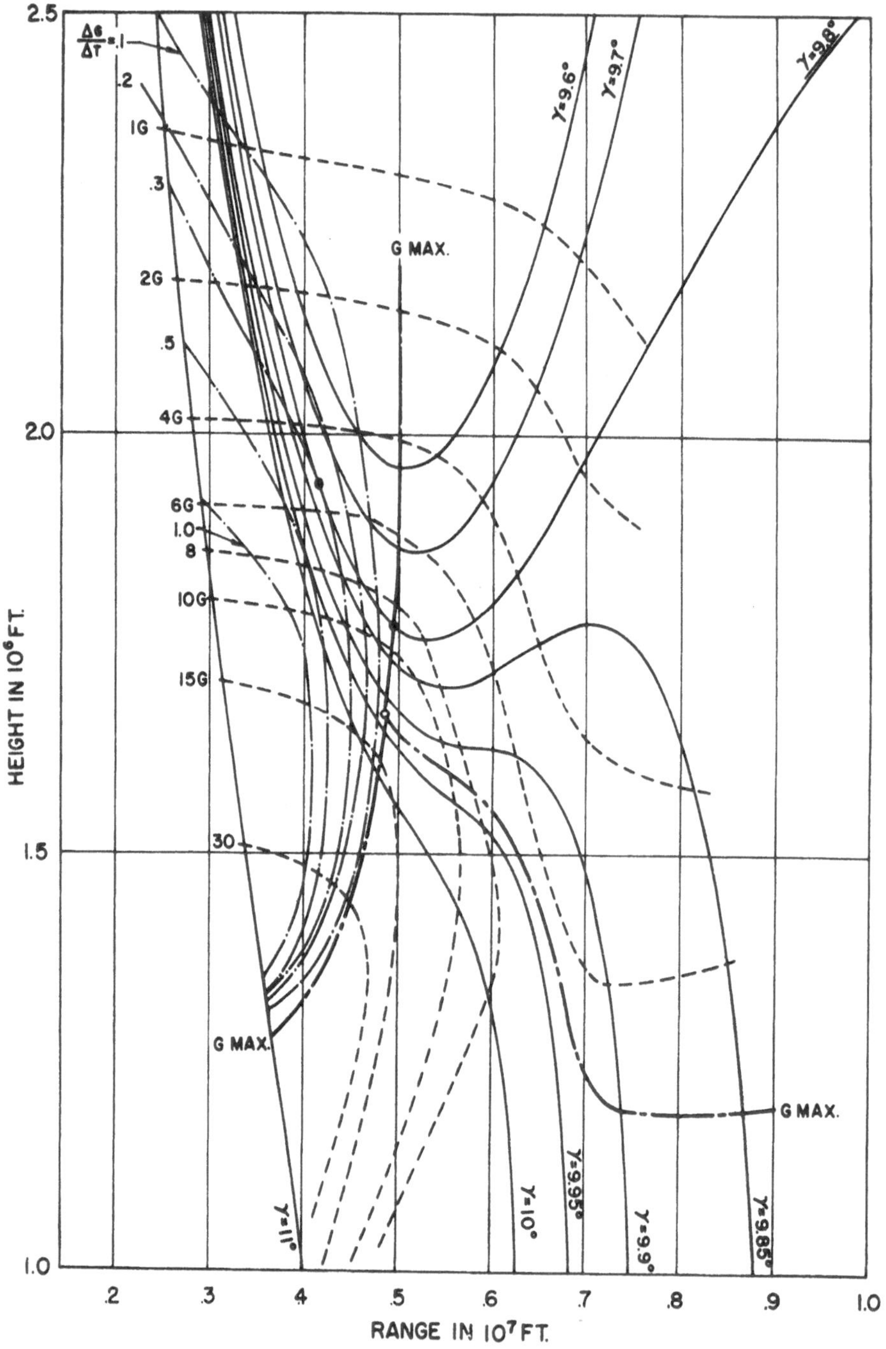

Fig. 12. Rate of change of deceleration in re-entry trajectories with $V = 44{,}250$ ft/sec ($\bar{V} = 1.705$) at 200 km. $W/(C_{D_0}A) = 62$ lbs/ft², $C_L = 0$

to $\alpha = 20°$, and $30°$, respectively. These data indicate that the range can be easily changed between 5×10^6 and 12×10^6 ft.

Results in Fig. 11 can also demonstrate the relative effectiveness of the constant elevation maneuver in reducing the peak deceleration as compared to a skip trajectory. In the upper left figure, the trajectory with $C_L = 0.2$ or $\alpha = 10°$ without constant elevation maneuver has a peak G equal to 3.94. When α increases gradually from 0 to 10° to maintain the constant elevation, and then remains constant at 10°, the trajectory as shown in the lower left figure has a peak G equal to 2.06, which is smaller than the aforementioned value of 3.94 in the skip trajectory.

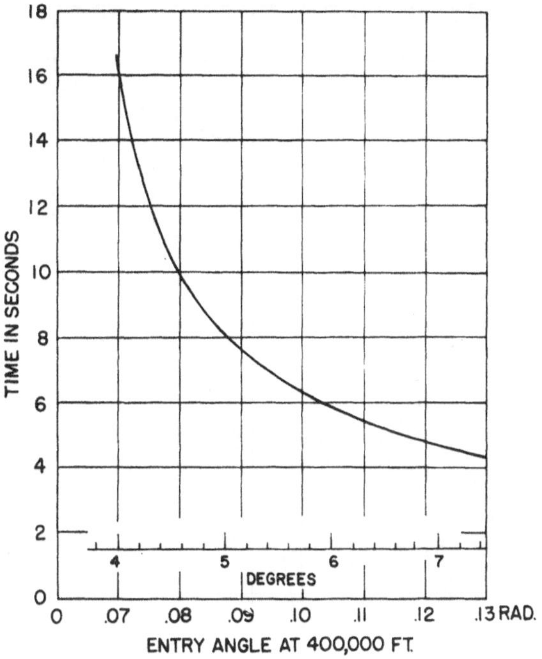

Fig. 13. Interval between 0.1 g and 0.2 g for $\bar{V} = 1.2$ at 400,000 ft. $W/(C_{D_0} A) = 62.5$, $C_L = 0$

VI. Significance of the Rate of Deceleration

It has been shown in [1 – 5] and also in this paper that the first peak deceleration is extremely sensitive to the entry angle. Therefore, it is necessary to determine the entry angle of a given trajectory in advance in order to use the lift necessary to reduce the peak deceleration below the maximum acceptable value. Instead of determining the entry angle directly with great accuracy, it is possible to determine in advance the entry angle or the expected first peak deceleration corresponding to the given entry trajectory indirectly by measuring the rate of change deceleration, dG/dt, or by measuring the decelerations and a time interval at the beginning of the re-entry trajectory.

To see the significance of dG/dt the region before the first peak G in Fig. 1 is enlarged in Fig. 12. The trajectory with entry angle 9.8° with the first peak $G = 9$ is tangential to the $dG/dt = .3$ curve. Therefore, along a trajectory where dG/dt reaches 0.3, while G is still small, we know that the first peak G will be greater than 9. This information will enable the pilot to use lift modulation or apply a control rocket to change the trajectory well in advance.

Fig. 13 shows the interval between 0.1 g and 0.2 g for non-lifting trajectories with various entry angles. Since the decelerations and time can be measured accurately inside the vehicle, the pilot will be able to determine indirectly the corresponding entry angle by measuring a time interval between two values of G and to perform in time the required lift modulation.

Fig. 14 shows the curve of maximum entry angle vs. entry velocity for first peak G equal to 5 with $C_L = 0$ for $G < 0.2$ and $C_L = 0.8$ for $G > 0.2$. This curve is only slightly below the corresponding curve with $C_L = 0.8$ for the entry trajectory. This indicates that it is possible to enter with zero lift until G reaches 0.2. The measurements of entry angle can be performed during this period and then the proper lift modulation can be applied. The reduction in the full effect of the lift modulation because of this delay is negligible.

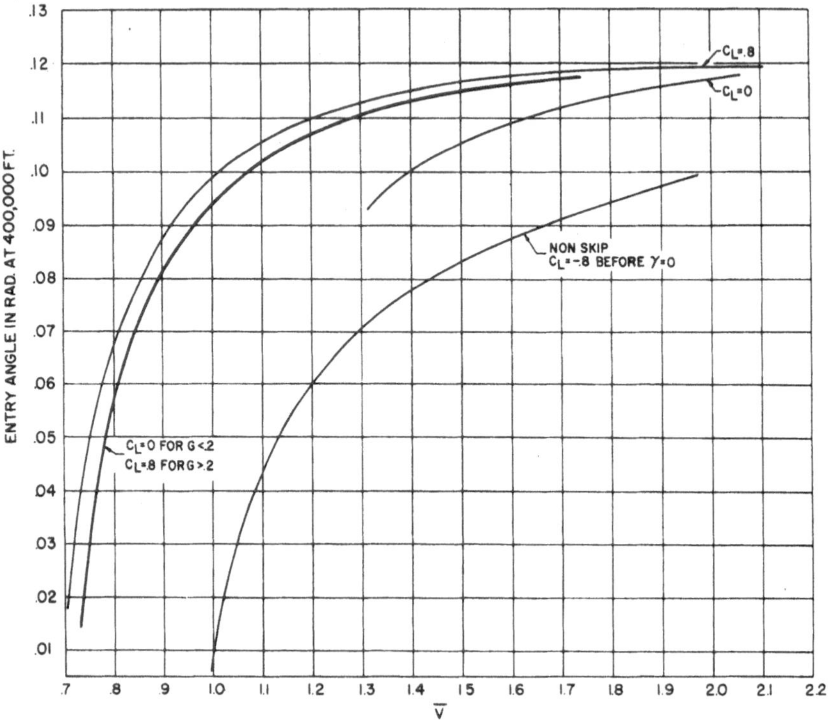

Fig. 14. Entry angle for the first peak $G = 5$ with $C_L = 0$ for $G < 0.2$, $C_L = 0.8$ for $G > 0.2$

References

1. D. CHAPMAN, An Approximate Analytic Method for Studying Entry into Planetary Atmospheres. NACA TN 4276, May 1958.
2. D. CHAPMAN, An Analysis of the Corridor and Guidance Requirements for Supercircular Entry to Planetary Atmospheres. NASA TR R-55, 1959.
3. L. LEES, F. W. HARTWIG, and C. B. COHEN, The Use of Aerodynamic Lift During Entry into the Earth's Atmosphere. ARS Journal **29**, 633–641 (1959).
4. J. D. C. CRISP, The Dynamics of Supercircular Multiple-Pass Atmospheric Braking. Polytechnic Institute of Brooklyn, PIBAL Report No. 582, October 1960; Astronaut. Acta **8**, 1–27 (1962).
5. J. D. C. CRISP, An Assèssment of Lift-Drag Modulation Techniques for Atmospheric Entry. Polytechnic Institute of Brooklyn, PIBAL Report No. 586, January 1961.
6. T. D. DUANE, E. L. BECKMAN, J. E. ZIEGLER, and H. N. HUNTER, Some Observations on Human Tolerance to Accelerative Stress. III. Human Studies of Fifteen Transverse G. J. Aviat. Med. **26**, 298–303 (1955).
7. R. EDELBERG, J. P. HENRY, J. A. MACIOLEK, E. N. SALZMAN, and G. D. ZUIDEMA, Comparison of Human Tolerance to Acceleration of Slow and Rapid Onset. J. Aviat. Med. **27**, 482–489 (1956).
8. H. J. ALLEN and A. J. EGGERS, JR., A Study of the Motion and Aerodynamic Heating of Ballistic Missiles Entering the Earth's Atmosphere at High Supersonic Speed. NACA Report 1381, 1958.

9. K. WANG and L. TING, An Approximate Analytic Solution of Re-Entry Trajectories with Aerodynamic Forces. ARS Journal **30**, 565–566 (1960).
10. K. WANG and L. TING, Approximate Solutions for Re-Entry Trajectories with Aerodynamic Forces. Polytechnic Institute of Brooklyn, PIBAL Report No. 647, AFOSR 684, May 1961.
11. L. BROGLIO, Similar Solutions in Re-Entry Lifting Trajectories. Space Research, Edited by HILDE KALLMANN BIJL, pp. 564–580. New York: Interscience Publishers, Inc. 1960.
12. A. FERRI and L. TING, Practical Aspects of Re-Entry Problems. Polytechnic Institute of Brooklyn, PIBAL Report No. 705, July 1961.

Discussion

Mr. NONWEILER: If I may first make a comment on the paper, though it is not intended as a criticism: it is worth pointing out that the variation of drag with lift of re-entry type vehicles can often be very strange. You have assumed that drag increases with C_L^2, which is fair enough as one has to make *some* assumption; but certain vehicles are designed to fly at such high incidences that increase of incidence implies increase of drag but *decrease* of lift. There is also often a large variation in drag obtained from the deflection of the trimming surface. Secondly, I would like to ask a question: can you tell me how the heat transfer may be affected by this manoeuvre? I have in mind that in the re-entry of ballistic vehicles, variation of the frontal area to reduce the deceleration during descent can often lead to increased rates of heat transfer.

Mr. FERRI: As an answer to the first question, it should have been mentioned in our presentation that the lift and drag relationship is based on the experimental data of a realistic body type vehicle for C_L less than $C_{L\max}$.

As an answer to the second question, the problem of aerodynamic heating *along a given trajectory* can be reduced by engineering devices e.g. ablation, while the maximum value of peak deceleration that can be sustained by average human beings cannot be changed. In selecting a trajectory the peak deceleration must be kept below the given value, therefore, this is the deciding criterion.

A Generalized Study of Two-dimensional Trajectories of a Vehicle in Earth-Moon Space[1]

By

H. Hiller[2]

(With 16 Figures)

Abstract — Zusammenfassung — Résumé

A Generalized Study of Two-dimensional Trajectories of a Vehicle in Earth-Moon Space. A study has been made of the two-dimensional motion of a vehicle in a simplified Earth-Moon system, using the concept of a "sphere of influence". The trajectory, starting from the vicinity of the Earth, is considered to be divided into three phases — to, across and away from the Moon's sphere of influence. Graphical results are presented of the following: the geocentric and selenocentric velocity vectors and location of the vehicle for arrival at and departure from the Moon's sphere of influence; the periselene distance; the parameters defining the departure conic-section trajectory, so that subsequent motion is determined; and finally, the conditions for grazing and centrally hitting the Moon.

An additional study, using some of the above results, has shown that the tolerance in direction of the initial geocentric velocity, to miss the Moon's centre by its radius, has a mean value of approximately $\pm$ 0.2° about an initial value of 90°.

Allgemeine Untersuchungen über zweidimensionale Bahnen im Erde-Mond-Bereich. Die betrachteten Bahnen, die in unmittelbarer Nachbarschaft der Erde beginnen sollen, werden in drei Phasen behandelt, und zwar im Einflußbereich des Mondes, im Übergangsgebiet und in größerer Entfernung. Folgende graphische Resultate werden gezeigt: Die geozentrischen und die selenozentrischen Geschwindigkeitsvektoren sowie die Position des Fahrzeuges bei Ankunft und Verlassen des Einflußbereichs des Mondes, die periselenische Distanz und schließlich die Bedingungen für eine Umfahrung sowie für ein zentrales Auftreffen auf dem Mond.

In einer zusätzlichen Untersuchung wird unter Verwendung obiger Resultate gezeigt, daß die Toleranzen in der Richtung der anfänglichen geozentrischen Geschwindigkeit etwa $\pm$ 0,2° bei einem Anfangswinkel von 90° betragen können, wenn das Zentrum der Mondscheibe maximal um ihren Radius verfehlt werden darf.

Etude généralisée de trajectoires bidimensionnelles dans l'espace Terre-Lune. L'étude est basée sur le concept de la "sphère d'influence". La trajectoire est divisée dans les phases: vers-intérieure-hors de la sphère d'influence lunaire. Résultats graphiques présentés: vecteurs vitesse géocentriques et sélénocentriques et position du véhicule à l'arrivée et au départ de la sphère d'influence lunaire; distance perisélénique; paramètres de la conique de départ; conditions d'effleurement ou d'impact central sur la Lune.

Un prolongement de l'étude montre que la tolérance angulaire sur la vitesse géocentrique initiale est en moyenne de $\pm$ 0.2° autour d'une valeur initiale de 90° pour manquer le centre de la Lune d'une distance égale à son rayon.

[1] This paper is an abridged version of a Ministry of Aviation report of the same title.

[2] Royal Aircraft Establishment, Farnborough, Hants, England.

I. Introduction

The restricted three-body problem, for a vehicle moving under the gravitational attraction of an isolated Earth-Moon system, is intractable analytically but, by reducing the problem to a succession of two-body problems, it was considered that an approximate solution could be obtained which would give some insight into the solution of the original problem. The vehicle is considered to be moving under the gravitational attraction of either the Earth or the Moon, the changeover from one régime to the other being carried out using the concept of a "sphere of influence" of one of the celestial bodies (the smaller) in the presence of the other. The sphere of influence provides the boundary surface between the two gravitational fields and is defined for the Moon (in the Earth's presence) as follows — it is the region around the Moon where the ratio of the Earth's perturbing force (on the vehicle's selenocentric motion) to the Moon's attractive force is less than the ratio of the Moon's perturbing force (on the vehicle's geocentric motion) to the Earth's attractive force.

The sphere-of-influence concept is referred to in [1—6] and is discussed in some detail in the Ministry of Aviation report of which this paper is an abridged version. It is important to note that this concept can only be used here for a single transit around the Earth-Moon system; more than one perturbation by the Moon could cause prohibitive errors in the vehicle's motion.

In the first phase of vehicle motion, from the Earth to the Moon's sphere of influence, both Direct and Retrograde trajectories are considered, and these may be either Outgoing or Incoming with respect to the Earth's centre as focus (see Fig. 2). All trajectories however are assumed to have a fixed perigee height at 100 miles above the Earth's surface.

A simplified Earth-Moon model has been considered and the assumptions involved, together with other assumptions required are given in section II. The errors are discussed in section III and the method explained in section IV. The results are given in Figs. 4—16 and Table I and discussed in section V. Finally, conclusions are presented in section VI.

II. Assumptions

The simplified Earth-Moon model considered here consists of a spherical Moon of radius 1,080 miles moving in a circular orbit of radius 238,857 miles, about the centre of a spherical Earth of radius 3,960 miles, with a constant angular velocity of 0.549°/hour.

The trajectory of the vehicle is considered to be divided into three phases, each a conic-section. The first is from the vicinity of the Earth to arrival at the Moon's sphere of influence, which has a radius of 41,000 miles and whose centre coincides with that of the Moon. The perigee is assumed to be at a fixed height 100 miles above the Earth's surface and so atmospheric drag can be assumed negligible. The second phase is taken to be a hyperbolic orbit across the Moon's sphere of influence, and lying in the plane of the Moon's orbit. The third phase is assumed to begin with the vehicle's departure from the Moon's sphere of influence into a new geocentric orbit and to end should the vehicle be perturbed again by any influence.

III. Errors

Since the motion of the vehicle is confined to the Earth-Moon plane, the set of results presented here may not be of great practical use. It is intended rather to give a first insight into the true motion of the vehicle. The errors involved in this two-dimensional restriction have been discussed in [10, 13, 18, 19]. The reduction of the original three-body problem to several two-body problems, using the "sphere of influence" concept, is considered to be a reasonable approximation and it is thought that the biggest error probably occurs when an initially elliptic orbit is tangential to the Moon's sphere of influence, at apogee. These errors can be checked by comparison with results obtained by the numerical integration of the equations of motion of the original three-body problem.

Other assumptions which may lead to small relative errors include the orbit of the Moon being circular, the Earth being spherical, and the gradient of the Sun's gravitational attraction being zero. These errors are discussed in [10, 11, 15, 20, 21].

Other unknown errors, which are not expected to be significant, include the assumptions that the region of influence of the Moon (in the Earth's presence) is spherical, the radius of this sphere of influence is 41,000 miles, the centres of the Moon and its sphere of influence coincide, and the barycentre is taken to be at the centre of the Earth.

IV. Method

Using Figs. 1 and 2; the mathematical equations were developed as in the Appendix and then programmed, using the Autocode method, on a *Pegasus* digital computer. The basic initial parameters are the initial perigee velocity V_0 and the true anomaly ϕ_1 (the angle measured from perigee of the conic-section

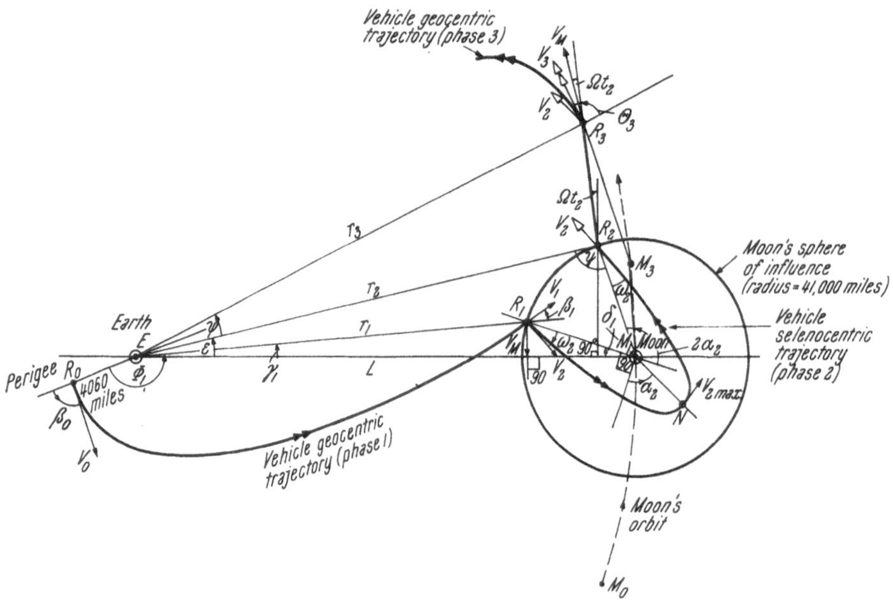

Fig. 1. Sketch of the Earth-Moon system, showing notation

L = 238,857 miles (Earth-Moon distance),

Ω = 0.549°/hour (angular velocity of Moon),

$T = 2\,t_2$ = time of flight, in hours, of vehicle from R_1 to R_2.

Suffixes: 0 — initial (all-burnt) perigee conditions for phase 1,
1 — final geocentric conditions for phase 1,
2 — selenocentric conditions for phase 2,
3 — initial geocentric conditions for phase 3

trajectory to the Earth-vehicle line at the instant of vehicle arrival at the Moon's sphere of influence), (see Fig. 1). However, it was found to be

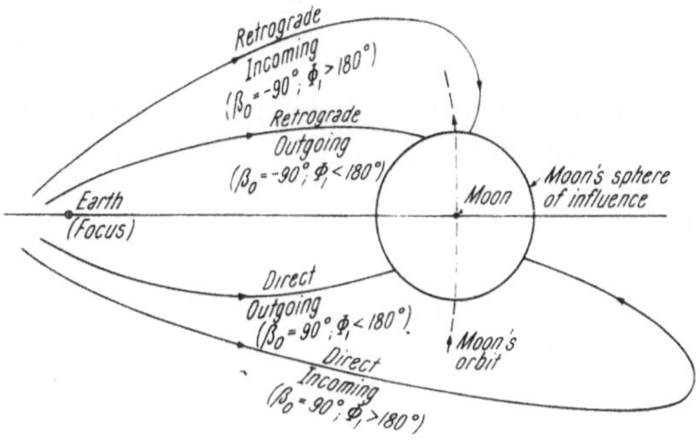

Fig. 2. Sketch of typical "phase 1" trajectories of the four types considered. The true anomaly ϕ_1 is measured anticlockwise for Direct and clockwise for Retrograde trajectories

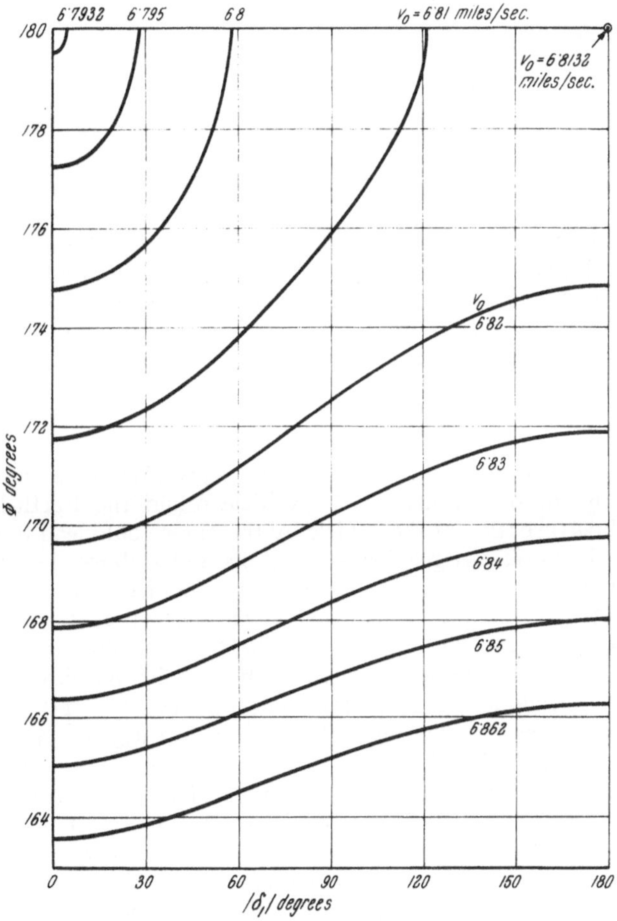

Fig. 3. All elliptic trajectories: variation of geocentric true anomaly, on arrival at the Moon's sphere of influence, ϕ_1, with δ_1. δ_1 is the angle between the vehicle-Moon and Earth-Moon lines at the instant of vehicle arrival at the Moon's sphere of influence. For Outgoing trajectories, $\phi_1 = \phi$ and is measured anticlockwise (from perigee) in Fig. 1 when Direct, and clockwise (from perigee) when Retrograde. For Incoming trajectories, $\phi_1 = (360 - \phi)^\circ$ and is measured as for Outgoing ones

more convenient to express the results in terms of δ_1 (the angle between the Moon-Earth and Moon-vehicle lines at the instant of vehicle arrival at the Moon's sphere of influence) rather than ϕ_1, the conversion curve being given in Fig. 3. A third parameter β_0, which is the angle between the direction of V_0 and the initial radial line from the Earth's centre, has been given the values $\pm 90°$ only, where $\beta_0 = 90°$ defines the Direct trajectory and $\beta_0 = -90°$ defines the Retrograde trajectory. Also $\phi_1 < 180°$ defines the Outgoing and $\phi_1 > 180°$ defines the Incoming trajectory.

For motion across the Moon's sphere of influence, the Moon is considered stationary and its velocity is added vectorially to the geocentric velocity V_1 (as shown in Fig. 1) to allow for the Moon's motion. Selenocentric hyperbolic motion then takes place and a distinction made between Direct and Retrograde motion as shown following eq. (18) of the Appendix.

On leaving the Moon's sphere of influence, the Moon's velocity is restored to the motion of the vehicle by adding it vectorially to the vehicle's selenocentric velocity, in the direction in which the Moon is travelling at the instant of departure of the vehicle from the Moon's sphere of influence. Further, since the Moon was assumed stationary during phase 2 motion, the position of the vehicle must also be adjusted to allow for the extra distance the vehicle should have moved during phase 2.

Having ascertained from eq. (2) of the Appendix that the minimum value of V_0, to reach to Moon's sphere of influence, is 6.7932 miles/sec. relative to the Earth's centre and that the parabolic value of V_0 is 6.8625 miles/sec., a maximum (hyperbolic) velocity V_0 of 8 miles/sec. was chosen and the whole range of values of V_0 sub-divided into intervals, as shown on the graphs. The range of values for δ_1 was also divided up, into 15° intervals generally and into smaller sub-divisions where greater accuracy was required.

V. Results

The results presented here in Figs. 4—16 depend on the assumptions made and apply only for one transit of the vehicle round the Earth-Moon system (i.e. a second perturbation of the vehicle by the Moon could lead to unacceptable errors) where the initial perigee distance is 4,060 miles above the Earth's centre.

Fig. 4 gives the geocentric arrival velocity and has been plotted in polar form to show the region into which no vehicle can enter, since the vehicle is always behind, and travelling slower than, the Moon, in this region. This graph is for phase 1 elliptic velocities only, and if hyperbolic velocities (up to infinite values) are also considered, the forbidden range is defined by $-115° < \delta_1 < -81.1°$. The graph also shows that for phase 1 elliptic velocities, the geocentric arrival velocity never exceeds 1 mile/sec.

Figs. 5—7 give results during selenocentric flight across the Moon's sphere of influence. Fig. 5 gives the selenocentric arrival velocity (at the Moon's sphere of influence). This is also the selenocentric departure velocity, except for a small region in each curve where impact on the Moon occurs. Fig. 6 gives the periselene distance (nearest approach to the Moon's centre). Since the Moon's radius is taken to be 1,080 miles, the Moon impact region is given by $1{,}080 > r_{P2} > 0$ miles. Fig. 7 gives the impact conditions for both grazing and central hits. Grazing conditions occur for both Direct and Retrograde selenocentric trajectories. This graph shows that for Direct Outgoing trajectories from the Earth, the minimum initial perigee velocity for a central hit on the Moon is 6.805 miles/sec. (35,930 ft./sec.) and for grazing hits, varies by $\pm$ 8 ft./sec. from this velocity.

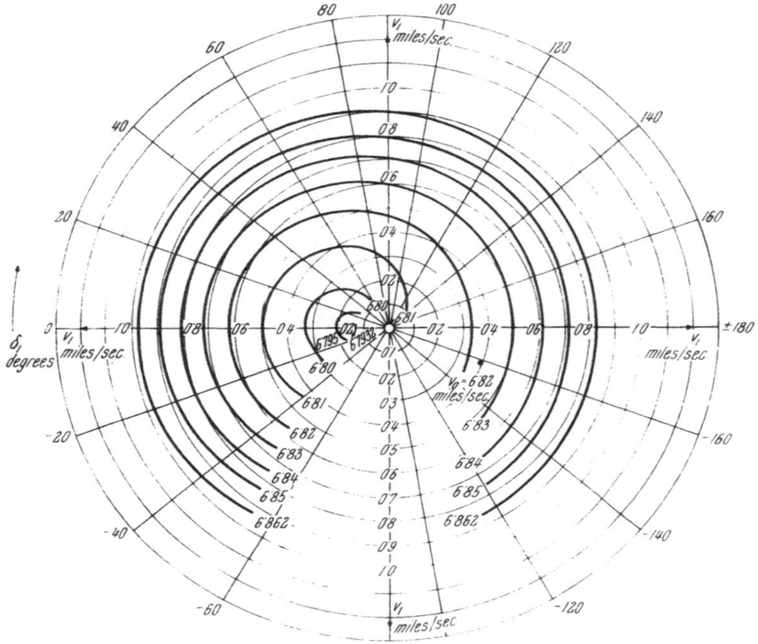

Fig. 4. All elliptic trajectories: variation of geocentric arrival velocity, V_1, with δ_1

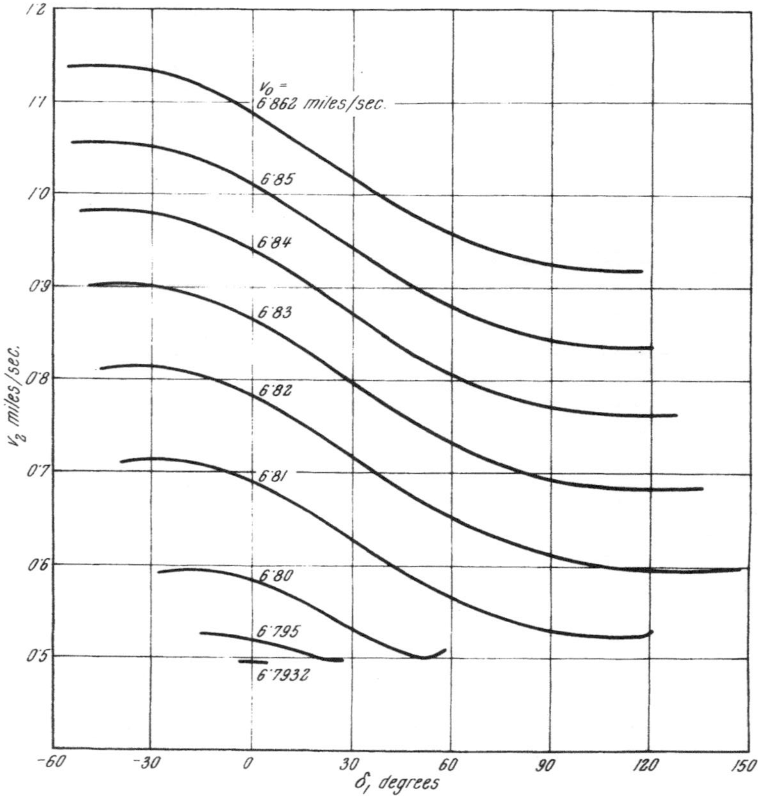

Fig. 5. Direct Outgoing elliptic trajectories: variation of selenocentric arrival (or departure) velocity, V_2, with δ_1

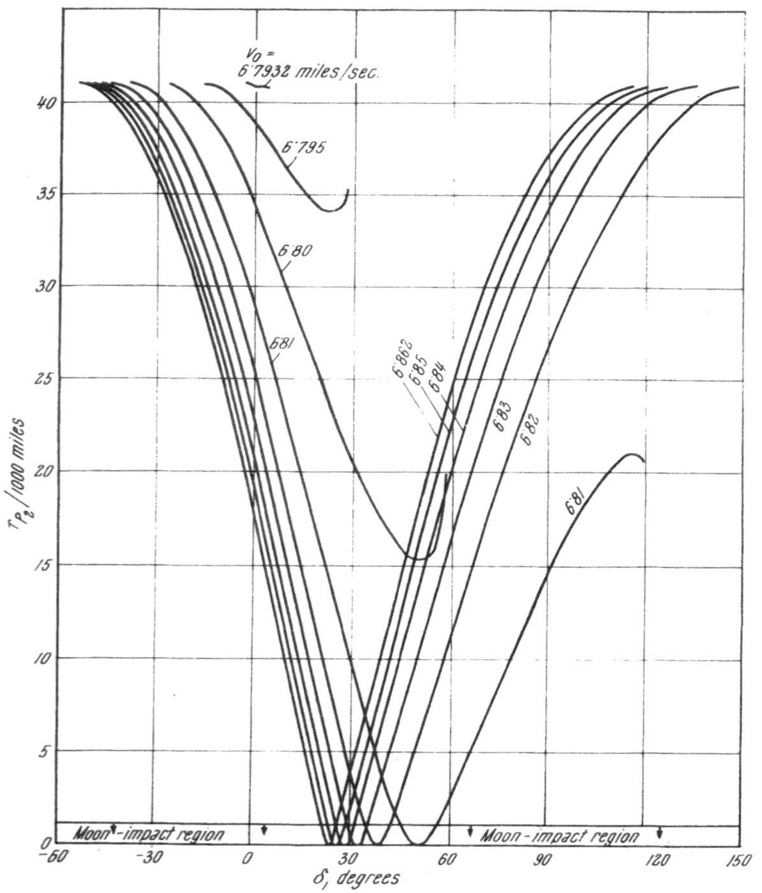

Fig. 6. Direct Outgoing elliptic trajectories: variation of periselene distance, r_{P2}, with δ_1

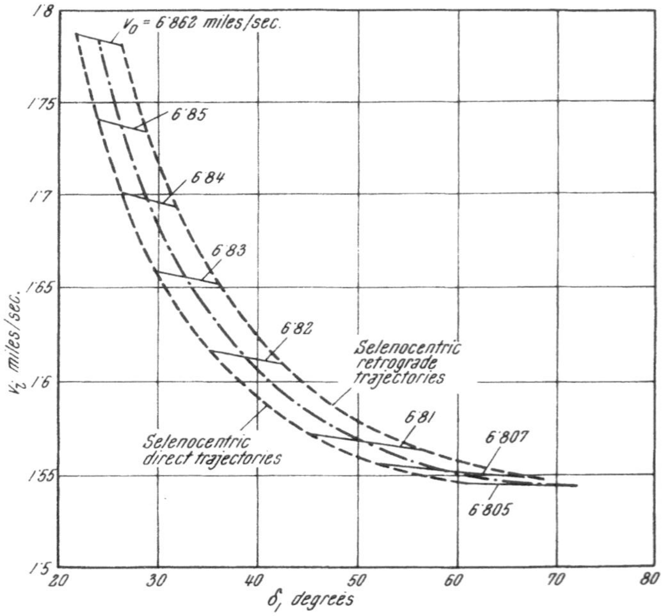

Fig. 7. Direct Outgoing elliptic trajectories: variation of selenocentric impact velocity on the Moon, V_i, with δ_1

— · — · — impact at centre of Moon's disc

— — — — grazing impact on Moon. δ_1 is the angle between the vehicle-Moon and Earth-Moon lines at the instant of vehicle arrival at the Moon's sphere of influence. V_0 is the initial perigee velocity

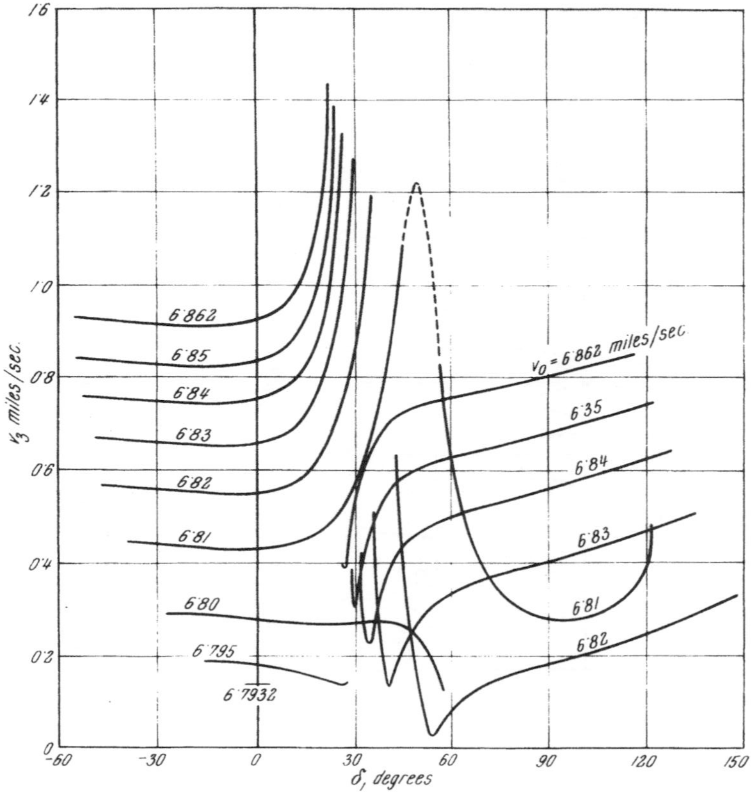

Fig. 8. Direct Outgoing elliptic trajectories: variation of geocentric departure velocity, V_3, with δ_1
– – – – typical curve for Moon-impact region

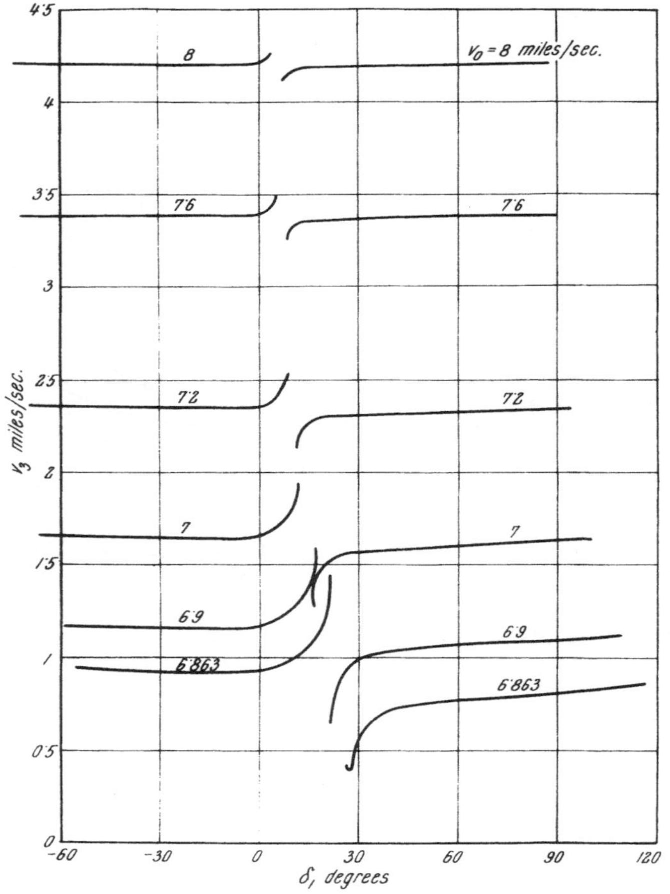

Fig. 9. Direct hyperbolic trajectories: variation of geocentric departure velocity, V_3, with δ_1. Gaps in curves indicate Moon-impact regions

The selenocentric impact velocity for a central Moon hit varies between 1.545 miles/sec. (8,160 ft./sec.) and 1.785 miles/sec. (9,420 ft./sec.) for initial elliptic trajectories. (Note that the time of launch varies with the initial perigee velocity V_0.)

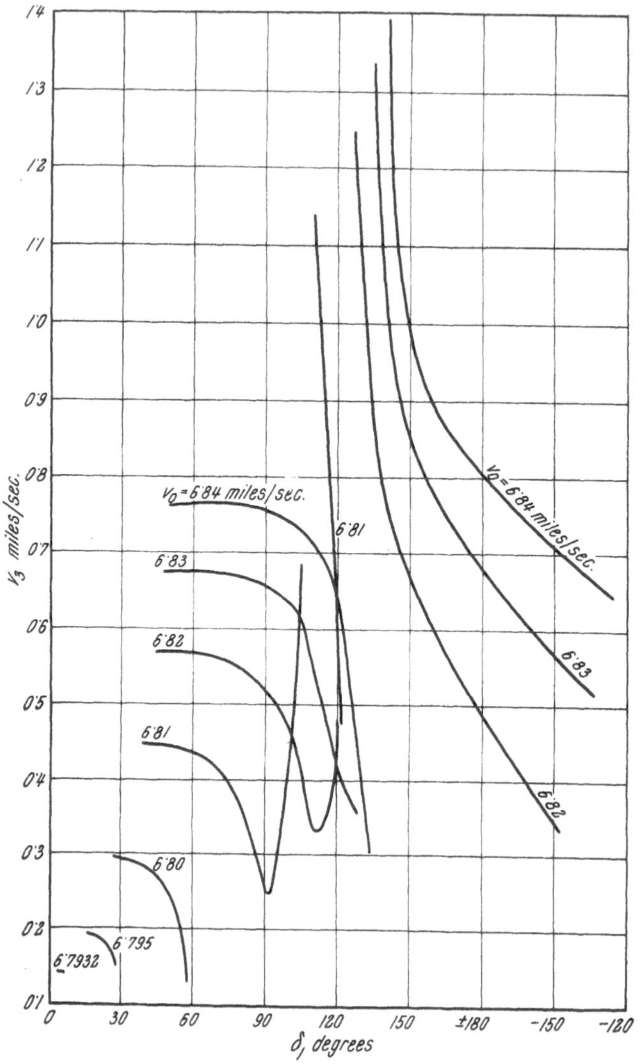

Fig. 10. Direct Incoming trajectories: variation of geocentric departure velocity, V_3, with δ_1. Gaps in curves indicate Moon-impact regions

Errors permissible in initial velocity or direction cannot be deduced directly from the results presented here, since changes in either of these parameters cause changes in initial (launch) time of the vehicle from perigee. Hence the errors can be found only by further extensive analysis to allow for the vehicle's flight time variations.

Figs. 8—16 give the results for geocentric departure trajectories from the Moon's sphere of influence. Figs. 8—11 give the velocity for various types of

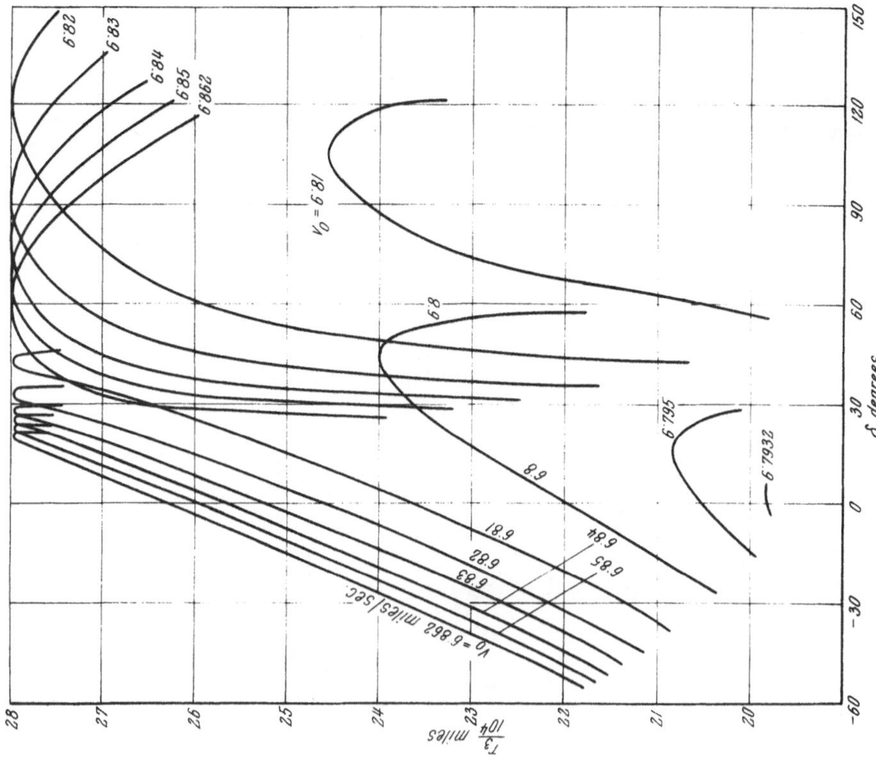

Fig. 12. Direct Outgoing elliptic trajectories: variation of geocentric distance, r_3, on departure from the Moon's sphere of influence, with δ_1. Gaps in curves indicate Moon-impact regions

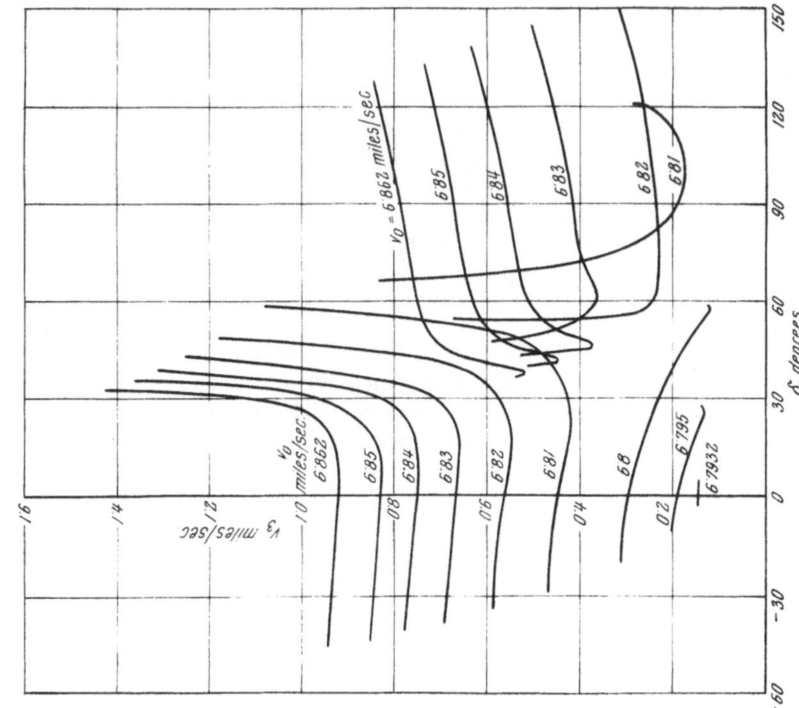

Fig. 11. Retrograde Outgoing elliptic trajectories: variation of geocentric departure velocity, V_3, with δ_1. Gaps in curves indicate Moon-impact regions

trajectory; Fig. 12 gives the location of the vehicle; Fig. 13 gives the direction of the velocity relative to the radial line from the Earth's centre to the vehicle; Figs. 14 and 15 give the conic-section parameters and finally, Fig. 16 gives the true anomaly which determines the direction of the major axis of the trajectory.

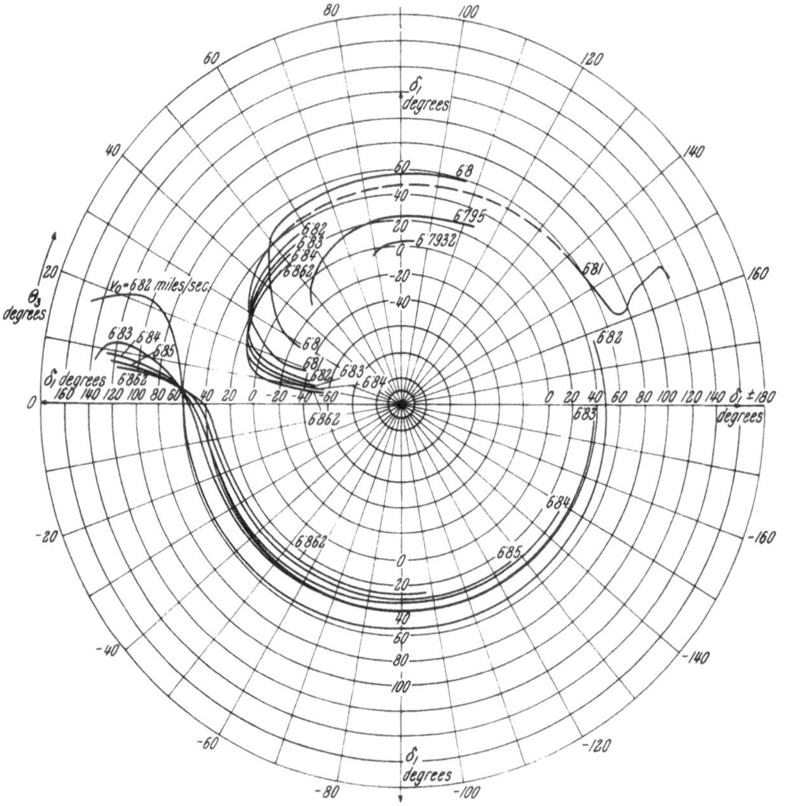

Fig. 13. Direct Outgoing elliptic trajectories: variation of direction of geocentric departure velocity (from geocentric departure radius), θ_3, with δ_1

– – – – – typical curve for Moon-impact region

Comparing Figs. 4 and 8, shows that the gravitational attraction of the Moon increases the geocentric velocity of the vehicle when it is close to but behind the moon and generally decreases the vehicle's velocity when it is close to but ahead of the Moon. The exceptions to this decrease occur when the initial perigee velocity is in the region of 6.81 to 6.82 miles/sec. Fig. 9 extends the range of Fig. 8 into initial hyperbolic perigee velocities and is drawn separately for greater accuracy. Fig. 10 is for Incoming trajectories and is given for initial perigee velocities up to 6.84 miles/sec. For velocities higher than this, the perturbations due to the Sun would have to be taken into account. These trajectories have similar properties to the Outgoing trajectories of Fig. 8, but the Retrograde selenocentric trajectories pass behind the Moon here whereas in Fig. 8 the Direct selenocentric trajectories pass behind the Moon. Fig. 11 and Fig. 8 show the differences between Direct and Retrograde trajectories. Generally the left-hand set of curves show higher velocities for Retrograde phase 1 trajectories (these represent Direct selenocentric trajectories) whereas the right-hand curves (Retrograde selenocentric) show higher velocities for Direct phase 1 trajectories.

Fig. 12 gives the distance of the vehicle from the Earth's centre (focus) at the instant of departure from the Moon's sphere of influence. The maximum distance is 279,857 miles (the Earth-Moon distance plus the radius of the Moon's sphere of influence) while the minimum distance is 197,857 miles (the Earth-Moon distance minus the radius of the sphere of influence).

Fig. 13 gives the angle θ_3 of the geocentric departure velocity measured from the radius vector to the Earth's centre (as focus). After departing from the Moon's sphere to influence, the trajectory is Direct for positive θ_3 and Retrograde for negative θ_3. When $|\theta_3|$ is 90°, the trajectory begins its phase 3 motion at apogee. When $\theta_3 = 0$, the vehicle is moving away from the Earth's centre in a straight line (the eccentricity $e_3 = 1$) and for all initial elliptic velocities the vehicle will reach apogee and then return in a straight line towards the Earth's centre. When $|\theta_3| = 180°$, the vehicle is moving directly towards the Earth's centre. In these two instances, where the vehicle is moving in a straight line, we have special cases of the solution. These curves are all for phase 1 Direct trajectories but whereas the upper group of curves (between $\theta_3 = 8°$ and $\theta_3 = 113°$ and ignoring the broken line) are phase 2 Direct trajectories, the lower group (between $\theta_3 = 21°$ and 134°) are phase 2 Retrograde trajectories. Also for the region $0 < \theta_3 < 180°$ (which includes all the upper curves) all motion is phase 3 Direct, while for $-180° < \theta_3 < 0$ (which includes most of the lower curves), we get phase 3 Retrograde motion. A slightly different set of conditions apply to Incoming trajectories. The types of trajectory for the various phases are summed up in Table I. For phase 1 Outgoing trajectories, Direct phase 2 trajectories on reaching the Moon's orbit pass behind the Moon while Retrograde phase 2 trajectories pass ahead of the Moon. The reverse is true for phase 1 Incoming trajectories, where the Direct phase 2 trajectories pass ahead of the Moon and Retrograde phase 2 trajectories pass behind the Moon (in the Moon's orbit).

Table I. *Classification of Phases into Direct* (*D*) *or Retrograde* (*R*) *Motion*

Initial Trajectory	Phase		
	1	2	3
Outgoing	*D*	*D*	*D*
	D	*R*	Mostly *R* (Some *D*)
	R	*D*	Mostly *D* (Some *R*)
	R	*R*	Mostly *R* (Some *D*)
Incoming	*D*	*D*	Mostly *R* (Some *D*)
	D	*R*	*D*
	R	*D*	Mostly *R* (Some *D*)
	R	*R*	Mostly *D* (Some *R*)

The length of the semi-axis a_3 ("major" for an ellipse and "transverse" for a hyperbola) of the phase 3 (geocentric) trajectory is given in Fig. 14. This graph also gives a measure of the total energy (kinetic plus potential) of the vehicle for any phase 3 trajectory with respect to infinity (where the total energy, which is proportional to the reciprocal of the semi-axis, is zero). The end-points of each curve, for negative δ_1, also give the total energy measure for phase 1 since these points represent trajectories which are unaffected by the Moon (having never passed within the Moon's sphere of influence); for these trajectories $a_1 = a_3$,

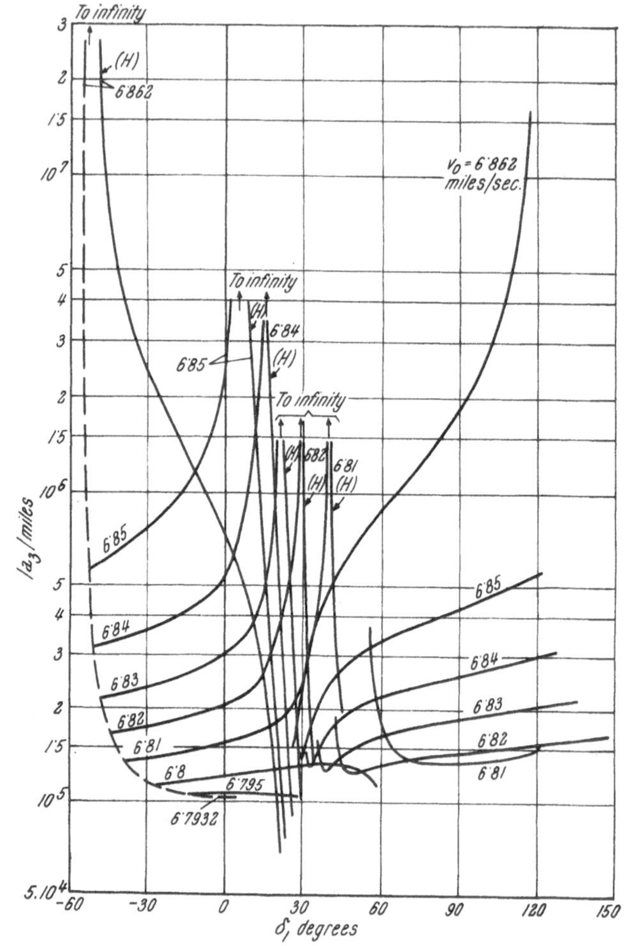

Fig. 14. Direct Outgoing elliptic trajectories: variation of semi-axis of departure geocentric trajectories, a_3, with δ_1

Semi-axis is semi major axis for ellipse and semi transverse axis for hyperbola. (H) on curve refers to final hyperbolic (a_3 + ve) trajectory. All other curves are elliptic (a_3 − ve)

— — — value of $|a_1|$

as given by the broken line. The variation of energy along any curve can be considered since each point represents a different trajectory. For example, for $V_0 = 6.85$ miles/sec., when $\delta_1 = -53°$, $a_3 = a_1$ and as δ_1 increases, the total energy of the vehicle (proportional to $-1/a_3$ for elliptic trajectories) increases at the Moon's expense from some negative value to zero at $\delta_1 = 5°$. The energy then increases along hyperbolic trajectories where a_3 is positive (since energy is now proportional to $+1/a_3$) until $\delta_1 = 23.84°$ where the trajectory grazes the Moon's surface after passing behind the Moon. The gap in the curve is due to

Moon impact conditions and the curve continues again at grazing conditions, the vehicle passing ahead of the Moon, and there is now a loss in energy. Along the curve the energy gradually increases until the end when a_3 is again equal to a_1 when the trajectory represented by the end point is once again unaffected by the Moon. This state of affairs is similar for all values of $V_0 \geqslant 6.8132$ miles/sec. but below this value the initial ellipse is not long enough to reach all points round the Moon's sphere of influence and so the end points of the curves for positive δ_1 are not quite at the same total energy-level as they are at the end points for negative δ_1. Hence the curves show that for vehicles passing behind the Moon, there is always a gain in total energy of the vehicle at the expense of the Moon; but for vehicles passing ahead of the Moon, although there is generally a loss in energy of the vehicle, there are exceptional cases in the region of $V_0=6.81$ and 6.82 miles/sec. when passing very closely ahead of the Moon.

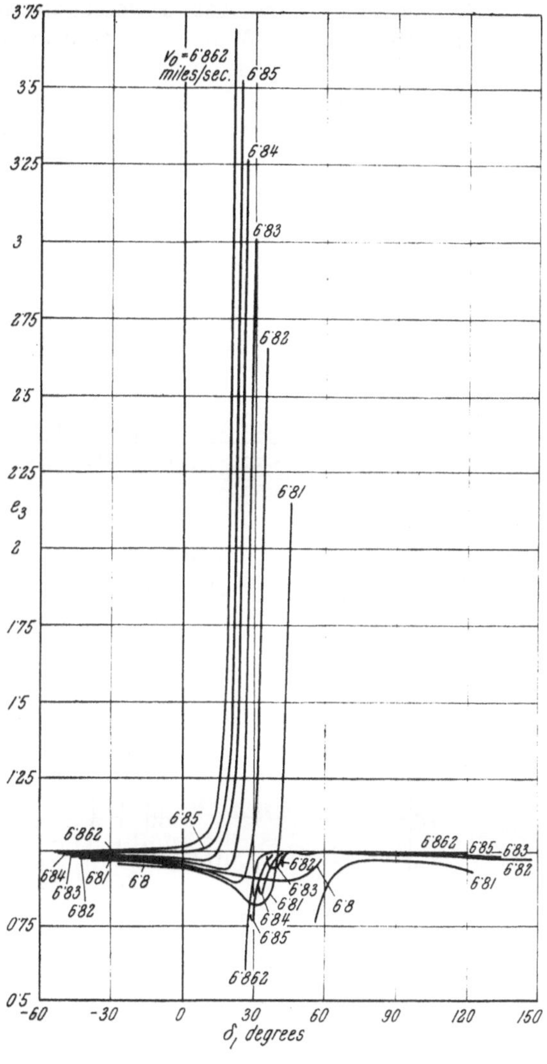

Fig. 15. Direct Outgoing elliptic trajectories: variation of eccentricity of departure geocentric trajectories, e_3, with δ_1

The geocentric eccentricity e_3 for departure trajectories is given in Fig. 15. Since this graph is for Direct Outgoing trajectories, the left-hand group of curves are for those trajectories passing behind the Moon. It can be seen that when close to the Moon and associated with the large increase in energy, there is a corresponding large increase in eccentricity from phase 1 (given by the end points of the horizontal parts of the curves where δ_1 is negative) to phase 3. No curves are shown below $V_0 = 6.8$ miles/sec. as they coincide with the $V_0 = 6.8$ miles/sec. curve at $e_3 = 0.96$. For the other group of curves (which correspond to trajectories passing ahead of the Moon) there are no hyperbolic trajectories although for each curve above $V_0 = 6.81$ miles/sec. there is a trajectory ($e_3 = 1$) corresponding to $\theta_3 = 0$. This is a special case where the vehicle is moving in a straight line directly away from the Earth's centre, at the beginning of phase 3. The corresponding initial Direct Incoming trajectories (not given here) exhibit the other special point corresponding to $|\theta_3| = 180°$.

Finally Fig. 16 gives the phase 3 geocentric true anomaly ϕ_3 at the instant of vehicle departure from the Moon's sphere of influence. This gives the direction of the major axis of the phase 3 conic-section trajectory. Again, the left-hand

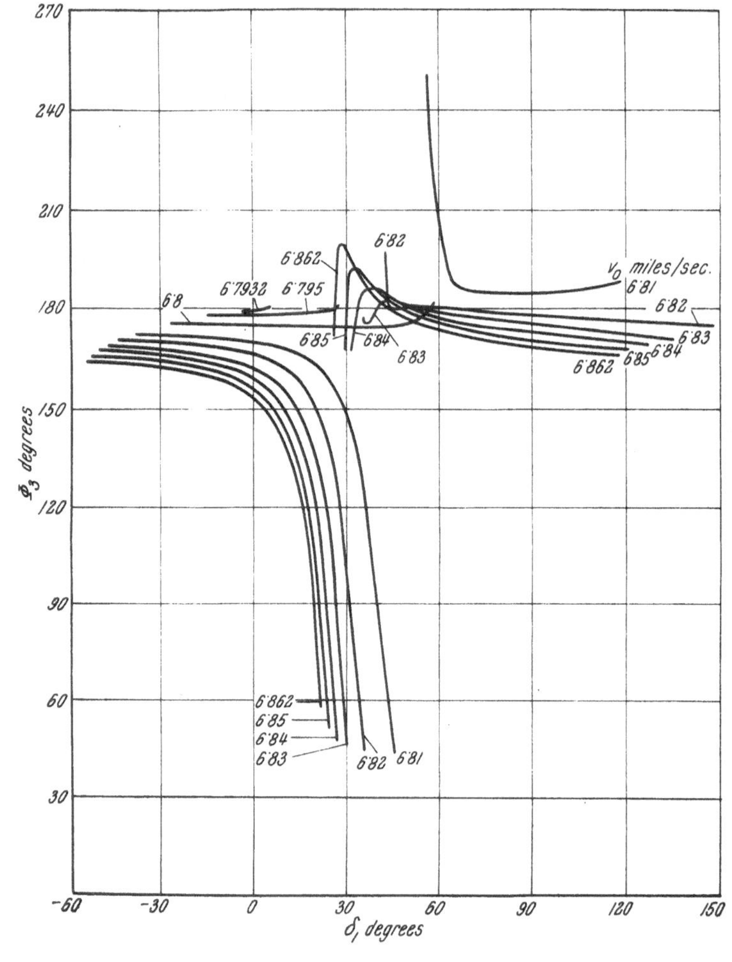

Fig. 16. Direct Outgoing elliptic trajectories: variation of geocentric true anomaly, ϕ_3, on departure from the Moon's sphere of influence, with δ_1. ϕ_3 is measured anticlockwise from perigee

group (of nine curves) represents trajectories passing behind the Moon and the right-hand group (of six curves) ahead of the Moon. The right-hand group is for elliptic trajectories only, given by Fig. 15, (where $\phi_3 = 180°$ gives the apogee of the ellipse) and is for both Direct and Retrograde phase 3 trajectories. The left-hand group is for both elliptic and hyperbolic trajectories which are all Direct for phase 3 motion.

The curve for $V_0 = 6.82$ miles/sec., in the right-hand group of curves passes through $\phi_3 = 180°$ three times (at $\delta_1 = 60°$, $\delta_1 = 53.8°$ and $\delta_1 = 45°$). When $\delta_1 = 60°$, $\theta_3 = 0$ and we have a special case of the motion where the vehicle is moving directly away from the Earth's centre (but for initial elliptic trajectories, the vehicle will always come back again towards the Earth's centre). Again when $\delta_1 = 45°$, $|\theta_3| = 180°$ and a second special case occurs where the vehicle is moving directly towards the Earth's centre. For $\delta_1 = 53.8°$, $\theta_3 = -90°$ and the vehicle is at apogee when beginning phase 3 Retrograde motion.

An additional study was carried out for phase 1 Direct Outgoing trajectories to find the tolerance in direction β_0 of the initial geocentric velocity, to miss the Moon's centre by its radius. A number of standard trajectories, for central hits on the Moon, were selected from the previous work and β_0 varied in small steps, the corresponding changes in periselene distance being noted, the flight time variations being allowed for. It was found that the tolerance has a mean value of approximately $\pm 0.2°$ about an initial value of 90°.

VI. Conclusions

Graphical results have been produced for the two-dimensional flight of a vehicle from the vicinity of the Earth to departure from the Moon's sphere of influence. The absolute values are not expected to be realistic in practice but a reasonably true overall picture is hoped for. Some of the more important features of the results are:

(1) The minimum initial perigee velocity to hit the Moon centrally is about 35,930 ft./sec., and to graze the Moon about 8 ft./sec. less. The resulting selenocentric velocity for a central hit is also a minimum, and equal to 8,160 ft./sec.

(2) Table I gives the classification of phases into Direct or Retrograde motion and shows, for example, that a phase 1 Direct trajectory which passes behind the Moon will always remain Direct for phase 3; but a Direct phase 1 trajectory which passes ahead of the Moon is more likely to change to Retrograde, for phase 3.

(3) For all vehicles which pass behind the Moon (in the direction of the Moon's motion), there is an increase in the vehicle's total energy (kinetic plus potential) at the expense of the Moon. For vehicles passing ahead of the Moon, however, there is generally an energy decrease but a few exceptional cases occur for initial perigee velocities near to 6.81 miles/sec.

(4) For phase 1 Direct Outgoing elliptic trajectories, the selenocentric Moon-impact velocity range is 8160 to 9,440 ft./sec. and is about 450 ft./sec. higher for phase 1 Retrograde Outgoing elliptic trajectories.

(5) An additional study, using some of the previous results, has shown that for phase 1 Direct Outgoing trajectories, the tolerance in direction of the initial geocentric velocity, to miss the Moon's centre by its radius, has a mean value of approximately $\pm 0.2°$ about an initial value of 90°.

Acknowledgements

The Author wishes to express appreciation to Mr. D. G. King-Hele for some helpful suggestions and to Miss E. Harding and Miss P. Hilton for programming the equations on a *Pegasus* digital computer. Crown Copyright is reserved on text and diagrams.

Appendix

The Equations Programmed for the Pegasus Digital Computer

The vehicle moving under the gravitational attraction of the Earth only, whose centre is the origin, traces out a conic-section trajectory given by the polar equation

$$r = \frac{a_1(e_1^2 - 1)}{1 + e_1 \cos \varphi}$$

for polar coordinates (r, φ) where φ (the true anomaly) is measured from perigee; e_1 is the eccentricity and a_1, the semi-axis, is positive for hyperbolic and negative for elliptic trajectories. Hence

$$\frac{r}{r_0} = \frac{1 + e_1}{1 + e_1 \cos \varphi} \tag{1}$$

where suffix 0 refers to initial (phase 1) perigee conditions where $\varphi_0 = 0$, φ being measured positively in an anticlockwise direction in Fig. 1.

Under the inverse square law of attraction with respect to the Earth's centre as focus, the conservation of energy gives at perigee

$$V_0^2 = \mu_E\left(\frac{2}{r_0} + \frac{1}{a_1}\right) \tag{2}$$

where V_0 is the initial (all-burnt) perigee velocity relative to the Earth's centre; the Earth's gravitational constant $\mu_E (= g_E R_E^2) = 95\,600$ miles^3 sec^{-2}, where g_E is the gravitational acceleration at the Earth's equator which is at a distance R_E from the Earth's centre.

At perigee, $r_0 = a_1(e_1 - 1)$, giving, with eq. (2),

$$V_0^2 = \mu_E(1 + e_1)/r_0. \tag{3}$$

We have chosen $r_0 = 4060$ miles (100 miles above the Earth's surface) and so from eq. (3) we get

$$e_1 = 0.04247\, V_0^2 - 1. \tag{4}$$

At the end of phase 1 (when the vehicle arrives at the Moon's sphere of influence) we have $r = r_1$ and $\varphi = \varphi_1$ (see Fig. 1) and so from eq. (1) we get

$$\cos \varphi_1 = \frac{1}{e_1}\left[\frac{4060}{r_1}(1 + e_1) - 1\right]. \tag{5}$$

Taking the Earth-Moon line ($E\,M_1$ in Fig. 1) at the instant of vehicle arrival at the Moon's sphere of influence, as datum line then triangle $E\,M_1\,R_1$ gives

$$(r_{1/10^4})^2 = 23.8857^2 + 4.1^2 - 8.2 \times 23.8857 \cos \delta_1 \ (\text{miles})^2$$

where δ_1 is the angle between the Moon-vehicle and datum lines.

Hence we get

$$\frac{r_1}{10^4} = 13.9951 \sqrt{[2.998\,72 - \cos \delta_1]} \text{ miles}, \tag{6}$$

and eqs. (5) and (6) give a relationship between φ_1 and δ_1.

Although V_0 and φ_1 are the basic initial parameters, δ_1 is much more convenient than φ_1 to use and the relationship between φ_1 and δ_1 enables the latter parameter to be used.

The geocentric arrival velocity V_1 (at the Moon's sphere of influence) can be found using the conservation of energy applied to the initial and final points of the vehicle in phase 1 (R_0 and R_1 in Fig. 1) to give

$$V_1^2 - V_0^2 = 2\,\mu_E \left[\frac{1}{r_1} - \frac{1}{r_0}\right]$$

and therefore

$$V_1 = \sqrt{[V_0^2 - 19.12\,(2.463\,05 - 10^4/r_1)]} \text{ miles/sec.} \qquad (7)$$

The direction, β_1, of V_1 is given, relative to the Earth-vehicle line, by the conservation of angular momentum about the Earth's centre, i.e.

$$r_0\, V_0 \sin\beta_0 = r_1\, V_1 \sin\beta_1, \qquad 0 < \beta_1{}^0 < 90 \qquad (8)$$

for Outgoing phase 1 trajectories (those reaching the Moon's sphere of influence before the vehicle arrives at apogee). For Incoming trajectories (those reaching the Moon's sphere of influence after the vehicle has passed its apogee), the equation becomes

$$r_0\, V_0 \sin\beta_0 = r_1\, V_1 \sin(180 - \beta_1), \qquad 0 < \beta_1{}^0 < 90. \qquad (8\text{ A})$$

Since $\beta_0 = 90°$ at perigee for Direct phase 1 trajectories (those having angular momentum vectors in the same direction as that of the Moon) and $\beta_0 = -90°$ for Retrograde phase 1 trajectories (having angular momentum vectors in opposite direction to that of Moon), eqs. (8) and (8 A) give positive β_1 for Direct and negative β_1 for Retrograde trajectories.

For further analysis, γ_1 (the angle between the Earth-vehicle and datum lines) is required and is given, from triangle $E\,M_1\,R_1$ of Fig. 1, by

$$\sin\gamma_1 = \frac{41{,}000}{r_1} \sin\delta_1 \qquad (9)$$

where γ_1 is negative when δ_1 is negative.

Although it is not essential that the initial (all-burnt) conditions be at perigee, the initial conditions must relate to a conic-section whose perigee is at a fixed distance 4060 miles above the Earth's centre, i.e. the perigee is fixed, but the initial point can be anywhere along any trajectory having this fixed perigee. If the initial velocity V is at a distance r from the Earth's centre and at an angle β to the radius vector from the Earth's centre, then the conservations of energy and angular momentum respectively give

$$V_0^2 - V^2 = 2\,\mu_E \left(\frac{1}{r_0} - \frac{1}{r}\right)$$

and

$$V_0\, r_0 = V\, r \sin\beta,$$

where suffix 0 is for perigee conditions. Hence given an initial point (V, r, β) the perigee conditions V_0 and r_0 can be found. If r_0 is near 4060 miles the results should not be very much in error.

For motion within the Moon's sphere of influence, i.e. phase 2 selenocentric motion, it is necessary to assume that the trajectory lies in the plane of the Moon's orbit, and further motion is considered by bringing the Moon to rest. A velocity vector, equal in magnitude and opposite in direction to that of the Moon (at the instant of the beginning of phase 2 motion), is compounded vectorially with the geocentric arrival velocity V_1 to give the selenocentric arrival velocity V_2 from

$$V_2^2 = V_1^2 + V_M^2 - 2\,V_1\,V_M \sin(\beta_1 + \gamma_1) \qquad (10)$$

where V_M the Moon's velocity is 0.636 miles/sec.

It has been established [5] that selenocentric motion is always hyperbolic for a single circuit of the vehicle from Earth to Moon and so the conservation of energy gives

$$V_2^2 = \mu_M \left(\frac{2}{\varrho_s} + \frac{1}{a_2}\right)$$

where the Moon's gravitational constant $\mu_M = \mu_E/(M_E/M_M)$; $\mu_E = 95\,600$ miles^3 sec^{-2}; M_E/M_M, the Earth to Moon mass ratio, is 81.45; and ϱ_s the radius of the Moon's sphere of influence is 41,000 miles. Hence we can find a_2, the semi-transverse axis of the hyperbola, from the reduced equation

$$a_2 = \frac{1174}{V_2^2 - 0.057\,27} \text{ miles.} \tag{11}$$

For Moon impact, conservation of energy between the surface of the Moon and the boundary of the Moon's sphere of influence gives

$$V_i^2 - V_2^2 = 2\,\mu_M \left(\frac{1}{\varrho_M} - \frac{1}{\varrho_s}\right)$$

where ϱ_M and ϱ_s are the respective radii of the Moon (1,080 miles) and the Moon's sphere of influence (41,000 miles) and V_i is the selenocentric impact velocity on the Moon. This equation reduces to

$$V_i = \sqrt{[V_2^2 + 2.1168]} \text{ miles/sec.}$$

A knowledge of ω_2, the inclination of the selenocentric velocity vector V_2 to the vehicle-Moon line, is required and is obtained by resolving velocities, at R_1 in Fig. 1, in a direction perpendicular to $R_1 M_1$, giving

$$\sin \omega_2 = \frac{V_M}{V_2} \cos \delta_1 - \frac{V_1}{V_2} \sin (\beta_1 + \gamma_1 + \delta_1) \tag{12}$$

If $V_{2\max}$ is the maximum selenocentric velocity at the periselene (the point of closest approach to Moon's centre — denoted by N in Fig. 1), the equations for the conservation of energy and angular momentum about the Moon's centre are respectively

$$V_{2\max}^2 = \frac{\mu_M}{a_2} \frac{e_2 + 1}{e_2 - 1} \tag{13}$$

and

$$a_2 (e_2 - 1)\, V_{2\max} = \varrho_s V_2 \sin \omega_2. \tag{14}$$

Eliminating $V_{2\max}$ from eqs. (13) and (14) we get the hyperbolic eccentricity e_2 from

$$e_2 = \sqrt{\left[1 + \frac{\varrho_s^2}{\mu_M} \frac{V_2^2 \sin^2 \omega_2}{a_2}\right]}. \tag{15}$$

where $\varrho_s = 41{,}000$ miles and $\mu_M = 1{,}174$ miles^3/sec^2.

Hence the periselene distance $r_{P2} = a_2 (e_2 - 1)$.

The selenocentric flight-time, T, between the beginning and end of flight across the Moon's sphere of influence, is given by

$$T = 2\,t_2 = 2 \sqrt{\frac{a_2^3}{\mu_M}} [e_2 \sinh \lambda_2 - \lambda_2] \text{ sec.} \tag{16}$$

where t_2 is the flight-time from periselene to the boundary of the Moon's sphere of influence and λ_2 is a convenient parameter given by the polar equation

$$\varrho_s = a_2 (e_2 \cosh \lambda_2 - 1). \tag{17}$$

The selenocentric true anomaly φ_2, of R_1 (or R_2) in Fig. 1, is measured from the periselene of the hyperbola and is given by

$$\varrho_s = \frac{a_2 (e_2^2 - 1)}{1 - e_2 \sin \alpha_2}, \tag{18}$$

since $\varphi_2 = (90 + \alpha_2)^0$ for positive ω_2 (Direct selenocentric motion) and $\varphi_2 = (270 - \alpha_2)^0$ for negative ω_2 (Retrograde selenocentric motion).

It follows that $\alpha_2{}^0$ must be replaced by $(180 - \alpha_2)^0$ whenever ω_2 is negative, where $\alpha_2 < 90^0$.

The vehicle, departing from the Moon's sphere of influence, returns to geocentric motion (phase 3) which continues until the vehicle is eventually perturbed by some other influence.

Since the selenocentric arrival and departure velocities V_2 are equal in magnitude, by energy considerations, the geocentric departure velocity can be found, at R_2 in Fig. 1, from

$$V_3 = \sqrt{[V_2^2 + V_M^2 - 2\, V_2\, V_M \cos \theta]}$$

where $\theta = \omega_2 + 2\,\alpha_2 + 90 - \delta_1 - \Omega\, T$ degrees, the Earth-Moon line turning through Ω (= 0.549) degrees/hour about the Earth's centre (anticlockwise in Fig. 1). Hence

$$V_3 = \sqrt{[V_2^2 + V_M^2 - 2\, V_2\, V_M \sin (\delta_1 - \omega_2 - 2\,\alpha_2 + \Omega\, T)]}. \tag{19}$$

Eq. (10) could, alternatively, have been

$$V_1 = \sqrt{[V_2^2 + V_M^2 - 2\, V_2\, V_M \sin (\delta_1 + \omega_2)]}. \tag{20}$$

This is of interest, if conditions are required to find the change in geocentric velocity due to the Moon's perturbation.

The angle ν between V_2 and V_3 the respective selenocentric and geocentric velocities is given by

$$\sin \nu = \frac{V_M}{V_3} \sin \theta, \tag{21}$$

from Fig. 1 where ν is positive for $\theta < 180°$ and negative for $\theta > 180°$. It is also necessary to find $\cos \nu$ from

$$\cos \nu = \frac{V_3^2 + V_2^2 - V_M^2}{2\, V_2\, V_3}$$

to determine if $|\nu| > 90°$ or $|\nu| < 90°$.

The apparent exit position of the vehicle, at R_2 in Fig. 1, is given in polar form by (r_2, ε) where

$$\frac{r_2}{10^4} = 13.9951 \sqrt{[2.998\,72 + \cos (2\,\alpha_2 - \delta_1)]} \text{ miles}, \tag{22}$$

obtained similarly to eq. (6), and

$$\sin \varepsilon = \frac{\varrho_s}{r_2} \sin (2\,\alpha_2 - \delta_1). \tag{23}$$

Eqs. (22) and (23) give the coordinates (r_2, ε) assuming a stationary Moon and so, allowing for the Moon's actual motion, the true exit position (relative to the Earth), at R_3 in Fig. 1, is given by $(r_3, [\psi + \varepsilon])$. $R_2 R_3$ is parallel to $M_1 M_3$ where M_3 is the true position of the Moon at the instant of vehicle departure and is given by

$$R_2 R_3 = 2\, L \sin \Omega\, t_2, \tag{24}$$

where (Ωt_2) is the inclination of $R_2 R_3$ (or $M_1 M_3$) to the line perpendicular to the datum line $(E M_1)$.

The true geocentric distance r_3 is given by

$$r_3^2 = r_2^2 + 4 L^2 \sin^2 \Omega t_2 - 4 L r_2 \sin \Omega t_2 \sin (\Omega t_2 - \varepsilon)$$

which can be written

$$r_3^2 = r_2^2 + 4 L \sin \Omega t_2 [L \sin \Omega t_2 - r_2 \sin (\Omega t_2 - \varepsilon)]. \tag{25}$$

ε is given by eq. (23) and so to find $(\psi + \varepsilon)$, we get ψ from

$$\sin \psi = \frac{2L}{r_3} \sin \Omega t_2 \cos (\Omega t_2 - \varepsilon). \tag{26}$$

The angle ζ between V_1 and V_3 (i.e. turned through by the geocentric velocity vector, due to the Moon's gravitational attraction on the vehicle) is given by

$$\zeta = \omega_2 + 2 \alpha_2 + \nu - (\beta_1 + \gamma_1 + \delta_1). \tag{27}$$

The final geocentric conic-section trajectory must be fully determined, so that subsequent motion is determined. V_3 and r_3 have already been found and the inclination of V_3, from the radial line through the Earth's centre, is θ_3, given by

$$\theta_3 = \nu + \omega_2 + y - \psi, \tag{28}$$

where y is given by

$$\sin y = \frac{L}{\varrho_s} \sin \varepsilon = 5.825\,78 \sin \varepsilon. \tag{29}$$

When $|y| > 90^0$, y^0 must be replaced by $(180 - y)^0$ in eq. (29). This occurs when

$$r_2^2 < L^2 - \varrho_s^2 \qquad \text{i.e.} \qquad r_2 < 235{,}312 \text{ miles.}$$

Also when θ_3 is positive, the geocentric phase 3 motion is Direct and for θ_3 negative, the motion is Retrograde.

The semi-axis ("major" for ellipse and "transverse" for hyperbola) a_3 is given by the conservation of energy i.e. $V_3^2 = \mu_E(2/r_3 + 1/a_3)$ and therefore

$$a_3 = \frac{\mu_E r_3}{r_3 V_3^2 - 2 \mu_E}, \tag{30}$$

where, as before, a_3 is negative for an ellipse. This parameter is also used to determine changes in total energy of the vehicle, due to the Moon.

To determine subsequent motion completely, it remains to find the eccentricity e_3 or the true anomaly φ_3 (which determines the direction of the major axis of the conic section). It is useful to know both these parameters.

Combining eqs. (11), (15), (18) for phase 2 motion, we get the equation

$$\cot \omega_2 = \frac{e_2 \cos \alpha_2}{1 - e_2 \sin \alpha_2},$$

and applying this equation to phase 3, we get

$$\cot \theta_3 = \frac{e_3 \sin \varphi_3}{1 + e_3 \cos \varphi_3}. \tag{31}$$

This is a standard equation which can be obtained easily from

$$\frac{1}{r} \frac{dr}{d\varphi} = \cot \theta$$

and

$$r = \frac{a(e^2 - 1)}{1 + e \cos \varphi} \text{ (the polar equation).}$$

The polar equation for phase 3 motion is

$$r_3 = \frac{a_3 (e_3^2 - 1)}{1 + e_3 \cos \varphi_3}. \tag{32}$$

Combining eqs. (31) and (32) gives

$$\tan \varphi_3 = \frac{[(r_3/a_3) + 2] \cot \theta_3}{(r_3/a_3) + 1 - \cot^2 \theta_3}, \tag{33}$$

and

$$e_3 = \frac{\cos \theta_3}{-\cos (\varphi_3 + \theta_3)}, \tag{34}$$

For all conic sections, we have the perigee distance given by

$$r_{P3} = a_3 (e_3 - 1)$$

and for elliptic motion only, the apogee distance is given by

$$r_{A3} = - a_3 (e_3 + 1)$$

since a_3 is negative for elliptic motion).

Before φ_3 can be used in eq. (34), its correct quadrant must be found from $\tan \varphi_3$, given by eq. (33), as shown in Table II. In fact, for Retrograde trajectories, φ_3 is negative (since θ_3 is negative) and so 360° has been added, as shown in Table II, to make φ_3 positive always and has been plotted this way in Fig. 16. The quadrants were obtained by considering the simple geometrical properties of an ellipse.

Table II. *The Quadrant for* φ_3

θ_3°	$\tan \varphi_3$	φ_3°
0 to 90	+	0 — 90
0 to 90	—	90 — 180
90 to 180	+	180 — 270
90 to 180	—	270 — 360
−90 to −180	+	0 — 90
−90 to −180	—	90 — 180
0 to −90	+	180 — 270
0 to −90	—	270 — 360

References

1. P. S. Laplace, Mécanique Céleste, Vol. 9, Ch. 2. Paris 1799.
2. F. F. Tisserand, Traité de Mécanique Céleste, Vol. 4, Ch. 12, p. 198. Paris 1889.
3. J. C. Watson, Theoretical Astronomy, para. 207, p. 546. Philadelphia: Lippincott Co., 1900.
4. W. E. Moeckel, Trajectories with Constant Tangential Thrust in Central Gravitational Fields. NASA Techn. Rep. R-53, 1960, 7.
5. V. A. Egorov, Problems on the Dynamics of Flight to the Moon. The Russian Literature of Satellites, Part 1. New York 1958.
6. K. A. Ehricke, Environment and Celestial Mechanics, Space Flight, Vol. 1, p. 477. London: D. Van Nostrand Co. Inc., 1959.
7. F. T. Sun, Launching Conditions and the Geometry of Orbits in a Central Gravity Field. Proceedings of the Xth International Astronautical Congress, London 1959, p. 537. Vienna: Springer, 1960.

8. R. W. Buchheim, Motion of a Small Body in Earth-Moon Space. RAND RM-1726, June, 1956.
9. L. G. Walters, Flight Mechanics of a Lunar Vehicle. Advances in Astron. Sciences, Vol. 3. London: Chapman and Hall Ltd., 1958.
10. R. W. Buchheim, Lunar Flight Trajectories. RAND P-1268, Jan., 1958.
11. R. W. Buchheim and H. A. Lieske, Lunar Flight Dynamics. RAND P-1453, Aug., 1958.
12. H. A. Lieske, Accuracy Requirements for Trajectories in the Earth-Moon System. RAND P-1022, Feb., 1957.
13. A. S. Boksenbom, Graphical Trajectory Analysis. NASA, TN D-4, Dec., 1959.
14. Trajectory Problems in Cislunar Space. AFOSR-TN-59-1284, Westinghouse, Dec., 1959.
15. R. F. Hoelker and N. J. Braud, Lunar Probe Trajectories: Characteristics and Tolerances for Coplanar Flights. A.B.M.A. Report DA-TR-7-59, June, 1959.
16. J. M. J. Kooy and J. Berghuis, On the Numerical Computation of Free Trajectories of a Lunar Space Vehicle. Astronaut. Acta 6, 115 (1960).
17. E. T. Benedikt, Collision Trajectories in the Three-body Problem. Engineering Paper No. 605A, Aug., 1958. Douglas Aircraft Co., California.
18. G. C. Goldbaum and R. J. Gunkel, Comparison of Two-dimensional and Three-dimensional Analyses of Earth-Moon Flight. Engineering Paper No. 634, Aug., 1958. Douglas Aircraft Co., California.
19. A. B. Mickelwait and R. C. Booton, Jr., Analytical and Numerical Studies of Three-dimensional Trajectories to the Moon. I.A.S. Paper No. 59-90, California, June, 1959.
20. W. H. Michael, Jr., and R. H. Tolson, Effect of Eccentricity of the Lunar Orbit, Oblateness of the Earth, and Solar Gravitational Field on Lunar Trajectories. NASA TN D-227, June, 1960.
21. H. Hiller and Miss J. Hughes, Flight Time to the Sphere of Influence of the Moon. Unpublished Ministry of Aviation Report, 1960.
22. H. O. Ruppe and C. L. Barber, Jr., Lunar Circumnavigation. I.R.E. Trans. on Military Electronics. Vol. MIL-4, April-July, 1960.

Discussion

Mr. Stern: I would like some information concerning a point in one of your graphs (Fig. 1). It appears that the vehicle, on arriving at the Moon's sphere of influence, suddenly changes direction through a large angle. Is this so or is it a mathematical concept to assist in the calculations?

Mr. Hiller: It is simply a change from one reference system to another, at the boundary of the Moon's sphere of influence. There is, therefore, no question of a sudden change in direction of the trajectory but merely of looking at the trajectory first as an observer on Earth and then as an observer on the Moon.

Mr. Ting: You appear to have neglected the attraction of the Sun. Is this justified?

Mr. Hiller: Since the sphere of influence of the Earth in the presence of the Sun has a radius of about 600,000 miles, the use of the concept of a sphere of influence means I can neglect the gradient of the Sun's gravitational attraction for the vehicle within 600,000 miles of the Earth's centre.

Mr. Kovalevsky: I calculated the trajectory of Lunik III by two methods, using the three-body problem and also the concepts of a sphere of influence. I obtained quite good agreement for the whole of the trajectory between the Earth and Moon. As soon as I considered the part of the trajectory beyond the Moon, I obtained large discrepancies. Taking into account the Sun and the Moon, I arrived at a difference for the apogee height of about 100,000 km.

Mr. Hiller: I have explained in the lecture that I do not expect high numerical accuracy so much as high relative accuracy of the results. However, I have mentioned that similar studies to your Lunik III study have been made at the Royal Aircraft Establishment and fairly good agreement obtained. I am surprised your discrepancy is so large.

Mr. BONEFF: Le travail de M. HILLER me semble très interessant. Pourtant il me semble que le fond de ce travail est un peu simplifié? Ce n'est pas qu'il prend la lune comme sphérique, car elle est sphérique, mais il prend la trajectoire de la lune autour de la terre comme circulaire.

Mr. KOVALEVSKY: Ce n'est jamais qu'une première approximation et un début de manière à avoir une idée approximative de la forme des trajectoires. Cependant il ne faut certainement pas s'en tenir là.

Mr. HILLER: I agree. The error involved in this approximation has been shown by other workers to have a maximum value of about 50 ft./sec. in an initial perigee velocity of 35,000 ft./sec. However the relative errors are much smaller.

. Mr. BONEFF: J'ai suivi les travaux de GRÖBNER depuis trois ans, depuis Amsterdam et Londres. J'ai l'impression qu'au fond ces travaux sont très sérieux; j'ai l'impression qu'en appliquant les méthodes de GRÖBNER et de son école, on obtiendrait de meilleurs résultats.

Mr. HILLER: GRÖBNER and CAP have produced a solution to the three-body problem but not in the form of a closed analytical solution. The form is that of an infinite series which is time-dependent and so use can be made of this solution only with the aid of a digital computer. In this case, I can see no advantage over the more usual step-by-step numerical integration of the equations of motion of the three-body problem.

Mr. KOVALEVSKY: Mais puis-je répondre à ce que vous venez de dire au sujet du travail de GRÖBNER et de CAP que j'ai regardé de très près. En général, cela converge beaucoup moins bien que toutes les méthodes d'intégration numérique comme celle de COWELL par exemple, qui sont les seuls méthodes de Mécanique Céleste avec lesquels on puisse comparer ce travail.

Mr. BONEFF: Pourtant je voudrais objecter qu'avec ces méthodes analytiques on est forcé de rester au voisinage du système Terre-Lune.

Mr. FRAEIJS DE VEUBEKE: Je voulais faire la même remarque, c'est peut-être le moment de le faire. Je crois qu'il pourrait y avoir confusion sur les travaux de GRÖBNER et CAP. A Amsterdam, la méthode qu'ils ont présentée est une extrêmement belle formulation analytique des séries de puissance dans le temps, mais on sait depuis 150 ans que ces méthodes ne convergent pas, en tous cas d'un point de pratique. Par conséquent, il s'agit d'une beauté qui est purement formelle et qui n'a aucune valeur pratique. Mais depuis lors, je crois, que des modifications sont intervenues, des quelles j'avoue ne pas être au courant.

Mr. SCHÜTTE: I would like to call your attention to the fact that someone has published some orbits, two of which have the same initial conditions but one with the Sun and one without the Sun. One of these orbits went round the Moon and the other did not. You see, we cannot neglect the Sun.

Mr. HILLER: Other workers have shown that the effect of the Sun is very small, about 10 ft./sec., in an initial perigee velocity of 35,000 ft./sec.

Optimal Intermediate-Thrust Arcs in a Gravitational Field

By

Derek F. Lawden[1]

(With 2 Figures)

Abstract — Zusammenfassung — Résumé

Optimal Intermediate-Thrust Arcs in a Gravitational Field. An earlier general theory of optimal rocket trajectories is extended to allow for the possibility that powered arcs, over which the motor thrust is not a maximum, may be included in the trajectory. The spiral form taken by such arcs in the field due to a single centre of inverse square law attraction is examined in detail and the question of whether a transfer between two KEPLERian orbits via such a spiral can ever be more economical than a two-impulse transfer is considered but not finally resolved.

Optimale Bahnen mit Zwischenschub im Gravitationsfeld. Eine frühere allgemeine Theorie über optimale Raketenbahnen wird auf den Fall erweitert, daß auch Bahnstücke, entlang denen der Schub kein Maximum ist, in die Bahn einbezogen werden. Die Spiralform der hier auftretenden Bahnen wird genauer untersucht und es wird die Frage behandelt, ob eine solche Bahn als Übergang zwischen zwei KEPLER-Ellipsen günstiger sein kann als ein Zwei-Impuls-Übergang.

Arcs optima correspondant à des poussées intermédiaires dans un champ de gravitation. Une précédente théorie générale des trajectoires optimales est étendue pour permettre d'inclure, dans la trajectoire, des arcs propulsés par une poussée qui n'a plus sa valeur maximale. La forme en spirale prise par de tels arcs, dans un champ dû à un centre unique d'attraction inversement proportionnel au carré de la distance, est examinée en détail, et la question de savoir si un transfert entre deux orbites KEPLÉRiennes par une telle spirale peut être parfois plus économique qu'un transfert classique au moyen de deux impulsions est considérée, mais non finalement résolute.

I. Introduction

The problem of joining two given terminals in a known gravitational field by a trajectory along which a rocket may be navigated with minimum expenditure of propellant has been considered in [1]. A general *in vacuo* theory was developed and applied to the particular case, of greatest importance for astronautics, where the field is an inverse square law of attraction to a point. In the general case, an optimal trajectory was taken to comprise a number of arcs meeting at junction points, the arcs being of two types, (a) arcs of null thrust along which the rocket coasts under the influence of the gravitational field alone and (b) arcs of maximum thrust along which the rocket motor is operative under conditions of maximum propellant expenditure. Conditions necessarily satisfied at the junction and terminal points by an optimal trajectory were obtained and it was demonstrated that these determined the optimal trajectory. Intermediate thrust arcs, along

[1] Professor of Mathematics, University of Canterbury, Christchurch, New Zealand.

which the motor is operative at less than maximum thrust, were thought, except in very special fields, to be excluded by the stringent nature of the condition (eq. (52), [1]) it was found such arcs must satisfy. However, after doubts had been expressed to the author [2] regarding the validity of this conclusion, the matter was investigated more closely and a class of such arcs, satisfying the required condition, was found to exist in a geneıal field; it was also found that the form taken by these arcs in the particular case of the inverse square law field was describable in terms of elementary functions.

The object of this paper is to complete the earlier theory by taking account of the possibility that an optimal trajectory may include intermediate thrust arcs and also to investigate the properties of these arcs in an inverse square law field. The results of the earlier theory are still valid if intermediate thrust arcs are specifically excluded (e.g. because navigation along such arcs is inconveniently complex) and, in particular, the theory of optimal transfer between orbits employing impulsive thıusts, which was based on the general theory, is not affected. However, it is possible that by utilising such arcs, significant propellant economies can be effected in certain cases and the arcs may have some significance for space vehicles propelled by micro-thrust motors. For these reasons, the matter deserves a thorough investigation.

II. General Theory — Types of Optimal Arc

$Ox_1\, x_2\, x_3$ is a rectangular cartesian inertial frame in the given gravitational field and f_i ($i = 1, 2, 3$) are the components of attraction in the directions of the arcs. It will be assumed that the f_i are known functions of the x_i and the time t, i.e.

$$f_i = f_i(x_1, x_2, x_3, t). \tag{1}$$

D is a terminal point of departure having given coordinates d_i and A is a terminal point of arrival having given coordinates a_i. The problem is to navigate a rocket between D and A in such a manner that it leaves D with a specified velocity having components D_i and arrives at A with a specified velocity whose components are A_i and the characteristic velocity for the manoeuvre is as small as possible. The times of departure and arrival T_0, T_1, may, or may not, be predetermined.

Let x_i be the coordinates of the rocket, v_i its velocity components, M its mass and l_i the direction cosines of the direction of its motor thrust, all at time t. Then its equations of motion are

$$\dot{v}_i - \frac{c}{M}\beta\, l_i - f_i = 0, \tag{2}$$

$$\dot{x}_i - v_i = 0, \tag{3}$$

where dots denote differentiations with respect to t, $i = 1, 2, 3$, c is the jet velocity and β is the rate of propellant consumption. M and β are obviously related by the equation

$$\dot{M} + \beta = 0. \tag{4}$$

The l_i are also related thus:

$$l_1^2 + l_2^2 + l_3^2 - 1 = 0. \tag{5}$$

β will satisfy inequalities of the form,

$$0 \leqslant \beta \leqslant B, \tag{6}$$

where B depends upon the design of the motor. Account will be taken of these restrictions by making β depend upon a parameter α, i.e.

$$\beta = \beta(\alpha), \tag{7}$$

the functional relationship being chosen so that β increases monotonically from 0 to B as α increases from $-\infty$ to $+\infty$ and $d\beta/d\alpha \to 0$ as $\alpha \to \pm\infty$. It follows that $d\beta/d\alpha = 0$ implies that $\beta = 0$ or B.

The mathematical problem is to express the eleven quantities x_i, v_i, l_i, M, α as functions of t in such a way that the eqs. (2)—(5) and the terminal conditions at D, A are satisfied and the characteristic velocity given by

$$V = c \log \frac{M_0}{M_1} \tag{8}$$

is minimized (M_0, M_1 are the masses of the rocket at D, A respectively. M_0 will be regarded as a given quantity).

We first define a LAGRANGE function

$$F = \lambda_i\left(\dot{v}_i - \frac{c}{M}\beta\, l_i - f_i\right) + \lambda_{i+3}(\dot{x}_i - v_i) + \lambda_7(\dot{M} + \beta) + \lambda_8(l_1^2 + l_2^2 + l_3^2 - 1), \tag{9}$$

where the multipliers $\lambda_i (i = 1, 2, \ldots 8)$ are functions of t to be determined later and the repeated subscript summation convention is operative in respect of an index such as i, for $i = 1, 2, 3$, from this point onward. Then the unknown functions x_i, etc. and the λ_i of necessity satisfy the EULER characteristic equations

$$\frac{d}{dt}\left(\frac{\partial F}{\partial \dot{x}_i}\right) = \frac{\partial F}{\partial x_i}, \qquad \text{etc.} \tag{10}$$

These equations take the form

$$\dot{\lambda}_i = -\lambda_{i+3}, \tag{11}$$

$$\dot{\lambda}_{i+3} = -\lambda_j \frac{\partial f_j}{\partial x_i}, \tag{12}$$

$$0 = -\frac{c}{M}\beta\,\lambda_i + 2\,\lambda_8\, l_i, \tag{13}$$

$$\dot{\lambda}_7 = \frac{c}{M^2}\beta\,\lambda_i\, l_i, \tag{14}$$

$$0 = \left(-\frac{c}{M}\lambda_i\, l_i + \lambda_7\right)\frac{d\beta}{d\alpha}. \tag{15}$$

It follows from eqs. (11), (12) that the quantities $\lambda_i (i = 1, 2, 3)$ satisfy the equations

$$\ddot{\lambda}_i = \lambda_j \frac{\partial f_j}{\partial x_i} \tag{16}$$

and are components of a vector relative to rectangular cartesian frames. This is the *primer* vector. Eqs. (13) imply that the thrust and primer vectors are parallel, their components being related thus:

$$\lambda_i = \frac{2\,M}{c\,\beta}\lambda_8\, l_i. \tag{17}$$

It follows from eq. (15) that either $d\beta/d\alpha = 0$ and the thrust is a maximum or zero, or

$$\lambda_7 = \frac{c}{M}\lambda_i\, l_i. \tag{18}$$

In this latter event, eq. (14) shows that

$$\dot{\lambda}_7 = \frac{\beta}{M}\lambda_7 = -\frac{\dot{M}}{M}\lambda_7 \tag{19}$$

and this equation can be integrated to yield

$$\lambda_7 = \frac{c\,k}{M}, \tag{20}$$

where k is a constant. Substituting for λ_7 from eq. (20) into eq. (18), it is then found that

$$\lambda_i\, l_i = k. \tag{21}$$

But the vectors l_i, λ_i are parallel and hence

$$l_i = \lambda_i / \sqrt{\lambda_1^2 + \lambda_2^2 + \lambda_3^2}. \tag{22}$$

It now follows from eq. (21) that

$$\sqrt{\lambda_1^2 + \lambda_2^2 + \lambda_3^2} = k, \tag{23}$$

i.e. the primer is of constant magnitude. Eq. (22) now gives

$$l_i = \lambda_i / k. \tag{24}$$

Substituting for l_i from eq. (24) in eq. (17), it will then be found that

$$\lambda_8 = \frac{c\,k\,\beta}{2\,M}. \tag{25}$$

If, however, the thrust is zero, then $\beta = 0$ and eqs. (13) and (14) show that $\lambda_8 = 0$ and $\lambda_7 =$ constant.

If the thrust is a maximum, then $\beta = B$. Also, since the thrust and primer are aligned, eq. (22) is valid and it follows from eqs. (13) and (14) that

$$\lambda_8 = \frac{c\,B}{2\,M}\sqrt{\lambda_1^2 + \lambda_2^2 + \lambda_3^2}, \tag{26}$$

$$\dot{\lambda}_7 = \frac{c\,B}{M^2}\sqrt{\lambda_1^2 + \lambda_2^2 + \lambda_3^2}. \tag{27}$$

To summarize, the Euler characteristic equations can be satisfied along arcs of three types:

a) *Arcs of null thrust* along which

$$\lambda_7 = \text{constant}, \qquad \lambda_8 = 0. \tag{28}$$

b) *Arcs of maximum thrust along which* λ_7, λ_8 are governed by eqs. (26) and (27).

c) *Arcs of intermediate thrust* along which

$$\sqrt{\lambda_1^2 + \lambda_2^2 + \lambda_3^2} = k, \tag{29}$$

$$\lambda_7 = \frac{c\,k}{M}, \tag{30}$$

$$\lambda_8 = \frac{c\,k\,\beta}{2\,M}. \tag{31}$$

Along all such arcs the primer components must satisfy the eqs. (16) and the thrust (if non-zero) and primer vectors must be aligned.

III. General Theory — Junctions and Terminals

At a junction between two arcs belonging to the different types considered in the previous section, the Weierstrass-Erdmann corner conditions are applicable. These require that the following quantities shall be continuous across the junction:

$$\frac{\partial F}{\partial \dot{x}_i}, \frac{\partial F}{\partial \dot{v}_i}, \frac{\partial F}{\partial l_i}, \frac{\partial F}{\partial \dot{M}}, \frac{\partial F}{\partial \dot{\alpha}},$$

$$F - \dot{x}_i \frac{\partial F}{\partial \dot{x}_i} - \dot{v}_i \frac{\partial F}{\partial \dot{v}_i} - l_i \frac{\partial F}{\partial l_i} - \dot{M} \frac{\partial F}{\partial \dot{M}} - \dot{\alpha} \frac{\partial F}{\partial \dot{\alpha}}. \tag{32}$$

With F defined by eq. (9), these conditions require that the $\lambda_i (i = 1, 2, 3, 4, 5, 6, 7)$ and

$$-\lambda_i \left(\frac{c}{M} \beta l_i + f_i \right) - \lambda_{i+3} v_i + \lambda_7 \beta \tag{33}$$

should be continuous. But the f_i and v_i are certainly continuous across such a junction and hence the conditions reduce to the requirements that (a) the primer and its first derivative, (b) λ_7 and (c) the quantity

$$\beta \left(\frac{c}{M} \lambda_i l_i - \lambda_7 \right), \tag{34}$$

are to be continuous.

Consider the continuity of the quantity (34). Since

$$\frac{c}{M} \lambda_i l_i - \lambda_7 \tag{35}$$

is, in any case, continuous by the other corner conditions, the quantity (34) will be continuous if β is continuous, (as, for example, when moving from a null-thrust regime into an intermediate-thrust regime where the thrust increases from an initial value of zero or when moving from a maximum-thrust regime into an intermediate-thrust regime where the thrust decreases from an initial maximum value). If, however, β is discontinuous at a junction, it will be necessary that

$$\frac{c}{M} \lambda_i l_i - \lambda_7 = 0 \tag{36}$$

and hence that

$$\lambda_7 = \frac{c}{M} \lambda_i l_i = \frac{c}{M} \sqrt{\lambda_1^2 + \lambda_2^2 + \lambda_3^2}. \tag{37}$$

To summarize, at every junction the primer, its first time derivative and λ_7 must be continuous and, in addition, at any junction where β is discontinuous the condition (37) must be satisfied.

Conditions to be satisfied at the end points of an optimal trajectory are obtained as in [1] (p. 13) (slight modification is necessary since we are taking M_0 = constant). They prove to require that

$$\lambda_7 = \frac{c}{M} \tag{38}$$

at the terminal A and, if the time of departure and/or of arrival is not predetermined, then

$$\lambda_i \dot{v}_i - \dot{\lambda}_i v_i - \lambda_7 \beta = 0 \tag{39}$$

at the corresponding end point.

IV. Impulsive Thrusts

For an orthodox rocket motor, B is so large that the duration of any period of maximum thrust is usually short by comparison with the total time of transit between the terminals. If such is the case, the motion of the vehicle during a period of maximum thrust may often be disregarded and the thrust treated as if it were impulsive. It then follows from the form of the equations satisfied by the components of the primer (eq. (16)), that the change in this quantity during the period of maximum thrust will also be negligible and hence, since the primer and its first derivative are continuous at both the commencement and conclusion of this period, these quantities will be continuous across the whole impulsive thrust.

Further, as shown in [1] (p. 15), if eq. (27) is to be satisfied by λ_7 throughout a period of maximum thrust and if eq. (37) is to be satisfied at the commencement and termination of this period, then

$$\lambda_i \dot{\lambda}_i = 0, \tag{40}$$

if the thrust is treated as impulsive. Since condition (37) is not necessarily satisfied at the commencement or termination of a period of maximum thrust if this is the commencement or termination of the complete trajectory, condition (40) also need not be satisfied at any impulse which occurs at either end point.

If the equation

$$\lambda_7 = \frac{c}{M} \tag{41}$$

is true at the commencement of an arc of null-thrust, then it is true at its termination. This follows since both λ_7 and M are constant over such an arc.

Again, if eq. (41) is valid at the commencement of an arc of intermediate-thrust, then it is also true at its termination; for, by eq. (30), $k = 1$ at the beginning of the arc and hence over the whole arc. Hence eq. (41) is valid over the whole arc. In this case

$$\sqrt{\lambda_1^2 + \lambda_2^2 + \lambda_3^2} = 1, \tag{42}$$

i.e. the primer is of unit magnitude over the arc.

Finally, if eq. (41) is true at the onset of a short period of impulsive thrust, it is true at its termination. For, assuming the λ_i do not change over the short duration of the impulse, by integrating eq. (27) over this time interval, it is found that

$$\Delta\lambda_7 = \sqrt{\lambda_1^2 + \lambda_2^2 + \lambda_3^2}\, \Delta\left(\frac{c}{M}\right). \tag{43}$$

where $\Delta\lambda_7$, $\Delta(c/M)$ are the increments in λ_7, c/M respectively caused by the impulse. But, eq. (37) must be satisfied at the beginning of the thrust period and, since we are supposing that $\lambda_7 = c/M$ at this instant, it follows that

$$\sqrt{\lambda_1^2 + \lambda_2^2 + \lambda_3^2} = 1 \tag{44}$$

at the impulse. Thus,

$$\Delta\lambda_7 = \Delta\left(\frac{c}{M}\right) \tag{45}$$

and eq. (41) will accordingly remain true at the end of the impulse.

But eq. (41) is certainly true at A on the trajectory. It is therefore valid over the whole trajectory.

To summarize, if impulsive thrusts are permitted, over an optimal trajectory it is necessary that the primer should satisfy the following conditions:

(i) the primer and its first derivative must be continuous everywhere;

(ii) whenever the motor is operative, the primer must be a unit vector in the direction of the thrust;

(iii) at an impulse $\lambda_i \dot{\lambda}_i = 0$, i.e. the primer and its first derivative are orthogonal.

The components of the primer must, of course, satisfy the eqs. (16).

V. First Integral of the Euler Equations

It is known that, provided the gravitational field does not vary with the time and the functions f_i are not therefore explicitly dependent upon t, a first integral of the EULER characteristic equations for the LAGRANGE function F (eq. (9)) exists in the form

$$F - \dot{x}_i \frac{\partial F}{\partial \dot{x}_i} - \dot{v}_i \frac{\partial F}{\partial \dot{v}_i} - \dot{M} \frac{\partial F}{\partial \dot{M}} = \text{constant.} \tag{46}$$

Since, by the WEIERSTRASS-ERDMANN corner conditions, the left-hand member of this equation is continuous across all discontinuities, the constant right-hand member must take the same value on all sub-arcs of the optimal trajectory.

Substituting for F from eq. (9) and employing eq. (5), the first integral is expressed in the form

$$-\lambda_i \left(\frac{c}{M} \beta l_i + f_i\right) - \lambda_{i+3} v_i + \lambda_7 \beta = \text{constant.} \tag{47}$$

By eq. (11), this is equivalent to the equation

$$\lambda_i f_i - \dot{\lambda}_i v_i + \beta \left(\frac{c}{M} \lambda_i l_i - \lambda_7\right) = \text{constant.} \tag{48}$$

If impulsive thrusts are permitted then, as proved in Section IV, either $\beta = 0$ or $\lambda_i = l_i$. In the latter case,

$$\frac{c}{M} \lambda_i l_i - \lambda_7 = \frac{c}{M} - \lambda_7 = 0, \tag{49}$$

since eq. (41) is valid at all points on the optimal trajectory. It follows that

$$\beta \left(\frac{c}{M} \lambda_i l_i - \lambda_7\right) = 0 \tag{50}$$

everywhere and eq. (48) reduces to

$$\lambda_i f_i - \dot{\lambda}_i v_i = C. \tag{51}$$

At a terminal where the time of arrival or departure is not specified, the condition (39) must be satisfied. Employing eq. (2), this may be expressed in the form

$$\lambda_i f_i - \dot{\lambda}_i v_i + \beta \left(\frac{c}{M} \lambda_i l_i - \lambda_7\right) = 0, \tag{52}$$

i.e. the constant is zero in eq. (48) and, if impulsive thrusts are permitted, eq. (51) takes the form

$$\lambda_i f_i - \dot{\lambda}_i v_i = 0. \tag{53}$$

Thus, if the time of transit between two points in a time-invariant field is not predetermined, a first integral of the EULER equations exists in the form

$$\lambda_i f_i - \dot{\lambda}_i v_i = 0. \tag{54}$$

It may also be shown, as in [1] (p. 17), that the existence of the first integral (51) with a value for C which is the same for all sub-arcs, implies that $\lambda_i \dot{\lambda}_i = 0$ at every impulse.

VI. Weierstrass Condition

This is a further condition necessarily satisfied by a trajectory which minimizes the characteristic velocity V. The form taken by this condition for a very general problem from the calculus of variations will be found given in [3]. However, BREAKWELL [4] has reduced the condition to a simplified form, applicable to problems of the type we are considering. Applying the condition in the form given by BREAKWELL, it will be found to require that

$$\beta\left(\frac{c}{M}\lambda_i l_i - \lambda_7\right) \geqslant \beta^*\left(\frac{c}{M}\lambda_i l_i^* - \lambda_7\right) \tag{55}$$

at all points on the trajectory and for all possible values of β^*, l_i^* satisfying the conditions (5), (6). Over an arc of null-thrust, $\beta = 0$, $\beta^* \geqslant 0$ and the condition is that

$$\frac{c}{M}\lambda_i l_i^* - \lambda_7 \leqslant 0. \tag{56}$$

But $\lambda_i l_i^*$ is the scalar product of the primer and a unit vector and hence, for variable l_i^*, takes its maximum value of $\sqrt{(\lambda_1^2 + \lambda_2^2 + \lambda_3^2)}$ when the primer and unit vector are aligned. Condition (56) is accordingly satisfied for all l_i^* satisfying eq. (5) if, and only if,

$$\lambda_7 \geqslant \frac{c}{M}\sqrt{(\lambda_1^2 + \lambda_2^2 + \lambda_3^2)}. \tag{57}$$

If impulsive thrusts are permitted, it has already been shown that $\lambda_7 = c/M$ over the whole trajectory. In this case, therefore, condition (57) reduces to

$$\sqrt{(\lambda_1^2 + \lambda_2^2 + \lambda_3^2)} \leqslant 1, \tag{58}$$

i.e. the primer must be a vector of magnitude less than unity over a null-thrust arc. This condition was first obtained in [5].

Over an arc of maximum thrust, $\beta = B$, $0 \leqslant \beta^* \leqslant B$ and condition (55) will only be satisfied for all β^* satisfying these inequalities if

$$\left.\begin{aligned} &\frac{c}{M}\lambda_i l_i - \lambda_7 \geqslant 0, \\ \text{and}\quad &\frac{c}{M}\lambda_i l_i \geqslant \frac{c}{M}\lambda_i l_i^*. \end{aligned}\right\} \tag{59}$$

Since the primer and thrust directions are aligned over an arc of maximum thrust and the scalar product $\lambda_i l_i^*$ takes its maximum value when the vectors λ_i, l_i^* are parallel, the second of these inequalities is certainly satisfied. Also, since the vectors λ_i, l_i are parallel, the first inequality requires that

$$\lambda_7 \leqslant \frac{c}{M}\sqrt{(\lambda_1^2 + \lambda_2^2 + \lambda_3^2)}. \tag{60}$$

If impulses are permitted, at a point of application the primer has unit magnitude and $\lambda_7 = c/M$ over the whole trajectory. Condition (60) is then satisfied as an equality.

If, however, B is finite, condition (60) with eq. (27) together imply that

$$\dot{\lambda}_7 \geqslant -\frac{\dot{M}}{M}\lambda_7,$$

i.e.

$$\frac{d}{dt}(\log \lambda_7) \geqslant \frac{d}{dt}\left(\log\frac{k\,c}{M}\right),$$

where k is an arbitrary constant. Thus, if $\lambda_7 = k\,c/M$ at the commencement of an arc of maximum thrust,

$$\lambda_7 \geqslant \frac{k\,c}{M} \tag{61}$$

at the end of the arc. Since $\lambda_7 = k\,c/M$ over the other types of optimal arc, it follows that the effect of a period of maximum thrust must always be to increase the value of k. By condition (38), the final value of k at A must be unity.

Finally, over an arc of intermediate thrust eq. (18) is valid and condition (55) becomes

$$\beta^* \left(\frac{c}{M} \lambda_i l_i^* - \lambda_7 \right) \leqslant 0.$$

But β^* is positive and hence

$$\lambda_7 \geqslant \frac{c}{M} \lambda_i l_i^* \tag{62}$$

for all admissible l_i^*. Since $\lambda_7 = c\,k/M$ (eq. (30)) over this type of arc, condition (62) requires that

$$\lambda_i l_i^* \leqslant k. \tag{63}$$

But the vectors λ_i, l_i are parallel and hence, by eq. (29), $\lambda_i l_i = k$. Since $\lambda_i l_i^* \leqslant \lambda_i l_i$, this shows that the condition (63) is satisfied over an arc of intermediate thrust.

To summarize, therefore, if impulsive thrusts are permitted the WEIERSTRASS condition leads to one additional requirement only, viz. that the magnitude of the primer shall never exceed unity at any point on the optimal trajectory.

VII. Case of the Inverse Square Law Field

The eqs. (16) governing the primer components have been integrated in terms of known functions for motion along a KEPLERian null-thrust arc in an inverse square law field due to a single attracting body. The complete solution will be found in Section VI of [1]. If impulsive thrusts are permitted, the theory of optimal rocket trajectories in a field of this type can be completed when the optimal arcs of intermediate thrust and the form taken by the primer on such arcs have been determined. This will be done in this and the following Sections under the assumption that the motion is confined to a plane through the centre of attraction.

Let (r, θ) be the polar coordinates, in the plane of motion, of the rocket P at time t relative to the centre of attraction O as pole (Fig. 1). Let f be the vehicle acceleration due to the motor thrust and let χ be the angle made by the direction of this thrust with the line $OA\,(\theta = 0)$. Let w be the component of the rocket's velocity in a direction perpendicular to the thrust and in the sense indicated in Fig. 1. Let $\mathfrak{a}, \mathfrak{b}$ be unit vectors in the directions of f, w respectively. Then, over an intermediate-thrust arc, the primer $\mathfrak{p}$ is given by

$$\mathfrak{p} = \mathfrak{a}. \tag{64}$$

But, since $\mathfrak{a}, \mathfrak{b}$ are unit vectors,

$$\dot{\mathfrak{a}} = \dot{\chi}\,\mathfrak{b}, \qquad \dot{\mathfrak{b}} = -\dot{\chi}\,\mathfrak{a}. \tag{65}$$

Differentiating eq. (64) twice and employing these results, it will be found that

$$\ddot{\mathfrak{p}} = -\dot{\chi}^2\,\mathfrak{a} + \ddot{\chi}\,\mathfrak{b}. \tag{66}$$

If γ/r^2 is the gravitational attraction towards O, taking rectangular cartesian axes Oxy through O the components of the gravitational field at the point (x, y) in the directions of the axes are easily found to be

$$f_x = -\frac{\gamma x}{(x^2+y^2)^{3/2}}, \qquad f_y = -\frac{\gamma y}{(x^2+y^2)^{3/2}}. \tag{67}$$

It follows that

$$\left.\begin{aligned} \frac{\partial f_x}{\partial x} = \gamma \frac{2\,x^2 - y^2}{(x^2+y^2)^{5/2}}, \qquad \frac{\partial f_y}{\partial y} = \gamma \frac{2\,y^2 - x^2}{(x^2+y^2)^{5/2}}, \\ \frac{\partial f_x}{\partial y} = \frac{\partial f_y}{\partial x} = \frac{3\,\gamma\, x\, y}{(x^2+y^2)^{5/2}}. \end{aligned}\right\} \tag{68}$$

If, now, $O\,x$ is taken to be parallel to $\mathfrak{a}$ and $O\,y$ parallel to $\mathfrak{b}$, then P has coordinates $(r \sin\psi, -r\cos\psi)$ relative to these axes, where ψ is the angle between $P\,O$ and $\mathfrak{b}$ shown in Fig. 1. Eqs. (68) then yield

$$\left.\begin{aligned} \frac{\partial f_x}{\partial x} = -\frac{\gamma}{r^3}(1 - 3\sin^2\psi), \qquad \frac{\partial f_y}{\partial y} = -\frac{\gamma}{r^3}(1 - 3\cos^2\psi), \\ \frac{\partial f_x}{\partial y} = \frac{\partial f_y}{\partial x} = -\frac{3\,\gamma}{r^3}\sin\psi\cos\psi. \end{aligned}\right\} \tag{69}$$

The components of the primer relative to these axes are simply (1, 0). It follows that the components of the right-hand member of the vector eq. (16) are

$$-\frac{\gamma}{r^3}(1 - 3\sin^2\psi), \qquad -\frac{3\,\gamma}{r^3}\sin\psi\cos\psi. \tag{70}$$

But, eq. (66) indicates that the corresponding components of $\ddot{\mathfrak{p}}$ are

$$-\dot{\chi}^2, \qquad \ddot{\chi}. \tag{71}$$

Eqs. (16) can accordingly be written in the form

$$\dot{\chi}^2 = \frac{\gamma}{r^3}(1 - 3\sin^2\psi), \tag{72}$$

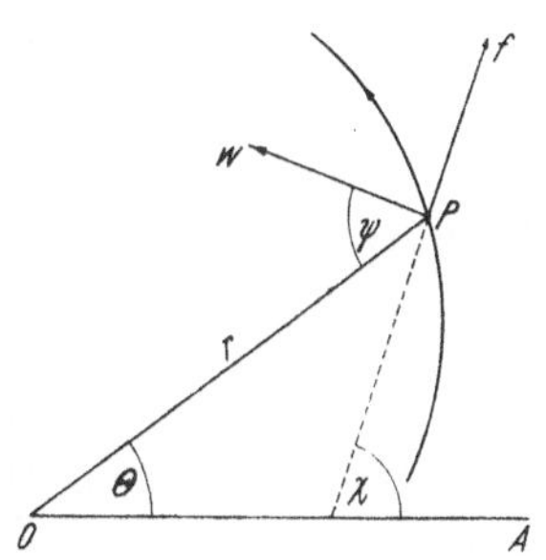

Fig. 1. Coordinates of vehicle

$$\ddot{\chi} = -\frac{3\,\gamma}{r^3}\sin\psi\cos\psi. \tag{73}$$

Let u be the component of the vehicle's velocity in the direction of the motor thrust. Then, if $\mathfrak{v}$ is the rocket velocity,

$$\mathfrak{v} = u\,\mathfrak{a} + w\,\mathfrak{b} \tag{74}$$

and hence, employing eqs. (65), the rocket acceleration is

$$\dot{\mathfrak{v}} = (\dot{u} - w\,\dot{\chi})\,\mathfrak{a} + (\dot{w} + u\,\dot{\chi})\,\mathfrak{b}. \tag{75}$$

But the components of rocket acceleration in directions along and perpendicular to the thrust are

$$f - \frac{\gamma}{r^2}\sin\psi, \qquad \frac{\gamma}{r^2}\cos\psi. \tag{76}$$

The equations of motion of the rocket are accordingly

$$\dot{u} - w\,\dot{\chi} = f - \frac{\gamma}{r^2}\sin\psi, \tag{77}$$

$$\dot{w} + u\,\dot{\chi} = \frac{\gamma}{r^2}\cos\psi. \tag{78}$$

Since the radial and transverse components of rocket velocity are $(\dot{r}, r\,\dot{\theta})$ respectively, it follows that

$$u = \dot{r}\sin\psi + r\,\dot{\theta}\cos\psi, \tag{79}$$

$$w = r\,\dot{\theta}\sin\psi - \dot{r}\cos\psi. \tag{80}$$

Finally, by reference to Fig. 1, it will be seen that

$$\chi = \frac{1}{2}\pi + \theta - \psi. \tag{81}$$

Eqs. (72), (73), (77)—(81) provide seven conditions to be satisfied by seven unknown functions of t, viz. r, θ, u, w, ψ, χ, f. These equations accordingly determine the family of intermediate-thrust arcs in the given plane of motion.

VIII. Solution of Equations for Intermediate Thrust Arcs

From eqs. (72), (73) it will be found that

$$\dot{\chi}^2 \cos\psi - \ddot{\chi}\sin\psi = \frac{\gamma}{r^3}\cos\psi. \tag{82}$$

Whence, from eqs. (78), (82), it follows that

$$\dot{w} + u\,\dot{\chi} - r\,\dot{\chi}^2\cos\psi + r\,\ddot{\chi}\sin\psi = 0. \tag{83}$$

Taking account of eqs. (79), (81) this last equation can be written in the form

$$\frac{d}{dt}(w + r\,\dot{\chi}\sin\psi) = 0. \tag{84}$$

Integration yields immediately

$$w + r\,\dot{\chi}\sin\psi = A. \tag{85}$$

Since the field is not time-dependent, a first integral exists in the form of eq. (51). Now $\lambda_i f_i$ is the scalar product of the primer with the field intensity. The primer has components (1, 0) in the directions of $\mathfrak{a}$, $\mathfrak{b}$ respectively and the field intensity has corresponding components $(-\gamma\sin\psi/r^2, \gamma\cos\psi/r^2)$. Thus

$$\lambda_i f_i = -\frac{\gamma}{r^2}\sin\psi. \tag{86}$$

Also, by the first of eqs. (65), $\dot{\mathfrak{p}}$ has components $(0, \dot{\chi})$. Since $\mathfrak{v}$ has corresponding components (u, w), it then follows that

$$\dot{\lambda}_i v_i = w\,\dot{\chi}. \tag{87}$$

The first integral is accordingly

$$w\,\dot{\chi} = C - \frac{\gamma}{r^2}\sin\psi. \tag{88}$$

Eliminating $\dot{\chi}$ between eqs. (72) and (88) and then between eqs. (85) and (88), the following equations are obtained:

$$w^2(1 - 3\sin^2\psi) = \frac{r}{\gamma}\left(C\,r - \frac{\gamma}{r}\sin\psi\right)^2, \tag{89}$$

$$\left(C\,r - \frac{\gamma}{r}\sin\psi\right)\sin\psi = w(A - w). \tag{90}$$

These determine w and r as functions of the parameter ψ without further integration.

Substituting for χ from eq. (81) into eq. (88), this latter equation can be written

$$w(\theta' - 1)\dot{\psi} = C - \frac{\gamma}{r^2}\sin\psi \tag{91}$$

where the prime denotes a differentiation with respect to ψ. Also, eq. (80) is equivalent to the equation

$$w = (r\,\theta'\sin\psi - r'\cos\psi)\,\dot{\psi}. \tag{92}$$

Eliminating $\dot{\psi}$ between eqs. (91), (92) and solving for θ', it is found that

$$\theta' = \frac{w^2 - r'\left(C - \frac{\gamma}{r^2}\sin\psi\right)\cos\psi}{w^2 - \left(C\,r - \frac{\gamma}{r}\sin\psi\right)\sin\psi}. \tag{93}$$

The right-hand member of this equation can be expressed as a function of ψ and then a single integration yields θ in terms of this parameter.

If θ' is eliminated between eqs. (91), (92) and $dt/d\psi$ solved for, the result proves to be

$$t' = \frac{w(r\sin\psi - r'\cos\psi)}{w^2 - \left(C\,r - \frac{\gamma}{r}\sin\psi\right)\sin\psi}. \tag{94}$$

An integration expresses t as a function of ψ.

u, f, χ can now be calculated as functions of ψ by appropriate substitutions in eqs. (79), (77) and (81) respectively.

IX. Transit Time Not Predetermined

As proved in Section V, if the time of transit between the terminals is not predetermined, but is also to be optimized, then $C = 0$. In this case it is easy to verify that the eqs. (89), (90) possess the simple solution

$$r = \frac{a\,s^6}{1 - 3\,s^2}, \tag{95}$$

$$w = \pm\left(\frac{\gamma}{a}\right)^{1/2} \cdot \frac{1}{s^2}, \tag{96}$$

where $s = \sin\psi$ and $a = 9\,\gamma/A^2$. The sign ambiguity in the expression for w will be resolved later in the argument.

From eq. (93), it can now be calculated that

$$\theta' = \frac{3}{s^2} - 4 \tag{97}$$

and hence that

$$\theta = \theta_0 - 4\,\psi - 3\cot\psi. \tag{98}$$

Eqs. (95), (98) constitute polar equations for an intermediate-thrust arc in parametric form. It will be observed that members of the family of such arcs differ from one another only in respect to orientation in the plane and to scale. As ψ increases from 0 to $\sin^{-1}(1/\sqrt{3})$, θ increases from $-\infty$ to $\{\theta_0 - 3\sqrt{2} - 4\sin^{-1}(1/\sqrt{3})\}$ and r increases from 0 to ∞. Further increase in ψ causes r to assume negative values; since, in the theory, r has been taken to be a quantity which is essentially positive, such values for ψ are not permissible. Similarly, as ψ increases from $-\sin^{-1}(1/\sqrt{3})$ to 0, θ increases from $\{\theta_0 + 3\sqrt{2} + 4\sin^{-1}(1/\sqrt{3})\}$ to $+\infty$ and r decreases from ∞ to 0. The positive range of variation of ψ accordingly generates a spiral which unwinds about the centre of attraction in the anti-clockwise sense, whereas the negative range of variation of ψ generates a spiral which winds into the centre in the anti-clockwise sense. Since we have taken the rocket motion to be in the anti-clockwise sense about the centre, the first spiral represents a vehicle receding from the centre and the second spiral a vehicle approaching the centre. A spiral of the first type has been drawn in Fig. 2.

The inequality $\sin^2 \psi < 1/3$ has already been deduced in [6].

Employing eq. (94), t' is determined in the form

$$t' = \pm \frac{a^{3/2}}{\gamma^{1/2}} \frac{s^7 (5 s^2 - 3)}{(1 - 3 s^2)^2}, \tag{99}$$

the sign ambiguity following that of eq. (96). But ψ and t increase together as the rocket moves along its trajectory and thus t' is always positive. In the case when the rocket is receding from the centre of attraction, s is positive and less than $1/\sqrt{3}$. Hence the negative sign must be taken in eq. (99) and

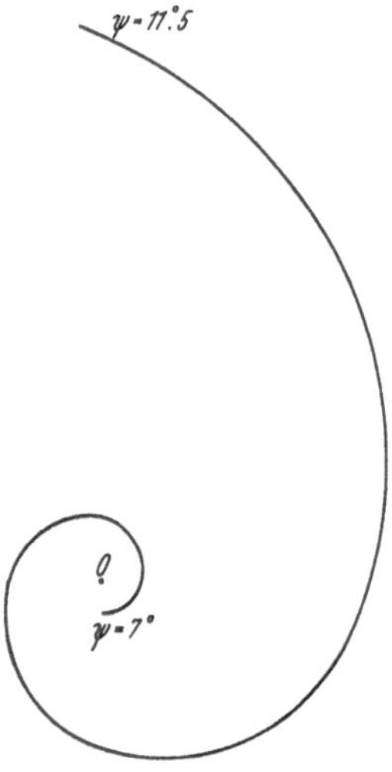

Fig. 2. Intermediate thrust spiral

$$t' = \frac{a^{3/2}}{\gamma^{1/2}} \cdot \frac{s^7 (3 - 5 s^2)}{(1 - 3 s^2)^2}. \tag{100}$$

The same sign is then appropriate in eq. (96), so that

$$w = -\left(\frac{\gamma}{a}\right)^{1/2} \frac{1}{s^2} \tag{101}$$

for a receding rocket. The signs of the right-hand members of eqs. (100), (101) are reversed in the case of a rocket approaching the centre of attraction.

Integrating eq. (100), it will be found that

$$\frac{\gamma^{1/2}}{a^{3/2}} t = \frac{1}{9} c^5 - \frac{31}{81} c^3 + \frac{5}{9} c + \frac{2}{81} \sqrt{\frac{2}{3}} \log \frac{\sqrt{3} c - \sqrt{2}}{\sqrt{3} c + \sqrt{2}} + $$
$$+ \frac{1}{81} \cdot \frac{c}{3 c^2 - 2} + \text{constant}, \tag{102}$$

where $c = \cos \psi$. Signs are reversed for a rocket approaching O.

Eq. (79) now leads to the result

$$u = \sqrt{\frac{\gamma}{a}} \cdot \frac{3 - 7 s^2}{s^3 (3 - 5 s^2)} \cos \psi; \tag{103}$$

if, therefore, $v = (u^2 + w^2)^{1/2}$ is the rocket velocity, eqs. (101), (103) show that

$$v = \sqrt{\frac{\gamma}{a}} \cdot \frac{(9 - 42 s^2 + 61 s^4 - 24 s^6)^{1/2}}{s^3 (3 - 5 s^2)}. \tag{104}$$

The right-hand member of eq. (103) must be prefixed by a minus sign if the rocket is approaching O.

From eqs. (81) and (98) it follows that

$$\chi = \tfrac{1}{2} \pi + \theta_0 - 5 \psi - 3 \cot \psi \tag{105}$$

and then, from eq. (77), that

$$f = \frac{\gamma}{a^2} \left(\frac{1 - 3 s^2}{3 - 5 s^2}\right)^3 \frac{1}{s^{11}} (27 - 75 s^2 + 60 s^4). \tag{106}$$

This equation is valid for both receding and approaching rockets.

The characteristic velocity for motion over any portion of an optimal spiral arc can be calculated from the equation

$$\int f \, dt = \text{constant} - \sqrt{\frac{\gamma}{a}} \cdot \frac{3 (1 - 2 s^2)(1 - 5 s^2)}{s^3 (3 - 5 s^2)} \cos \psi. \tag{107}$$

The right-hand member of this equation is reversed in sign for the case of an approaching rocket.

The radial and transverse components of rocket velocity are given by

$$v_r = \dot{r} = \sqrt{\frac{\gamma}{a}} \cdot \frac{6(1 - 2s^2)}{s^2(3 - 5s^2)} \cos\psi, \tag{108}$$

$$v_\theta = r\dot{\theta} = \sqrt{\frac{\gamma}{a}} \cdot \frac{(3 - 4s^2)(1 - 3s^2)}{s^3(3 - 5s^2)}, \tag{109}$$

for a receding rocket. Signs are reversed in both cases for an approaching rocket.

If ϕ is the angle made by the rocket velocity with the perpendicular to the radius vector r drawn in the sense of θ increasing, then

$$\tan\phi = \frac{v_r}{v_\theta} = \frac{6s(1 - 2s^2)}{(3 - 4s^2)(1 - 3s^2)} \cos\psi. \tag{110}$$

Over those portions of the trajectory where ψ is small, it is clear that we have approximately

$$\tan\phi = 2s = 2\sin\psi,$$

or (111)

$$\phi = 2\psi.$$

I.e. the direction of thrust bisects the angle between the direction of motion and the perpendicular to the radius vector. This agrees with a result obtained earlier in [7].

The total energy per unit mass of the vehicle at any point on this trajectory is given by the equation

$$E = \frac{1}{2}v^2 - \frac{\gamma}{r},$$

$$= -\frac{\gamma}{a} \cdot \frac{9 - 72s^2 + 169s^4 - 126s^6}{2s^6(3 - 5s^2)^2}. \tag{112}$$

As s increases towards the value $1/\sqrt{3}$, the rocket recedes from the centre of attraction and E increases through negative values, eventually becoming positive when the rocket attains a velocity exceeding that necessary for escape.

X. Junction of Intermediate Thrust and Keplerian Arcs

As shown in Section VI, the Weierstrass condition requires that the magnitude of the primer shall not exceed unity at any point on an optimal trajectory. Over an arc of intermediate thrust, the primer's magnitude is everywhere unity and the condition is clearly satisfied. On such an arc

$$\mathfrak{p}^2 = 1 \tag{113}$$

and hence, differentiating,

$$\mathfrak{p} \cdot \dot{\mathfrak{p}} = 0. \tag{114}$$

But, at a junction between an intermediate thrust arc and a null thrust arc, the primer and its first derivative are continuous (this will still be true if an impulse is applied at the junction) and it follows that eq. (114) is also valid on the null thrust arc at this junction point. This implies that the magnitude of the primer is stationary at this point on the null thrust arc and the Weierstrass condition then further requires that, in fact, the stationary value shall be a maximum. Thus, at a point on a null-thrust arc where it joins an arc of intermediate thrust, the magnitude of the primer increases to a maximum value of unity. In this section, we will check that this condition is satisfied at a junction between a Keplerian arc of null thrust in an inverse square law field and the spiral arc of intermediate thrust which was studied in the previous section. It will be assumed that no impulse is applied at this junction.

On a given spiral intermediate thrust arc, the primer is completely determined at every point as the unit vector in the direction making the angle ψ with the perpendicular to the radius vector r. In particular, $\mathfrak{p}$ and $\dot{\mathfrak{p}}$ are known at the junction point and, since these quantities are continuous, they will accordingly be known at this point on the KEPLERian arc. But, over this arc, the components of the primer satisfy the second order differential eqs. (16) and it follows that, if these components and their first derivatives are specified at one point on the arc, the primer is determined completely over the whole arc. In the case of transfer between two elliptical orbits via an intermediate thrust spiral, therefore, the primer will be determined at all points on both orbits and the WEIERSTRASS condition can be applied. Numerical calculation in particular cases suggests that this condition will be satisfied for such a transfer manoeuvre, but a general proof has not been obtained. It can, however, be shown that at the junctions the primer's magnitude is a maximum. This we proceed to do.

Referring to eq. (16) it will be seen that, since the quantities $\partial f_j/\partial x_i$ are continuous at the junction, the continuity of the primer components at this point implies that the $\ddot{\lambda}_i$ are also continuous. Again, differentiating eq. (16) with respect to t, it will be found that

$$\dddot{\lambda}_i = \dot{\lambda}_j \frac{\partial f_j}{\partial x_i} + \lambda_j \frac{\partial^2 f_j}{\partial x_i \, \partial x_k} \dot{x}_k. \tag{115}$$

No impulse being applied at the junction, the $\dot{x}_k$ are continuous at this point. We conclude that the $\dddot{\lambda}_i$ are also continuous at the junction. A further differentiation introduces "acceleration terms" $\ddot{x}_k$. These are not continuous at the junction, and neither are the quantities $\ddddot{\lambda}_i$ therefore. We have proved, then, that the primer and its time derivatives, up to and including the third order, are continuous at such a junction. This has been remarked in [8].

Let κ be the square of the magnitude of the primer. Thus

$$\kappa = \mathfrak{p}^2. \tag{116}$$

Differentiations with respect to t, yield the equations

$$\dot{\kappa} = 2\,\mathfrak{p} \cdot \dot{\mathfrak{p}}, \tag{117}$$

$$\ddot{\kappa} = 2\,\dot{\mathfrak{p}}^2 + 2\,\mathfrak{p} \cdot \ddot{\mathfrak{p}}, \tag{118}$$

$$\dddot{\kappa} = 6\,\dot{\mathfrak{p}} \cdot \ddot{\mathfrak{p}} + 2\,\mathfrak{p} \cdot \dddot{\mathfrak{p}}, \tag{119}$$

$$\ddddot{\kappa} = 8\,\dot{\mathfrak{p}} \cdot \dddot{\mathfrak{p}} + 6\,\ddot{\mathfrak{p}}^2 + 2\,\mathfrak{p} \cdot \ddddot{\mathfrak{p}}. \tag{120}$$

Hence, κ and its first three derivatives are continuous at the junction, whereas its fourth derivative is not. On the intermediate thrust arc $\kappa = 1$ identically and all its derivatives vanish. It follows that the first three derivatives of κ must vanish at the junction on the null thrust arc whereas, in general, the fourth will not. Thus κ (and with it the magnitude of $\mathfrak{p}$) will be a maximum at the junction on this arc, provided $\ddddot{\kappa}$ is negative at this point.

Let $|\ddddot{\kappa}|$ denote the magnitude of the discontinuity in $\ddddot{\kappa}$ at the junction in going from the arc of intermediate thrust to the arc of null-thrust. Then, since the first three derivatives of $\mathfrak{p}$ are continuous across the junction, eq. (120) indicates that

$$|\ddddot{\kappa}| = 2\,\mathfrak{p} \cdot |\ddddot{\mathfrak{p}}|. \tag{121}$$

But $\ddddot{\kappa} = 0$ on the intermediate thrust arc. Hence

$$\ddddot{\kappa} = 2\,\mathfrak{p} \cdot |\ddddot{\mathfrak{p}}| \tag{122}$$

at the junction on the null thrust arc.

Differentiating eq. (115) with respect to t, the components of $\ddddot{\mathfrak{p}}$ are found to be given by

$$\ddddot{\lambda}_i = \ddot{\lambda}_j \frac{\partial f_j}{\partial x_i} + 2\,\dot{\lambda}_j \frac{\partial^2 f_j}{\partial x_i\,\partial x_k}\,\dot{x}_k + \lambda_j \frac{\partial^3 f_i}{\partial x_i\,\partial x_k\,\partial x_l}\,\dot{x}_k\,\dot{x}_l + \lambda_j \frac{\partial^2 f_j}{\partial x_i\,\partial x_k}\,\ddot{x}_k\,. \tag{123}$$

In the right-hand member of this equation, only $\ddot{x}_k$ is discontinuous at the junction and hence

$$|\ddddot{\lambda}_i| = \lambda_j \frac{\partial^2 f_j}{\partial x_i\,\partial x_k}\,|\ddot{x}_k|. \tag{124}$$

From eq. (122) it now follows that

$$\ddddot{\kappa} = 2\,\lambda_i\,\lambda_j \frac{\partial^2 f_j}{\partial x_i\,\partial x_k}\,|\ddot{x}_k|. \tag{125}$$

The discontinuity in the acceleration is due to the cut off of motor thrust in passing from the powered arc to the null thrust arc. Since this thrust acts in the direction of the primer

$$|\ddot{x}_k| = -f\,\lambda_k. \tag{126}$$

Hence,

$$\ddddot{\kappa} = -2\,f\,\lambda_i\,\lambda_j\,\lambda_k \frac{\partial^2 f_j}{\partial x_i\,\partial x_k}\,. \tag{127}$$

For the case of plane motion in an inverse square law field, taking the origin O at the centre of attraction and x-axis through the junction point, the coordinates of this point will be $(r, 0)$. By further differentiation of eqs. (68) and then putting $x = r$, $y = 0$, it will be found that at the junction

$$\left.\begin{aligned} \frac{\partial^2 f_x}{\partial x^2} &= -\frac{6\gamma}{r^4}, & \frac{\partial^2 f_x}{\partial x\,\partial y} &= 0, & \frac{\partial^2 f_x}{\partial y^2} &= \frac{3\gamma}{r^4}, \\ \frac{\partial^2 f_y}{\partial x^2} &= 0, & \frac{\partial^2 f_y}{\partial x\,\partial y} &= \frac{3\gamma}{r^4}, & \frac{\partial^2 f_y}{\partial y^2} &= 0. \end{aligned}\right\} \tag{128}$$

Relative to these axes the primer makes an angle ψ with $O\,y$. Thus

$$\lambda_x = \sin\psi, \qquad \lambda_y = \cos\psi. \tag{129}$$

Substituting in eq. (127), it will now be found that

$$\ddddot{\kappa} = -\frac{3\gamma f}{r^4}\,s(3 - 5\,s^2). \tag{130}$$

But $s^2 < 1/3$ and thus $\ddddot{\kappa} < 0$. This proves that the primer's magnitude is a maximum at the junction.

XI. Conclusion

In the previous sections of this paper it has been demonstrated that powered arcs of intermediate thrust may form part of an optimal rocket trajectory *in vacuo* and that a number of necessary conditions for the propellant expenditure along the trajectory to be a minimum are, indeed, satisfied. However, it has not been proved that the satisfaction of these conditions ensures that the trajectory shall, in fact, be an absolute optimum and, from the nature of the argument employed, we can hope to prove no more than that a trajectory satisfying our conditions constitutes a local optimum. To calculate an absolute optimal trajectory by our methods, it would be necessary to determine *all* local optimal trajectories satisfying our conditions and then to compute the propellant expenditure appropriate to each. If this were done, it still might prove to be the case (as was supposed previously) that an absolute optimal trajectory comprises no arcs of intermediate thrust. Alternatively, this feature might be possessed by absolute optimal trajectories in an inverse square law field, but not in more complex fields. The methods of this paper seem to be insufficiently powerful to decide such questions. Other techniques which do not depend upon only *small* variations of the trajectory to establish that it is optimal, will probably have to be employed.

In the case of plane motion in an inverse square law field it is clear that, unless transfer between two KEPLERian orbits along an arc of intermediate thrust is, in some case, more economical of propellant than the orthodox mode of transfer employing two impulsive thrusts, then such powered arcs cannot form part of an absolute optimal trajectory; for any such arc could always then be replaced by two impulsive thrusts and the propellant expenditure further reduced. The author of this paper has not yet been able to arrive at any definite conclusions regarding this point. This is on account of the difficulty (a computational one) of calculating the optimal two-impulse transfer between two KEPLERian orbits whose axes are not aligned. However, some evidence has been obtained as follows:

At the point (r, θ) on the intermediate thrust spiral, "circular" velocity is given by

$$v_c = \sqrt{\frac{\gamma}{r}} = \sqrt{\frac{\gamma}{a}} \cdot \frac{(1 - 3\,s^2)^{1/2}}{s^3}. \tag{131}$$

Thus, from eqs. (108), (109), we deduce that the radial and transverse components of rocket velocity may be written in the form

$$v_r = \frac{6\,s\,(1 - 2\,s^2)\cos\psi}{(1 - 3\,s^2)^{1/2}\,(3 - 5\,s^2)}\,v_c, \tag{132}$$

$$v_\theta = \frac{(3 - 4\,s^2)\,(1 - 3\,s^2)^{1/2}}{3 - 5\,s^2}\,v_c. \tag{133}$$

It follows that, over an arc of this spiral along which ψ is small we have approximately

$$v_r = 0, \qquad v_\theta = v_c, \tag{134}$$

and this arc can accordingly be employed to effect transfer between two orbits which are approximately circular.

For small ψ, eq. (107) can be approximated to read

$$\int f\,dt = \text{constant} - \sqrt{\frac{\gamma}{a}} \cdot \frac{1}{s^3} \tag{135}$$

and thus, as s increases from $s = s_1$ to $s = s_2$, the characteristic velocity for the manoeuvre is given by

$$V = V_1 = \sqrt{\frac{\gamma}{a}} \left(\frac{1}{s_1{}^3} - \frac{1}{s_2{}^3}\right) \tag{136}$$

approximately.

To the same order of approximation, eq. (95) shows that during this manoeuvre r increases from $r = r_1$ to $r = r_2$, where

$$r_1 = a\, s_1^6, \qquad r_2 = a\, s_2^6. \tag{137}$$

But the characteristic velocity for HOHMANN transfer between two circular orbits having these radii is given by

$$V = V_2 = \sqrt{\frac{\gamma}{a}}\left[\sqrt{\frac{2}{s_1^6 + s_2^6}}\left\{\frac{s_2^3}{s_1^3} - \frac{s_1^3}{s_2^3}\right\} + \frac{1}{s_2^3} - \frac{1}{s_1^3}\right]. \tag{138}$$

Writing $(s_1/s_2)^3 = x$, it is clear that $V_1 > V_2$ provided

$$\sqrt{2}\,(1 - x) > \frac{1 - x^2}{(1 + x^2)^{1/2}}, \tag{139}$$

i.e. provided

$$(1 + x) < \sqrt{2 + 2\,x^2}, \tag{140}$$

since $0 < x < 1$. Squaring both sides of this inequality, it is easily seen to be equivalent to

$$x^2 - 2\,x + 1 > 0, \tag{141}$$

which is certainly true. We conclude that $V_1 > V_2$ and hence that transfer between two nearly circular orbits along an intermediate thrust spiral is not more economical than via the HOHMANN ellipse.

Nonetheless, it is still possible that transfer between two ellipses via the spiral is more economical than by the application of impulsive thrusts. This question deserves a close study.

Finally, we may remark that, even if it is proved that no part of a trajectory of absolute minimum propellant expenditure in an inverse square law (or other) field is an arc of intermediate thrust, it is still possible that such arcs may enter into trajectories which optimize other parameters or for which the transit time is predetermined. All these questions must be left open for future investigation.

References

1. D. F. LAWDEN, Advances in Space Science, Vol. I, p. 1. New York: Academic Press, 1959.
2. G. LEITMANN, Private Communication, 1960.
3. G. A. BLISS, Lectures on the Calculus of Variations, p. 220. Chicago: University of Chicago Press, 1946.
4. J. V. BREAKWELL, The Optimization of Trajectories. J. Soc. Indust. Appl. Math. 7, 215 (1959).
5. D. F. LAWDEN, Minimal Rocket Trajectories. J. Amer. Rocket Soc. 23, 360 (1953).
6. B. D. FRIED, Space Technology, p. 4—24. New York: Wiley, 1959.
7. D. F. LAWDEN, Optimal Escape from a Circular Orbit. Astronaut. Acta 4, 218 (1958).
8. G. LEITMANN, Extremal Rocket Trajectories in Position and Time Dependent Force Fields. Unpublished Memorandum, 1960.

Discussion

In reply to a question from Mr. STERN, Professor LAWDEN stated: I have calculated intermediate thrust arcs in an inverse square law field and the inverse square law field is, after all, a reasonably complex gravitational field, and if such a solution does exist in a gravitational field of that type then we can expect that it exists in the general field. Certainly, I am convinced myself that my previous argument against the existence of these arcs is invalid so that I am certain that there is an intermediate thrust arc in the square law field, and, I think, it follows from that that there will also be an intermediate thrust arc in the general field.

Factors Affecting the Accuracy of Impact Position of Satellites Re-Entered by Command

By

Terence R. F. Nonweiler[1]

(With 11 Figures)

Abstract — Zusammenfassung — Résumé

Factors Affecting the Accuracy of Impact Position of Satellites Re-Entered by Command. Choice of a reasonably large re-entry angle is necessary to effect good accuracy in line, and in practice the largest errors may arise due to unforeseen variations in the atmospheric properties. It is important to arrange that the rocket impulse (which precipitates re-entry) causes the *in-vacuo* orbit to intersect a specially defined "injection altitude" at the correct longitude, but accidental errors in the angle of descent of the vehicle's path through this altitude have little effect on impact accuracy. Errors in impact position due to misalignment of the impulse can be rendered very small by applying the impulse at an "optimum" point specially chosen relative to the impact longitude. All attempts to increase accuracy involve a cost in magnitude of the perturbing impulse, and so in rocket propellant. In considering accuracy in azimuth, it is shown that the relation between the latitude of the orbital node and that of the impact point is of primary importance.

Genauigkeitsfaktoren für den Auftreffpunkt auf Kommando zurückkehrender Satelliten. Große Eintauchwinkel sind nötig, um eine gute Genauigkeit zu erhalten, wobei im allgemeinen die größten Fehler durch unvorhergesehene Variationen in der Erdatmosphäre bewirkt werden dürften. Es ist wichtig, daß durch den Bremsimpuls, der die Rückkehrbahn einleitet, bewirkt wird, daß der außerhalb der Atmosphäre verlaufende Teil der Rückkehrbahn ein bestimmtes Niveau — das „Einschußniveau" — bei korrekter geographischer Länge erreicht. Zusätzliche Fehler im Winkel haben weniger Einfluß auf den Landepunkt. Fehler, die durch schlechte Dosierung des Bremsimpulses bewirkt werden, können durch Wahl eines optimalen Punktes, wo der Impuls eingeleitet wird, sehr klein gehalten werden. Alle weiteren Versuche, die Genauigkeit zu steigern, erfordern noch zusätzlichen Treibstoff. Bei Untersuchung der Azimut-Genauigkeit stellt sich heraus, daß die Beziehung zwischen der Breite des Bahnknotens und der des Landepunktes von primärer Bedeutung ist.

Facteurs affectant la précision d'impact des satellites à rentrée commandée. Un angle de rentrée suffisant est requis pour la précision d'impact en alignement et, en pratique, les erreurs les plus importantes peuvent être dues à des variations insoupçonnées des caractéristiques de l'atmosphère. La rétro-impulsion initiant la rentrée doit provoquer l'intersection de l'orbite "dans le vide" avec une altitude d'injection correcte à la longitude choisie. Les erreurs accidentelles d'inclinaison subséquentes de l'orbite ont peu d'importance. Les erreurs d'impact dues à un centrage incorrect de l'impulsion peuvent être minimisées par un choix approprié

---

[1] B.Sc., PhD., A.F.I.A.S., A.F.R.Ae.S., F.B.I.S.; Mechan Professor of Aeronautics and Fluid Mechanics at Glasgow University, Glasgow, W. 2, Scotland.

de l'initiation de l'impulsion. Toute tentative d'assurer une meilleure précision accroit l'impulsion requise et donc la consommation. Pour la précision transversale, la relation entre latitude du noeud de l'orbite et latitude du point d'impact est de la première importance.

This paper discusses in general terms some of the factors which affect the accuracy of prediction of the impact position on the earth's surface of a vehicle re-entered by command from an orbit of small eccentricity (close to the earth but effectively outside the atmosphere) and descending under the action of an aerodynamic drag force. The problem can be resolved into two parts: the effect of errors in the impulse of the rocket thrust which perturbs the orbit to bring it within the atmosphere, and the predictability of the aerodynamic effects when the vehicle is in contact with the atmosphere.

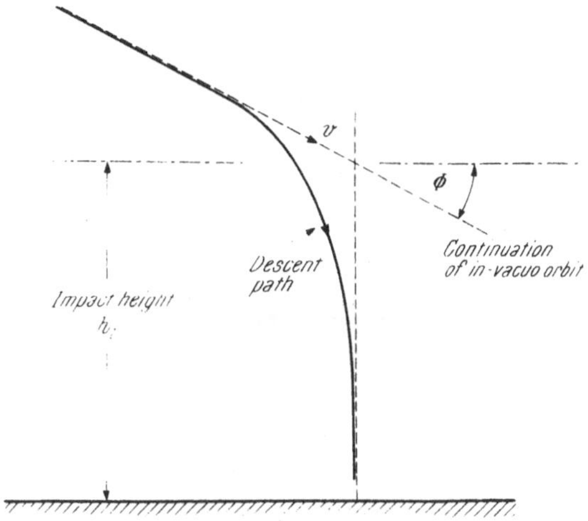

Fig. 1

Treating the latter problem first, it will be assumed that the required re-entry orbit is known, and then with assumed atmospheric properties and assumed variation of drag coefficient with MACH Number and REYNOLDS Number, it would be possible to calculate the descent path to an impact point on the earth's surface, which is the required impact point. The height of the *in-vacuo* orbit at the longitude of the impact point provides a reference altitude which we shall call the "impact height" (Fig. 1) — and the aim of the impulse applied to the orbit is therefore to produce an orbit which intersects the "impact height" at the longitude of the required impact point and with the required speed and angle of descent. The aerodynamic effects in the descent are not influenced by the accuracy in longitude with which this intersection is obtained — except so far as this may, in turn, influence the state of the atmosphere through which the vehicle descends — but they are certainly influenced by the properties of the *in-vacuo* orbit, determining what we may call the "re-entry speed" and "re-entry angle of descent" at the planned impact height.

If the vehicle has too steep an angle of descent or too high a speed, yet the *in-vacuo* orbit intersects the planned impact altitude at the correct longitude, then it will overshoot the planned impact point. In fact, it is the rate of descent at re-entry which is of primary importance in determining the *actual* impact height, and where this rate of descent is higher or lower than is planned, the actual impact altitude will be at a different height than that planned, where the air pressure is similarly higher or lower (Fig. 2). However, the effect of small errors in angle of descent can be eliminated if we regard the impulse applied to the orbit to be intended to produce not the correct longitude on the perturbed *in-vacuo* orbit at the planned impact altitude, but an appropriate intersection of the orbit with a specially chosen, higher, "injection" altitude (Fig. 3). Such an "injection" point and altitude may always be found, as indicated in the figure, so that if the *in-vacuo* orbit is steeper than planned (and its actual impact altitude is correspondingly lower) then the change in longitude between injection point and impact point will remain the same, due to the steeper descent.

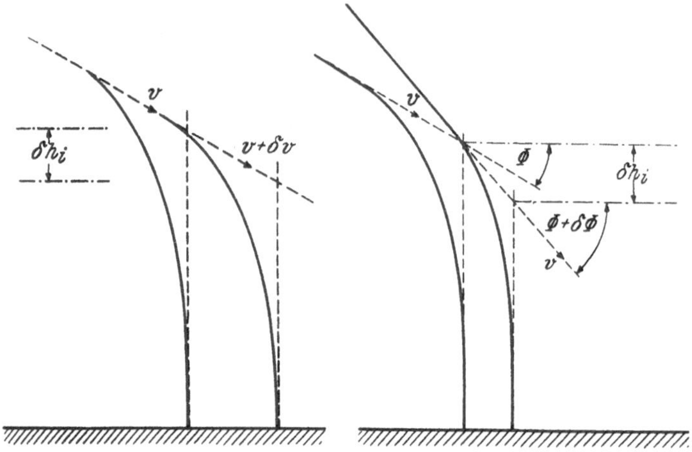

Fig. 2

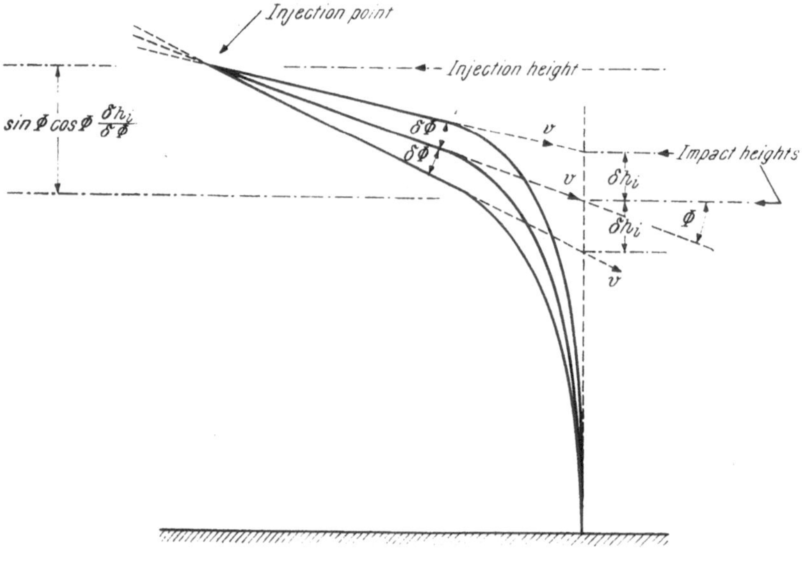

Fig. 3

This device is merely a way of looking at the problem of analysing accuracy, and is not of course, a means of improving the accuracy of the re-entry phase: it has the effect of showing that the angle of descent is relatively unimportant if correct positioning of the perturbed orbit to intersect the "injection" point can be ensured. Errors of speed at the injection point — or, what amounts to the same thing, errors in the semi-latus rectum (l) of the perturbed orbit —

still have an effect, even if such precise positioning is possible. However, calculations quickly suggest that — provided at least the speed of descent is not small compared with the speed of sound — the impact error from this cause would be quite negligible compared with the error in positioning the injection point implicit in the very existence of the wrong value of l.

But there are sources of error in the prediction of the atmospheric re-entry orbit other than those arising at injection. Errors in the prediction of scale-height (i.e. of air temperature) produce small effects like errors of the same proportion in l, and plainly the scale height is predictable within much coarser limits than those which are likely to encompass the relative variation of l. This emphasizes again the unimportance of error in speed at injection, viewed in the perspective of other errors.

As well as this, of course, the absolute value of the air pressure at the injection point is important. Thus even if the scale-height is as predicted, but the air pressure 1% low (say), then the sequence of aerodynamic effects is the same but is removed to an altitude 1% of the scale height lower, thus producing an error in the longitude of the impact point of about 2 cot ϕ seconds of arc where ϕ is the planned angle of descent at injection. Bearing in mind that variations of 10% in pressure — if not more — are known to exist at great altitudes it is easy to see that from this cause alone one may expect errors of several kilometers in the impact point position. Fig. 4 illustrates some values of the injection altitude, showing that it may indeed be well outside the range of normal meteorological prediction.

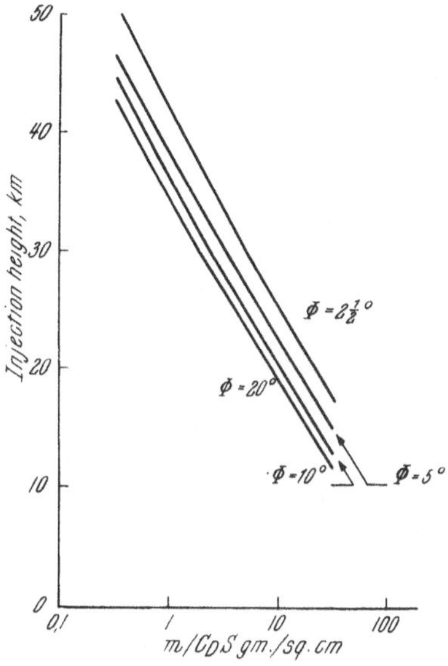

Fig. 4

Errors in the prediction of the drag coefficient have a similar relative magnitude to those in prediction of air pressure, but it is worth nothing that so far as the shape of the re-entry trajectory is concerned, unexpected variations in drag coefficient in a broad range of Mach Number about unity are of much more importance than variations at hypersonic speeds or at very low "incompressible" speeds, — provided at least that the re-entry angle is steep (say, greater than $2^1/_2{}^\circ$). This is because the trajectory shape does not differ significantly from the *in-vacuo* orbit until the Mach Number is below the hypersonic range, and of course at very low Mach Numbers the fall of the vehicle is nearly vertical. Errors in the alignment of the vehicle (relative to its direction of motion) on initial contact with the atmosphere — before the aerodynamic forces of stabilisation are fully effective — may of course, produce errors of a different order of magnitude, particularly as lift may be developed, but these are beyond the scope of the analysis cited here, and are clearly dependent primarily on the actual configuration of vehicle, and on the accuracy of its system of *in-vacuo* orientation control. In as much as errors in the application of rocket impulse may cause an unexpected change in mass of the re-entering vehicle, this may also affect the atmospheric re-entry path, since it is the drag to mass ratio which is kinematically important: however, errors from this source are likely to be unimportant compared with the others implied by such an eventuality.

Turning our attention to the errors introduced by the rocket impulse, we recall that it is the aim of this impulse to perturb the orbit to intersect what we have called the injection point, — irrespective of the direction of the orbit through this point. The angle of descent is certainly of importance in determining the

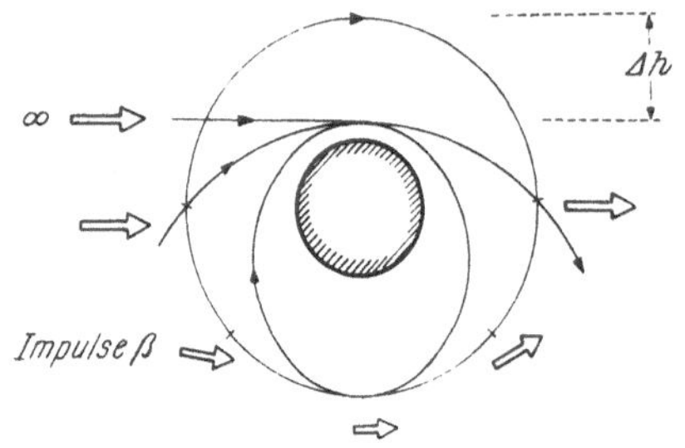

Fig. 5

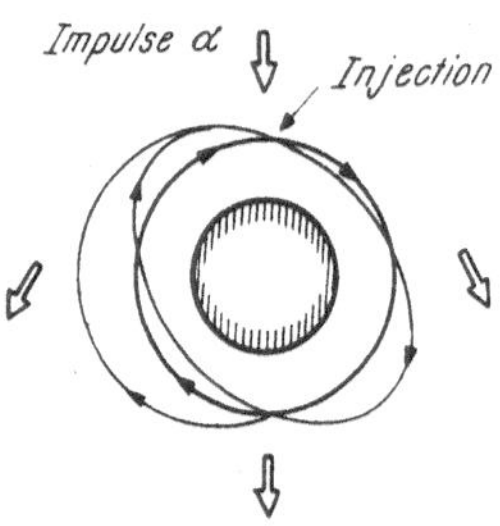

Fig. 6

retardation and aerodynamic heating rates during the re-entry phase, and so is a quantity whose required value might well have to be carefully selected, but none the less small variations about this chosen angle have, as we have seen, no effect on impact point accuracy — to the first order of approximation.

We shall first of all assume that the required impact point lies within the plane of the unperturbed orbit, so that the perturbing impulse will likewise be required to be directed within the plane of the orbit. We shall also ignore, for the moment, errors which contribute to an impulse perpendicular to the orbit, and we shall suppose that there is no error in the timing impulse (1 second of time might typically equal 3 or 4 minutes arc of longitude in travel), so that it may be regarded as being applied at the planned longitude on the orbit. At such a longitude, the required impulse to perturb the orbit to pass through the injection point, with the required angle of descent, will have a definite magnitude and direction. The perturbation will certainly be such as to reduce the altitude of the vehicle at the longitude of the injection point, and will probably also be such as to increase its angle of descent at this altitude, — obviously this would be so if the unperturbed orbit is closely circular as assumed in most of our quoted examples. To obtain the reduction in height Δh by itself, a tangential retardation is required on the opposite side of an orbit of low eccentricity or an acceleration if the impulse is applied on the same side as the injection point; this tangential impulse has to be combined with a downward impulse if the impulse is applied on the approaching side, or an upward impulse on the retreating side of the orbit (Fig. 5), producing a resultant impulse $\boldsymbol{\beta}$, say. A steeper angle of descent (by an amount $\Delta\phi$, say), with the orbit passing through the same injection point (i.e. $\Delta h = 0$), can be arranged by providing an upward impulse combined with a deceleration on the approaching side, or an acceleration on the retreating side of the orbit (Fig. 6), producing a resultant $\boldsymbol{\alpha}$, say.

An error in the impulse will cause not only a change in the angle of descent through the injection altitude — which may be unimportant to impact error as we have reasoned before — but the longitude of the intersection of the perturbed orbit with the injection altitude will be changed; in other words, there will be an error in the longitude of the injection point, which will cause an equal error in longitude of the impact point. An excessive retardation will cause the impact to occur earlier, particularly for impulses applied on the opposite side of the orbit: an excessive downward impulse has the same effect if applied on the approaching side of the orbit. However, if the error impulse is parallel to that impulse direction which produces no change in injection height but merely a change in injection angle (parallel to $\boldsymbol{\alpha}$ of Fig. 6), then there will be no first-order error in the longitude of injection.

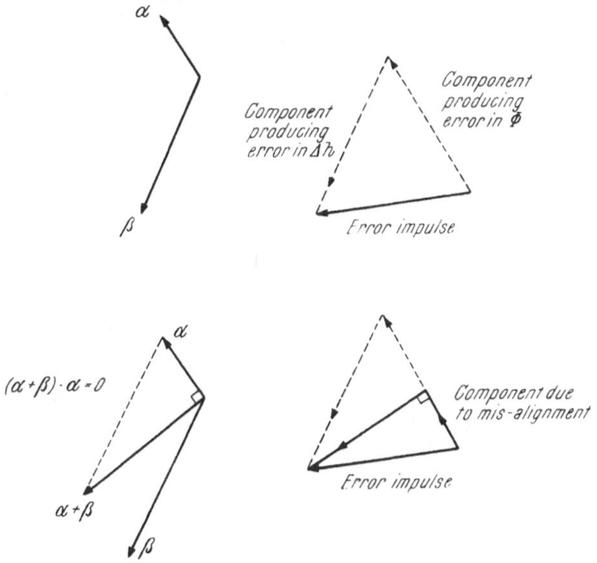

Fig. 7

This is an important, if simple, deduction. For it is evident that if the impulse is correctly directed, but its magnitude is wrong, there will be an error of similar proportion in both height-reduction and re-entry angle at the longitude of the injection point, no matter where the impulse is applied on the orbit. This source of error is inescapable, and the error in height-reduction will cause an error in injection point longitude. However, if the required resultant impulse is at right-angles to the impulse $\boldsymbol{\alpha}$ which changes the angle of descent but not the height at the injection longitude, it will be clear that an error in the direction of the total impulse (but not its magnitude) will have no first-order effect on the height-reduction or, consequently, on injection point longitude (Fig. 7). Since the directions of the impulses $\boldsymbol{\alpha}$ and $\boldsymbol{\beta}$ are not in general orthogonal, it is thus generally possible to find a particular angle of injection which will make any misalignment of the impulse unimportant, if the longitudes of the impulse and impact points are both known. (Such an angle will be the maximum or minimum obtainable with an impulse of the required magnitude applied at the impulse point.) Conversely, if the planned angle of injection is stipulated, as well the height-reduction at the injection longitude, it is always possible to find some position for the point of application of the impulse such that the resultant impulse is at right angles to $\boldsymbol{\alpha}$, i.e. such that

$$\boldsymbol{\alpha} \cdot (\boldsymbol{\alpha} + \boldsymbol{\beta}) = 0$$

Such "optimum" impulse points occur on the side of the orbit approaching the injection point if the injection angle is planned to be increased (Fig. 8), and if the unperturbed orbit is nearly circular, this impulse point is opposite to the impact point if there is to be no perturbation in angle of descent.

Having arranged in this way for the errors in thrust-direction to be unimportant, we are left with the inevitable error in injection point longitude caused by errors in total impulse *magnitude*. A 1% error in impulse will then be associated with a movement of the injection point by at most $\cot \phi\%$ of the height-reduction, where as before ϕ is the planned injection angle. Thus with a height reduction (Δh) of 200 km, a 1% error in impulse would lead to a probable maximum error of about 2 cot ϕ km. in injection (or about cot ϕ minutes error in injection longitude). Clearly, if the relative disposition of the orbital apse-line and injection longitude is open to choice, the error is least if the injection point is beneath the perigee of the unperturbed orbit.

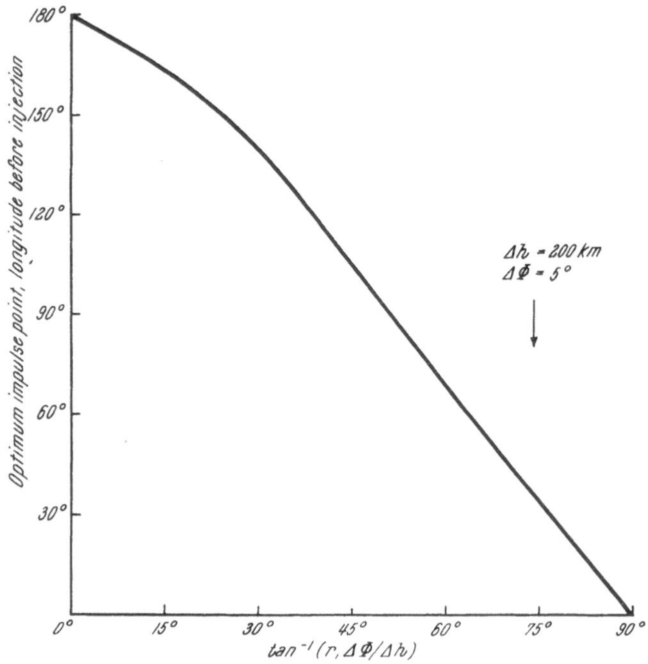

Fig. 8

The importance of this method of eliminating errors due to mis-alignment depends, of course, on the likelihood of such errors: where any direction of the error impulse is equally likely, and the impulse point is not chosen to correspond with the "optimum" so as to minimize the error — the probable error in impact point can be many times increased above the maximum figure we have just quoted, particularly if the injection angle is large. However there is good reason to expect that errors in alignment of impulse are much more likely than errors in magnitude, which further increases the probable error in impact position unless the impulse point is selected to be the obtimum. On the other hand, the total impulse needed may be appreciably larger when applied at the optimum point than the minimum impulse which can produce the appropriate injection conditions. For the perturbation to a nearly circular orbit, the impulse is minimum when applied a longitude ($\tan^{-1} \Delta h/r\, \Delta\phi$) ahead of injection, except for large $\Delta\phi$ (Fig. 9). There is no such disparity where $\Delta\phi = 0$, (i.e. the angle of descent at injection is that of the unperturbed orbit at the injection altitude); but the steeper the required descent, the greater the penalty in impulse — and so in rocket propellants — by applying the impulse at the optimum position (Fig. 10).

If the penalty is regarded as too severe, and a limit is set to the total impulse, either the injection angle must be reduced, or the impulse must be applied closer to the minimum impulse point. In practice the best compromise might be a mixture of both remedies, but as the angle of descent at injection affects the error

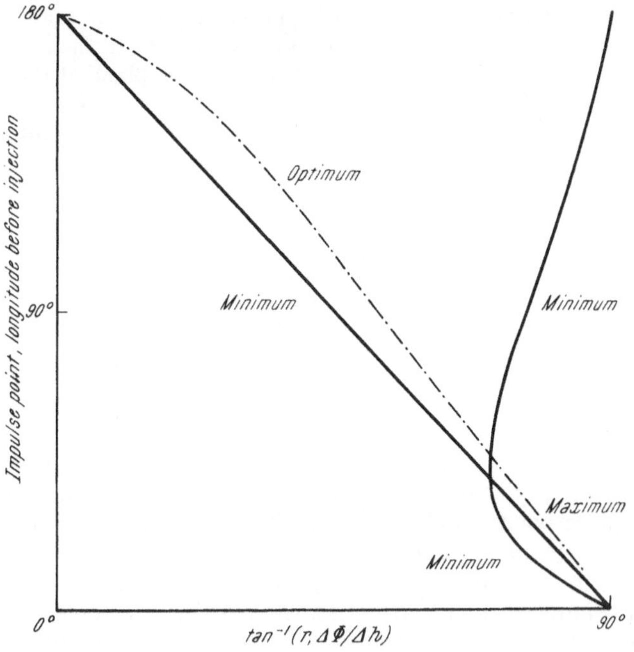

Fig. 9

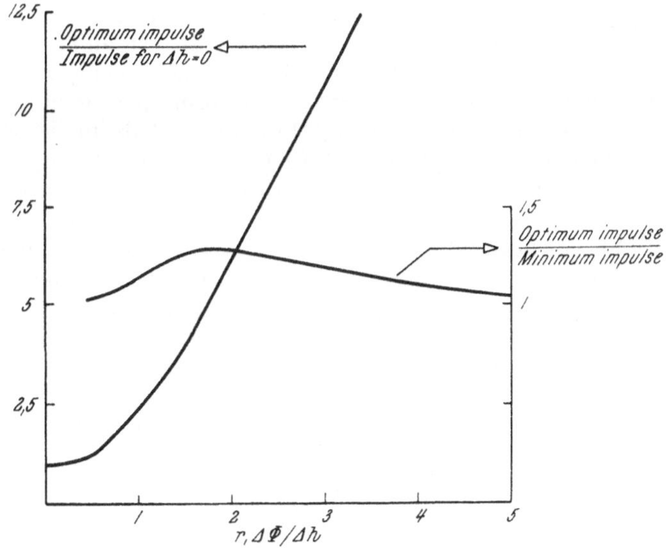

Fig. 10

in impact point due to atmospheric variations, examples quoted are too arbitrary to be of much value.

The matter is still further complicated if account is taken of errors in misalignment of the impulse producing a deflection of the orbital plane. Error impulse components at right angles to the orbital plane are possibly as likely to occur as those at right angles to the required impulse direction in the orbital plane. If the impulse were applied at the antipodal point to the impact point, the azimuthal error at injection would be zero, even if the plane of the orbit were to be inadvertently deflected; and although obviously no large azimuthal error would result if the impulse were applied close to the impact point, this effect is offset by the larger impulse then needed to perturb the orbit. The magnitude

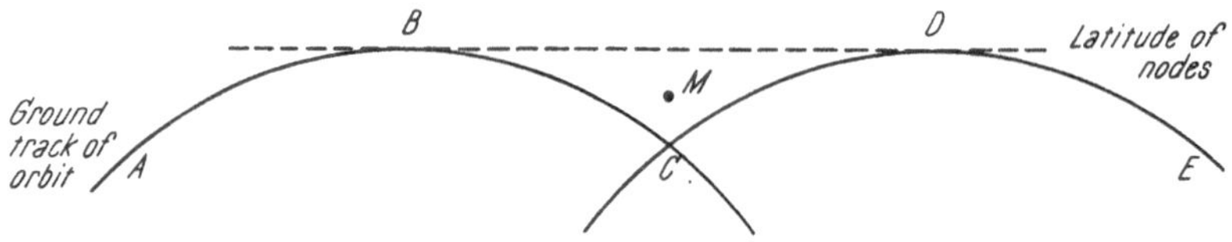

Fig. 11

of azimuthal errors is generally relatively small because their magnitude is not dependent on cot ϕ, but it will be clear that if they are taken into account, the optimum impulse point could be moved either towards the antipodal point, or towards the minimum impulse point.

Where the impact point does not lie in the plane of the orbit, the whole problem of the impulse and its errors assumes a different complexion. To consider the worst possible case: an impact point on the earth's equator could be up to 12° of earth longitude away from the closest approach of a polar-orbit satellite. The least impulse at right angles to the orbit (applied at the node) which will deflect the orbital plane to that of the great circle through the impact point would be sufficient, if applied within the orbital plane, to reduce the orbital height by about 5000 km. Errors in the alignment of such an impulse will have components arbitrarily directed in the orbital plane, and the only way of reducing the effect on accuracy in line of these errors is to bring the impulse point closer to the injection point and steepen the injection angle. This implies a still larger impulse, and there is obviously a limit to the enhancement of the accuracy thereby attained. We do not intend to pursue this problem. It is of such a different complexity that, where re-entry is envisaged, and delay of the re-entry until the required impact point is favourably placed relative to the orbit is for some reason undesirable, it would obviously enhance accuracy to choose the orbit to suit the impact point, or *vice-versa*. For instance, if an equatorial impact point is essential, the equatorial plane could be chosen as that of the orbit. If it is impractical to choose such an orbit, then the impact point could be situated on the latitude of the orbital nodes.

In the Fig. 11, if ABC and CDE represent successive ground tracks of the satellite with nodes at B and D, then the largest azimuthal deviation of the line of the orbit from the impact at M happens if M is, as shown, on the longitude $M\,C$ halfway between the nodes. The distance

$$M\,C = \frac{1}{32}\,R\,\Omega^2\,t_p^2 \sin 2\,\psi$$

approximately, where R is the earth radius, Ω is the rate of change of earth longitude of the nodes, t_p is the orbital period (so that $BD = \Omega\,t_p$), and ψ is the inclination of the orbit relative to the equator. For instance with $\Omega\,t_p = 24°$, and $\psi = 60°$, the value of MC is about 30 km, which is a figure which may not in practice differ very much from the probable impact error in line. If this were so, the impact accuracy could not be greatly improved by any attempt to correct such an azimuthal deviation by an impulse perpendicular to the orbit.

Discussion

Reply to Mr. STERN: The injection height is generally some 5 or 10 km above the impact height. The position of the optimum impulse point is illustrated in Fig. 8 for orbits of low eccentricity.

Reply to Dr. HILTON: The impact angle – or at least what I termed the re-entry angle at the impact height – can indeed be quite precisely defined in terms of the angle between the radius vector from the earth centre at the impact point and the tangent to the *in-vacuo* orbit at its intersection with this radius vector (Fig. 1). In practice where one is dealing with re-entry angles of $2^1/_2°$ or so, one does not have to be so very precise, since the angle of descent of the *in-vacuo* orbit changes only a little over the depth of the aerodynamically important atmosphere: the "perigee" of the orbit is well below the earth's surface. However, with the grazing contact considered in Dr. HILTON's own paper, it is much more satisfactory to speak of perigee height, rather than any arbitrarily defined angle.

Reply to Mr. FERRI: I think I ought to emphasize the need to make the re-entry angle reasonably large (however it may be defined!) if one is to avoid undue errors in line. In all expressions for this error, there exists a factor cot ϕ, and with an uncontrolled grazing contact (with $\phi = 0$, or nearly so) the error is indefinitely large. After all, with most earth satellites which are not intended for re-entry, it is impossible to predict the life-time with any accuracy, let alone the point of impact! Of course with manned vehicles it is necessary to limit the angle ϕ in order to limit deceleration during descent, so that there is an inescapable and inherent lack of accuracy in predicting their impact, unless they are controlled – by variable drag, or by use of lift (as pointed out in other papers presented this session).

Etude théorique des trajectoires optimales dans un champ de gravitation. Application au cas d'un centre d'attraction unique

Par

P. Contensou[1]

(Avec 10 Figures)

Résumé — Zusammenfassung — Abstract

Etude théorique des trajectoires optimales dans un champ de gravitation. Application au cas d'un centre d'attraction unique. Un système étant défini par n paramètres $x_1, x_2 \ldots x_n$ comprenant des positions, des vitesses, le temps et la masse de combustible dépensé, une trajectoire est une suite unidimensionnelle de valeurs des x_i. Elle est représentée par une courbe dans l'espace x_i. Si l'évolution du système n'est pas entièrement définie par des conditions initiales, on dira qu'il est doué de manoeuvrabilité. On peut alors distinguer dans l'espace x_i une zone accessible et une zone inaccessible. La frontière de la zone accessible est le lieu des points accessibles infiniment voisins de points inaccessibles. Une trajectoire optimale est une trajectoire qui passe par un point de cette frontière. La manoeuvrabilité de l'engin est dite canonique si le vecteur vitesse (c'est-à-dire les $n - 1$ dérivés des x_i par rapport à l'un d'entre eux) peut être choisi arbitrairement dans un domaine dépendant des x_i qui définit la manoeuvrabilité élémentaire du système. On indique les méthodes mathématiques permettant de déterminer, à partir de la manoeuvrabilité élémentaire, les trajectoires optimales (la méthode proposée est quelque peu différente de la méthode de Mayer).

Theoretische Betrachtungen über Optimalbahnen in einem durch ein Gravitationszentrum bestimmten Gravitationsfeld. In einem von n Parametern $x_1, x_2 \ldots x_n$, die Lagen, Geschwindigkeiten, Zeit und Masse des abgegebenen Treibstoffes bedeuten, definierten System wird die Bahn durch eine Kurve im x_i-Raum repräsentiert. Wenn die Entwicklung dieses Systems durch die Anfangsbedingungen nicht vollkommen festgelegt ist, soll dieses System als manövrierbar bezeichnet werden. Man kann den x_i-Raum in eine erreichbare und eine unerreichbare Zone einteilen. Eine Optimalbahn ist nun eine Bahn, die einen Punkt der Grenze zwischen den beiden Zonen berührt. Die Manövrierfähigkeit des Fahrzeuges wird als kanonisch bezeichnet, wenn der Geschwindigkeitsvektor (bzw. die $n - 1$ Ableitungen der x_i) auf einer Fläche in Abhängigkeit von den x_i beliebig gewählt werden kann, wodurch die elementare Manövrierfähigkeit des Systems definiert wird. Die mathematischen Methoden, die die Bestimmung der Optimalbahnen auf der Basis der elementaren Manövrierfähigkeit gestatten, werden angeführt. Die vorgeschlagene Methode unterscheidet sich etwas von der Mayerschen Methode.

Theoretical Study of Optimal Trajectories in a Gravitational Field. Application in the Case of a Single Centre of Attraction. In a system defined by n parameters $x_1, x_2 \ldots x_n$ comprising positions, speeds, time and mass of expended fuel, the trajectory is a series of one-dimensional values of x_i. It is represented by a curve in

[1] Directeur Technique Adjoint de l'O.N.E.R.A., Paris, France.

x_i space. If the evolution of the system is not completely defined by the initial conditions, it is said to be maneuverable. One can then distinguish an accessible zone and an inaccessible zone in x_i space. The border of the accessible zone is the locus of the accessible points, infinitely close to inaccessible points. An optimal trajectory is a trajectory which touches a point of this border. The maneuverability of the vehicle is said to be canonical if the speed vector (i.e. the $n - 1$ derivatives of the x_i in relation to any one of these) can be arbitrarily selected in an area dependent on the x_i which defines the elementary maneuverability of the system. The mathematical methods which permit the determination of the optimal trajectories on the basis of the elementary maneuverability are indicated (the proposed method is slightly different from the Mayer method).

I. Généralités sur les trajectoires optimales

1. Les notions développées ci-dessous ne sont pas particulières au cas du véhicule spatial. Elles s'appliquent à un système évolutif quelconque, doté d'un nombre fini de degrés de liberté, que nous désignerons par le terme général de mobile.

Supposons le mobile défini par les valeurs de $(n + 1)$ paramètres $x_1\, x_2 \ldots x_n + 1$ parmi lesquels peuvent figurer des grandeurs cinématiques (positions, vitesses, la masse, le temps, etc.). Ces valeurs définissent un point d'un espace ε_{n+1} à $n + 1$ dimensions.

Parmi ces $n + 1$ paramètres nous pouvons en choisir un, x_{n+1} par exemple, comme variable indépendante pour décrire le mouvement. Pour fixer le langage, nous donnerons ce rôle à la variable temps t, mais c'est un choix qui n'a rien d'indispensable. Le système est donc représenté, soit par un point P de l'espace $\mathscr{E}_n(x_1\, x_2 \ldots x_n)$, affecté d'une valeur de t, soit par un point de l'espace $\mathscr{E}_{n+1}(x_1\, x_2\, x_n\, t)$. Une "trajectoire $\mathscr{C}$" est une suite unidimensionnelle de points P de $\mathscr{E}_n$ ou $\mathscr{E}_n\, t$ qui peut se définir par n fonctions

$$x_i = f_i(t)$$

Le vecteur "vitesse" V au point $P(x_{i1}\, t)$ a par définition pour composantes, les n valeurs $d_{x1}/dt, \ldots d_{xn}/dt$. On peut lui associer un point de l'espace hodographe $\mathscr{H}$. Les mots trajectoires et vitesses sont pris dans un sens plus général que celui de la Mécanique.

2. Si le mouvement du mobile est déterminé, à un point initial $P_0(x_i^0, t)$ correspond une seule trajectoire.

Le cas qui va nous occuper est au contraire celui où le mouvement est partiellement indéterminé. Nous dirons alors que le mobile est doué de *manoeuvrabilité*. Les conditions initiales étant données une fois pour toutes, nous sommes amenés à distinguer:

a) un ensemble de trajectoires $\mathscr{C}$ permises, c'est-à-dire compatible avec la manoeuvrabilité, l'ensemble complémentaire étant celui des trajectoires interdites.

b) pour une valeur donnée de t, l'ensemble des points de $\mathscr{E}_n$ auxquels conduit au moins une trajectoire permise et que nous appellerons, $\mathscr{A}(t)$, *domaine accessible* au temps t. L'ensemble complémentaire constitue le domaine inaccessible.

c) pour chaque domaine accessible $\mathscr{A}(t)$, nous définirons sa *frontière* $\mathscr{F}(t)$, ensemble des points accessibles infiniment voisins de points inaccessibles. Les besoins du calcul infinitésimal nous amèneront à définir une *frontière au sens large* $\mathscr{F}'(t)$ qui inclut la précédente: c'est l'ensemble des points accessibles P infiniment voisins de points auxquels on ne peut accéder par aucune trajectoire permise infiniment voisine de la trajectoire conduisant à P (Si plusieurs trajectoires conduisent à P, la propriété énoncée doit s'appliquer au moins à l'une d'entre elles).

d) lorsque t varie, le domaine $\mathscr{A}(t)$ engendre un domaine $\mathscr{A}$ de l'espace $\mathscr{E}_{n+1}$ qui ne dépend que des conditions initiales et de la manoeuvrabilité du mobile et qui est, de façon absolue, le domaine accessible. La frontière $\mathscr{F}$ est évidemment engendrée par la frontière $\mathscr{F}(t)$.

e) le point initial P_0 fait évidemment partie de la frontière de $\mathscr{A}$ puisque $\mathscr{A}(t_0) \equiv \mathscr{F}(t_0) \equiv P_0$.

3. L'importance pratique de la frontière $\mathscr{F}$ réside dans le fait qu'elle inclut tous les points intéressants dans l'utilisation du mobile. Il suffit pour le montrer d'admettre qu'à l'un, au moins, des paramètres (x_i, t). x_1 par exemple, se rattache une notion de valeur, c'est-à-dire que l'on préfère, toutes choses égales d'ailleurs, le voir le plus petit ou le plus grand possible. Il est bien évident alors qu'à un point P intérieur à $\mathscr{A}$ on préfèrera un point P' voisin qui aura les mêmes coordonnées $(x_2 \ldots x_n\, t)$ avec un x_1 meilleur.

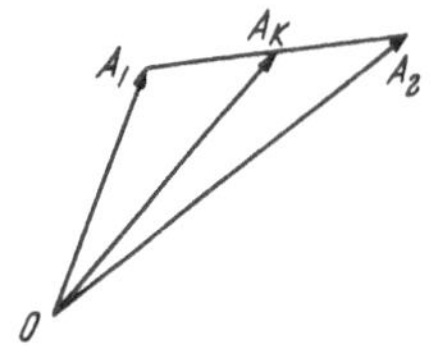

Fig. 1

Par définition, nous appellerons donc *trajectoire optimale* une trajectoire conduisant à un point de la frontière. Une trajectoire optimale peut ne contenir qu'un ou plusieurs points frontières. Elle est alors *localement optimale*. Elle peut aussi appartenir tout entière à la frontière. Elle est alors *globalement optimale*.

4. Pour aller plus loin, il est nécessaire de préciser nos idées sur la manière dont peut être définie la manoeuvrabilité du mobile.

Considérons toutes les trajectoires permises passant par un point $P(x_i\, t)$ de $\mathscr{E}_{n+1}$. A chacune est associée en ce point un vecteur vitesse V. L'ensemble de ces vecteurs définit un domaine de l'espace hodographe $\mathscr{H}_n$, fonction de P, que nous dénommerons $V(P)$. Nous disons que la *manoeuvrabilité du mobile est du type canonique, par rapport au système de variable $x_i\, t$, si le domaine $\vec{V}(P)$ est une donnée du problème et que le choix de V est entièrement libre à chaque instant parmi les éléments de $V(P)$.* La liberté de ce choix implique en particulier qu'il soit indépendant des choix antérieurs, donc que le vecteur $\vec{V}$ ne soit astreint à aucune espèce de continuité.

Il est bien évident que la propriété précédente est indépendante du rôle particulier que nous avons attribué à la variable t.

Une manoeuvrabilité canonique est entièrement définie par la donnée du domaine $V(P)$ ou $V(x_1 \ldots x_n\, t)$ que nous appellerons le *domaine de manoeuvrabilité.*

5. Une propriété importante du domaine de manoeuvrabilité est d'être obligatoirement *convexe*, c'est-à-dire de contenir tout segment de droite dont il contient les extrêmités. On s'en convaincra en remarquant que l'utilisation successive des vecteurs vitesses $\overrightarrow{O\,A}$ et $\overrightarrow{O\,A_2}$ pendant les instants infinement petits dt_1 et dt_2, est exactement équivalente à l'utilisation d'un certain vecteur $\overrightarrow{O\,A_K}$ pendant le temps $dt = dt_1 + dt_2$, A_K étant un point convenablement choisi du segment $A_1\,A_2$. Le point A_K doit donc être considéré comme appartenant à V si A_1 et A_2 lui appartiennent.

Si le domaine V tel qu'il résulte de l'analyse des possibilités mécaniques du système n'est pas naturellement convexe, il faudra donc le compléter en l'habillant du plus petit solide convexe circonscrit.

6. Le problème fondamental de la théorie des trajectoires optimales consiste à déterminer ces trajectoires connaissant le domaine de manoeuvrabilité. Le calcul infinitésimal va nous permettre de déterminer tout au moins les trajectoires optimales au sens large, une discussion ultérieure étant nécessaire dans chaque cas particulier, pour ne conserver parmi elles que les trajectoires optimales au sens strict.

Considérons donc une trajectoire $\mathscr{C}$ conduisant du point initial (P_0, t_0) à un point frontière (P_1, t_1). Nous devons exprimer qu'il existe des points voisins de P_1 non accessibles par des trajectoires permises voisines de τ.

De

$$\overrightarrow{OP} = \int_{t_0}^{t_1} \vec{V} \cdot dt$$

on tire d'abord

$$\overrightarrow{\delta P} = \int_{t_0}^{t_1} \overrightarrow{\delta V}\, dt$$

Cette formule montre qu'il suffit qu'un $\overrightarrow{\delta V}$ quelconque soit permis pendant un intervalle de temps quelconque $(t_2\, t_3)$ pour qu'on puisse obtenir un δP quelconque en choisissant

$$\overrightarrow{\delta V} = \frac{\overrightarrow{\delta P}}{t_3 - t_2}$$

dans l'intervalle t_2, t_3 et

$$\overrightarrow{\delta V} = 0$$

en dehors de cet intervalle.

7. *Une condition nécessaire pour qu'une trajectoire soit optimale est donc que $\vec{V}$ soit choisi à tout instant sur la frontière du domaine V.*

Cette condition étant supposée remplie, et en introduisant une représentation paramétrique de la frontière de V en fonction de p paramètres λ_j, nous pourrons poser

$$\dot{x}_i = g_i\,(x_1\, x_2 \ldots x_n\, t,\ \lambda_1\, \lambda_2\, \lambda_p) \tag{1}$$

les λ_j étant des fonctions entièrement arbitraires du temps. Le nombre p est au plus égal à $n - 1$. Il peut être plus petit si le domaine V est contenu dans un sous-espace linéaire de $\mathscr{H}_n$.

Affectons aux $\lambda_j(t)$ des variations $\delta\lambda_j(t)$ autour de la fonction λ_j qui correspond à la trajectoire τ. Les variations des x_i vont se trouver définies par le système linéaire

$$\frac{d}{dt}\,\delta x_i = \sum_K \frac{\partial g_i}{\partial x_K}\,\delta\, x_K + \sum_l \frac{\partial g_i}{\partial \lambda_j}\,\delta\lambda_j \tag{2}$$

Considérons d'autre part un vecteur $\vec{\xi}(t)$ dont les composantes ξ_i sont assujetties à vérifier le système adjoint de (2)

$$\frac{d\xi_i}{dt} = - \sum_K \frac{\partial g_K}{\partial x_i} \cdot \xi \tag{3}$$

Nous pouvons nous donner arbitrairement la valeur ξ^1 d'un tel vecteur au temps t_1.

Compte-tenu de (2) et (3), on établit immédiatement la relation

$$\frac{d}{dt}(\overrightarrow{\xi}\,\overrightarrow{\delta P}) = \sum_j \delta\lambda_j \sum_i \frac{\partial g_i}{\partial \lambda_j} \xi_i$$

qui, intègrée de t_0 à t, et compte-tenu de $\delta P_0 \equiv 0$ donne

$$\overrightarrow{\xi_1}\,\overrightarrow{\delta P_1} = \sum_l \int_{t_0}^{t_1} \sum_i \xi_i \frac{\partial g_i}{\partial \lambda_j} \times \delta\lambda_j \cdot dt$$

$\overrightarrow{\xi^1}$ pouvant être quelconque, cette formule montre la possibilité d'obtenir, par un choix convenable des $\delta\lambda_j$, un vecteur δP de projection donnée sur tout vecteur donné, c'est-à-dire finalement un vecteur δP arbitraire, excepté si pour un certain choix de $\overrightarrow{\xi}$, on a simultanément et pour toute valeur de t

$$\begin{aligned} &\sum_i \xi_i \frac{\partial g_i}{\partial \lambda_1} = 0 \\ &\sum_i \xi_i \frac{\partial g_i}{\partial \lambda_2} = 0 \\ &\cdots\cdots\cdots\cdots \\ &\sum_i \xi_i \frac{\partial g_i}{\partial \lambda_p} = 0 \end{aligned} \tag{4}$$

Finalement, les n fonctions inconnues x_i, les p fonctions inconnues λ_j, et les n fonctions inconnues auxiliaires ξ_i sont définies par:

$$\begin{aligned} &n \text{ équations:} \quad \frac{dx_i}{dt} = g_i(x_1\, x_2\, x_n\, t,\, \lambda_1\, \lambda_2\, \lambda_p) \\ &n \text{ équations:} \quad \frac{d\xi_1}{dt} = -\sum_K \frac{\partial g_K}{\partial x_i} \xi_K \\ &p \text{ équations:} \quad \sum_i \xi_i \frac{\partial g_i}{\partial \lambda_j} = 0 \end{aligned} \tag{5}$$

L'interprètation du vecteur $\overrightarrow{\xi}$ introduit comme inconnue auxiliaire est immédiate. Les équations (4) et le fait qu'elles entraînent $\overrightarrow{\xi_1}\,\overrightarrow{\delta P} = 0$ montrent que $\overrightarrow{\xi}$ est la normale commune, du point P à la frontière $\mathcal{F}'$ de $\mathcal{A}(t)$ et à la frontière de V.

Le système (5) ne faisant pas intervenir le temps t, il en résulte qu'une trajectoire optimale est optimale en tous points. *Lorsque la manoeuvrabilité est canonique, les trajectoires optimales sont globalement optimales.*

8. La définition de la manoeuvrabilité canonique peut paraître assez restrictive. En fait, on peut ramener à la manoeuvrabilité canonique le cas, beaucoup plus général où ce sont des dérivées supérieures des x_i qui jouent le rôle que nous avons attribué aux dérivées premières.

Supposons par exemple que la manoeuvrabilité soit définie par la possibilité de choisir le vecteur $\overrightarrow{\Gamma}(\ddot{x}_1\, \ddot{x}_2\, \ldots\, \ddot{x}_n)$ dans un domaine $g(x_i, \dot{x}_i', t)$. On se ramènera à un cas de maniabilité canonique en ajoutant au groupe de paramètre x_i leurs dérivées par rapport au temps. On posera donc

$$\dot{x}_i = u_i$$

Le vecteur vitesse a alors pour composantes

$$\dot{x}_1, \dot{x}_2, \dot{x}_3 \dots \dot{x}_n, \dot{u}_1 \dot{u}_2 \dots \dot{u}_n$$

Le domaine de manoeuvrabilité (ou plus exactement sa frontière), est défini par un système du type (1) qui s'écrit

$$\dot{x}_i = u_i$$
$$u_i = g_i (x_i \, u_i \, t \, \lambda_1 \, \lambda_2 \, \lambda_p)$$

et auquel on applique la méthode précédente.

Si on n'attache pas d'intérêt aux paramètres u_i, il faudra évidemment une fois le problème résolu, projeter le domaine accessible $\mathscr{A}$ de l'espace $(x_i \, u_i \, t)$ dans l'espace plus restreint $(x_i \, t)$. Les seules trajectoires optimales à retenir

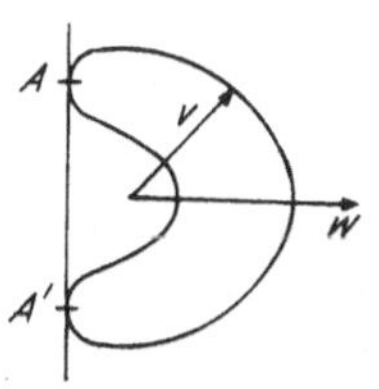

Fig. 2

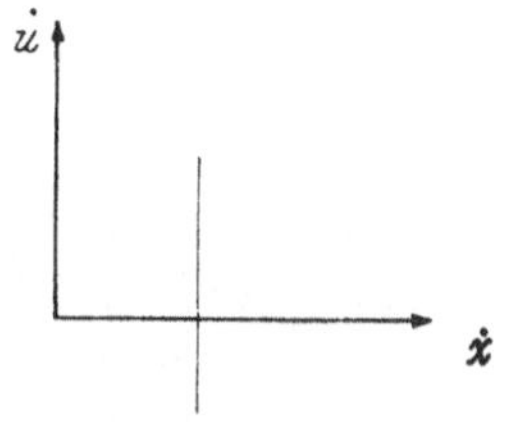

Fig. 3

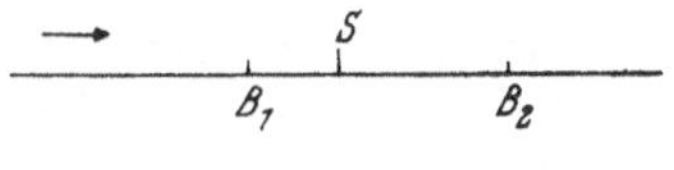

Fig. 4

sont celles qui conduisent à des points du contour apparent de $\mathscr{A}$, relatif à cette projection. Mais une telle propriété n'appartiendra en général qu'à des points particuliers de la trajectoire. Dans le champ des variables $(x_i \, t)$, les *trajectoires optimales ne seront donc que localement optimales.* Cette circonstance se produit chaque fois que l'on limite l'étude des trajectoires optimales à un sous-espace par rapport à celui qui correspond à la manoeuvrabilité canonique, exception faite du cas où la manoeuvrabilité serait encore canonique dans ce sous-espace.

9. Avant d'appliquer les notions précédentes à l'étude des trajectoires spatiales, éclairons-les de quelques exemples plus simples:

Un navire à voiles prend, pour une configuration de voilure déterminée et un vent déterminé W, un vecteur vitesse bien défini. En négligeant le temps d'établissement du régime, on peut lui affecter une manoeuvrabilité canonique dans l'espace de position $x \, y$. Le domaine V s'obtient en prenant l'enveloppe de tous les vecteurs V possibles, en tenant compte des diverses combinaisons de voilure. Mais un tel domaine doit être rendu convexe en le complètant de la bitangente $A \, A'$. Des combinaisons des vecteurs $O \, A$ et $O \, A'$ seront avantageusement utilisées pour remonter dans le vent, l'arc intérieur $A \, A'$ n'étant jamais utilisé.

Admettons que l'accélérateur et le frein d'une voiture permettent de choisir à chaque instant la valeur de l'accélération entre des limites pouvant dépendre de la vitesse, de l'espace et du temps. On aura donc:

$$\ddot{x} = f(\dot{x}, x, t, \lambda)$$

On se ramènera à une manoeuvrabilité canonique en introduisant le paramètre $\dot{x} = u$. Le domaine V est alors défini par:

$$\dot{x} = u$$
$$\dot{u} = f(x, u, t, \lambda)$$

Les trajectoires optimales sont faciles à déterminer. Globalement optimales dans l'espace x, u, t, elles ne sont que localement optimales dans l'espace x, t.

Il faut en conclure que si la voiture participe à une course, la tactique à suivre dépend de la position du poteau d'arrivée. On le vérifiera sur l'exemple suivant:

Supposons que la route soit barrée par un signal de circulation S, que la voiture trouve fermé. Si le but est en B_1, elle doit ralentir le plus tard possible pour arriver sans vitesse en S. Si le but est en B_2, elle doit s'arrêter à une distance suffisante de S, et repartir assez tôt avant l'ouverture du signal pour le franchir à vitesse maxima au moment précis de cette ouverture.

II. Application aux trajectoires spatiales

1. Nous allons appliquer les notions précédentes au véhicule spatial à propulsion chimique, dans le cadre des hypothèses ci-dessus.

Le mouvement est supposé plan. Les coordonnées du centre d'inertie de l'engin sont x et y, les composantes de sa vitesse u et v. Les seules forces appliquées sont une force de gravitation ne dépendant que du lieu et du temps, dont les composantes suivant les axes sont $m \cdot X(x, y, t)$ et $m \cdot Y(x, y, t)$ et la force propulsive arbitraire en grandeur et direction. La vitesse d'éjection w est parfaitement définie, mais elle peut être une fonction de la masse restante m, ce qui inclut le cas d'engins à étages successifs qui seront ainsi simplement caractérisés par le fait que certaines fractions de la masse sont éjectées à vitesse nulle. La manoeuvrabilité réside dans la possibilité d'éjecter la masse du propergol dans une direction arbitraire et avec un débit arbitraire, la vitesse d'éjection étant au contraire imposée. Ce schéma implique la possibilité d'émettre un débit infini, ce qui est naturellement irréalisable en pratique.

Si nous appelons λ l'angle de la poussée avec $O\,x$ et $\tau\, m$ son module (avec $\tau \geqslant 0$), les équations du mouvement s'écrivent:

$$\begin{aligned} \dot{x} &= u \\ \dot{y} &= v \\ \dot{u} &= X\,(x, y, t) + \tau \cos\lambda \\ \dot{v} &= Y\,(x, y, t) + \tau \sin\lambda \\ \dot{m} &= -\frac{m\,\tau}{w(m)} \end{aligned} \qquad (6)$$

Les dérivées sont prises par rapport au temps.

Les équations précédentes constituent la représentation paramétrique par λ et τ d'un domaine de manoeuvrabilité. La manoeuvrabilité est donc canonique dans l'espace x, y, u, v, m, t.

2. La dernière équation du système peut être simplifiée par le changement de variable

$$\varphi = \int_m^{m_0} w(m)\,\frac{dm}{m}$$

d'où

$$d\varphi = -\frac{w}{m}\,dm$$

Avec la nouvelle variable, le système s'écrit

$$\begin{aligned} \dot{x} &= u \\ \dot{y} &= v \\ \dot{u} &= X + \tau \cos \lambda \\ \dot{v} &= Y + \tau \sin \lambda \\ \dot{\varphi} &= \tau \end{aligned} \tag{7}$$

L'interprétation de φ est bien claire. Cette grandeur représente la vitesse qu'aurait acquise l'engin, par la même consommation de masse, en l'absence de tout champ de force. Du fait que φ est lié univoquement à la masse consommée, il peut servir à chiffrer la dépense de l'opération, indépendamment de la constitution et de la consommation spécifique des étages, dont la discussion sera entièrement indépendante de l'étude des trajectoires optimales. Il paraît utile de donnei un nom à la grandeur φ. Nous proposons celui de *vitesse latente*, et nous conviendrons de dire qu'un engin au départ a une réserve de vitesse latente qu'il épuise progressivement avec son propergol. Le signe adopté pour la définition de φ revient d'ailleurs à caractériser la situation à un instant donné par la *vitesse latente consommée depuis le départ*.

3. Les trajectoires optimales définies dans le cadre précédent, font intervenir à la fois la masse et le temps. Elles correspondent donc à la gamme des compromis possibles entre l'économie de masse et l'économie de temps. Si on ne s'intéresse qu'à l'économie de masse, on est conduit à rechercher les trajectoires optimales dans l'espace (x, y, u, v, φ) dont t est exclu. En divisant les quatre premières équations par la dernière, il vient:

$$\begin{aligned} \frac{dx}{d\varphi} &= \frac{u}{\tau} \\ \frac{dy}{d\varphi} &= \frac{v}{\tau} \\ \frac{du}{d\varphi} &= \frac{x}{\tau} + \cos \lambda \\ \frac{dv}{d\varphi} &= \frac{y}{\tau} + \sin \lambda \end{aligned} \tag{8}$$

Ce système montre que la maniabilité reste canonique *à condition que X et Y soient indépendants du temps*.

Les trajectoires optimales de l'espace x, y, u, v, φ sont les trajectoires économiques au sens habituel. Elles ne relèvent d'un problème autonome que dans le cas où la restriction précédente est satisfaite, ce que nous supposerons désormais.

Le problème qui sera traité dorénavant, sera celui des trajectoires optimales dans l'espace x, y, u, v, φ, la force de gravitation étant indépendante du temps (Le choix d'axes mobiles, assortis de forces massiques d'inertie venant s'ajouter aux forces massiques de gravitation, permet d'étendre la théorie à quelques cas simples de champs fonctions du temps).

4. Appliquons la théorie du § I.6. Soient ξ, η, ζ, χ, les quatre composantes du vecteur ξ suivant les directions $O\,x, O\,y, O\,u, O\,v$. Le système (5) prend la forme suivante, où les points désignent des *dérivées par rapport à* φ.

$$\left\{\begin{aligned} \dot{x} &= \frac{u}{\tau} \\ \dot{y} &= \frac{v}{\tau} \\ \dot{u} &= \frac{X}{\tau} + \cos\lambda \\ \dot{v} &= \frac{Y}{\tau} + \sin\lambda \end{aligned}\right. \tag{9}$$

$$\left\{\begin{aligned} -\tau\dot{\xi} &= X_x\zeta + Y_x\chi \\ -\tau\dot{\eta} &= X_y\zeta + Y_y\chi \\ -\tau\dot{\zeta} &= \xi \\ -\tau\dot{\chi} &= \eta \end{aligned}\right. \tag{10}$$

$$-\zeta\sin\lambda + \chi\cos\lambda = 0 \tag{11}$$

$$\text{et}\quad u\xi + v\eta + X\zeta + Y\chi = 0 \tag{12}$$

En dérivant (12) totalement par rapport à φ et compte tenu de (9) et (10), on trouve

$$\xi\cos\lambda + \eta\sin\lambda = 0 \tag{13}$$

Les relations (11) et (13), compte-tenu des deux dernières relations (10) définissent $\xi\,\eta\,\chi$ à un facteur près, soit

$$\begin{aligned} \xi &= \tau\dot{\lambda}\sin\lambda \\ \eta &= -\tau\dot{\lambda}\cos\lambda \\ \zeta &= \cos\lambda \\ \chi &= \sin\lambda \end{aligned} \tag{14}$$

Les valeurs (14) introduites dans l'expression (12) donnent

$$\tau\dot{\lambda}[u\sin\lambda + v\cos\lambda] + X\cos\lambda + Y\sin\lambda = 0 \tag{15}$$

Introduisons enfin les valeurs (14) dans les deux premières équations (10). Il vient

$$-(\tau\dot{\tau}\dot{\lambda} + \tau^2\ddot{\lambda})\sin\lambda - \tau^2\dot{\lambda}^2\cos\lambda = X_x\cos\lambda + Y_x\sin\lambda$$

$$(-\tau\dot{\tau}\dot{\lambda} + \tau^2\ddot{\lambda})\cos\lambda - \tau^2\dot{\lambda}^2\sin\lambda = X_y\cos\lambda + Y_y\sin\lambda$$

dont une combinaison s'écrit

$$-\tau^2\dot{\lambda}^2 = X_x\cos^2\lambda + (X_y + Y_x)\sin\lambda\cos\lambda + Y_y\sin^2\lambda \tag{16}$$

La quantité $\tau\dot{\lambda}$ se trouve donc définie par deux équations différentes qui n'ont aucune raison d'être compatibles. Sauf cas très particulier, dont l'étude devrait être approfondie, où les équations (14) et (15) seraient compatibles tout au long d'une trajectoire, on peut donc affirmer qu'il n'y a pas de trajectoire optimale correspondant à une valeur finie et non nulle de τ. La théorie générale qui suppose la liberté de variations $\delta\tau$ de signe contraire, ne s'applique pas aux valeurs zéro et l'infini de τ qui fourniront donc les trajectoires économiques. *Celles-ci se composeront donc d'arcs non propulsés et d'arcs correspondant à des poussées infinies*, c'est-à-dire à des discontinuités de u et v par rapport à la variable t.

5. Il reste évidemment à étudier la manière de choisir les points où la propulsion doit être mise en oeuvre et la direction de la poussée, de manière à obtenir un effet cinématique donné avec la consommation minima de vitesse latente.

Nous aborderons le problème en introduisant *un changement de système de coordonnées dans l'espace* (x, y, u, v).

Le nouveau système fera jouer un rôle privilégié *aux trajectoires libres* dans le champ de gravitation X, Y.

Les équations de la trajectoire libre s'écrivent:

$$\frac{dx}{dt} = u$$

$$\frac{dy}{dt} = v$$

$$\frac{du}{dt} = X$$

$$\frac{dv}{dt} = Y$$

D'après la théorie des équations différentielles, ce système admet trois intégrales premières indépendantes que l'on peut écrire:

$$\begin{aligned} A &= f(x, y, u, v) \\ B &= g(x, y, u, v) \\ C &= h(x, y, u, v) \end{aligned} \quad (17) \quad \text{avec} \quad \begin{aligned} u f_x + v f_y + X f_u + Y f_v &= 0 \\ u g_x + v g_y + X g_u + Y g_v &= 0 \\ u h_x + v h_y + X h_u + Y h_v &= 0 \end{aligned} \quad (18)$$

A, B, C sont des constantes pour une trajectoire déterminée. Ce sont donc trois grandeurs que l'on peut prendre comme variables de définition de la trajectoire elle-même.

Donnons-nous d'autre part une fonction quelconque de (x, y, u, v)

$$\mu = k(x, y, u, v) \quad (19)$$

assujettie à la seule condition

$$u \mu_x + v \mu_y + X \mu_u + Y \mu_v \neq 0$$

en tout point.

Les grandeurs A, B, C, μ constituent un nouveau système de coordonnées de l'espace x, y, u, v.

6. *Si les trajectoires libres sont fermées*, le paramètre μ doit être logiquement éliminé de la qualification du résultat cinématique recherché. Peu importe qu'une orbite soit atteinte en un point ou en un autre, puisque de toute façon, elle sera parcourue un nombre infini de fois. Nous sommes alors conduits à traiter le problème des trajectoires optimales dans le *sous-espace* (A, B, C, φ) de l'espace initial x, y, u, v, φ.

La manoeuvrabilité dans ce sous-espace est définie en dérivant les formules (17) par rapport à φ, soit:

$$\dot{A} = f_x \dot{x} + f_y \dot{y} + f_u \dot{u} + f_v \dot{v}$$

etc.

Compte-tenu de (18) et de (8), ces expressions se réduisent à

$$\begin{aligned} \dot{A} &= f_u \cos \lambda + f_v \sin \lambda \\ \dot{B} &= g_u \cos \lambda + g_v \sin \lambda \\ \dot{C} &= h_u \cos \lambda + h_v \sin \lambda \end{aligned} \quad (20)$$

Il faut imaginer que dans ces formules, x, y, u, v sont remplacés par leurs valeurs en fonction de A, B, C, μ que fournit l'inversion des formules (17) et (19). Les expressions (20) donnent donc $\dot{A}$, $\dot{B}$, $\dot{C}$, en fonction de A, B, C, λ et μ. Or,

μ peut au même titre que λ, être considéré comme un paramètre arbitraire puisqu'on est libre du point de la trajectoire $(A\ B\ C)$ où on exerce l'impulsion de direction λ. *Les formules* (20) *définissent donc un domaine de manoeuvrabilité canonique dans l'espace* $(A\ B\ C\ \varphi)$.

Il faut noter que le paramètre τ a disparu de cette définition. En fait, il est étroitement lié à μ comme on le voit en dérivant (19)

$$\frac{d\mu}{d\varphi} = (k_x u + k_y v + k_u X + k_v Y) \frac{1}{\tau} + k_u \cos\lambda + k_v \sin\lambda \qquad (21)$$

7. Si les trajectoires ne sont pas fermées, il n'est pas certain que l'utilisation se désintéressera de μ, qui fixe la portion d'une trajectoire qui sera effectivement parcourue. D'autre part, μ n'est plus absolument arbitraire: on ne peut utiliser que les points de la trajectoire situés sur cette même portion. Cette restriction se traduit d'ailleurs par une condition concernant $d\mu/d\varphi$ qui découle de (21) où l'on doit avoir $\tau > 0$.

Ces restrictions n'empêchent pas cependant d'utiliser la méthode précédente et d'en retenir les résultats, s'il se trouve que les trajectoires trouvées sont effectivement réalisables.

Nous nous bornerons, dans les applications qui vont suivre, à traiter des trajectoires économiques dans l'espace $(A\ B\ C\ \varphi)$.

Cet espace étant un sous-espace de (x, y, u, v, φ), les trajectoires économiques en question sont prélevées parmi l'ensemble des trajectoires optimales de (x, y, u, v, φ). Elles sont donc caractérisées par l'utilisation exclusive de poussées infinies.

III. Application au cas d'un centre d'attraction unique

A. Trajectoires elliptiques

1. Nous supposons la force d'attraction centripète et égale à K/μ^2.

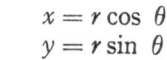

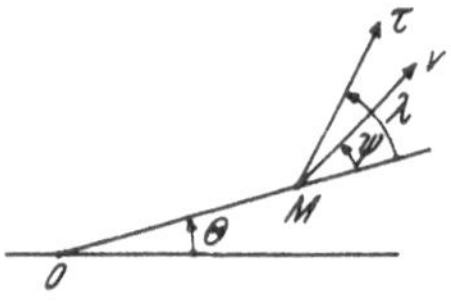

Fig. 5

Nous posons, M étant le point courant de la trajectoire:

$$x = r\cos\theta$$
$$y = r\sin\theta$$

La vitesse V fait un angle ψ avec $O\,M$.

Nous conviendrons désormais de noter λ *l'angle de la poussée avec le vecteur* $\overrightarrow{O\,M}$.

Comme grandeurs A, B, C caractéristiques de la trajectoire, nous choisirons les suivantes:

$$A = V\sin\psi = v\,x - u\,y$$

est le double de la *vitesse aréolaire*.

B caractérise l'energie et a les dimensions d'une vitesse. Nous poserons:

$$-\frac{B^2}{2} = \frac{V^2}{2} - \frac{K}{r}$$

C est l'angle que fait le grand axe de l'orbite avec Ox. On établit facilement:

$$C = \operatorname{Arc\,tg}\frac{y}{x} + \operatorname{Arc\,tg}\frac{(v\,x - u\,y)^2 - K\,r}{(u\,x + v\,y)\,(v\,x - u\,y)}$$

Pour appliquer les formules (20), nous avons à calculer les dérivées partielles de A, B, C par rapport à u et v. Un seul calcul est un peu compliqué, c'est celui qui est relatif à C. On le simplifie en remarquant qu'on peut placer les axes de manière à avoir $y = 0$ au point considéré. Dans ce cas:

$$C = \operatorname{Arc\,tg} \frac{v^2 - (K/r)}{u\,v}$$

L'expression de $\dot{A}$, $\dot{B}$, $\dot{C}$ se simplifie avec les conventions de notation suivantes:

a) On prend pour paramètre μ la quantité:

$$\mu = \frac{1}{r} \frac{A}{B}$$

b) On pose par ailleurs, pour simplifier l'écriture

$$\alpha = \frac{K}{A\,B} \qquad \text{(on a } \alpha \leqslant 1\text{)}$$

Avec ces notations on a, en particulier, les relations

$$V \sin \psi = \frac{A}{r} = B\,\mu$$

$$V \cos \psi = \varepsilon \sqrt{\frac{2\,K}{r} - B^2 - \frac{A^2}{r^2}} = B\,\varepsilon \sqrt{2\,\alpha\,\mu - \mu^2 - 1} \qquad \text{avec } \varepsilon = \pm 1$$

On trouve pour $\dot{A}\,\dot{B}\,\dot{C}$ les expressions suivantes:

$$\begin{aligned} X &= \dot{A} \frac{B}{A} = \frac{1}{\mu} \sin \lambda \\ Y &= -\dot{B} = \varepsilon \cos \lambda \sqrt{2\,\alpha\,\mu - \mu^2 - 1} + \mu \sin \lambda \\ Z &= \dot{C}\,B \sqrt{\alpha^2 - 1} = \cos \lambda\,(\alpha - \mu) + \varepsilon \sin \lambda \sqrt{2\,\alpha\,\mu - \mu^2 - 1} \left(1 + \frac{\alpha}{\mu}\right) \end{aligned} \tag{22}$$

(X, Y, n'étant naturellement pas à confondre avec les projections de la force, dont nous ne ferons plus usage).

Le domaine de manoeuvrabilité V_E défini par les formules (22) ne dépend, aux affinités près qui résultent des facteurs du premier membre, de la seule combinaison α de A, B. Il constitue donc une famille de surfaces à 1 paramètre α (λ et μ étant les paramètres servant à décrire la suiface).

2. Les propriétés de transfert d'une orbite elliptique à une orbite voisine sont entièrement commandées par la surface V_E.

La surface V_E admet le plan $X\,O\,Y$ comme plan de symétrie, et l'axe $O\,Z$ comme axe de symétrie. Elle affecte la forme générale d'un diabolo creux à paroi épaisse. Sa trace sur le plan $X\,O\,Y$ est une courbe à 4 points anguleux inscrite dans un rectangle.

Pour rendre le domaine V_E convexe, il est nécessaire de le compléter de deux surfaces réglées développables ε_1, et ε_2 (ε_2 étant répétée par symétrie).

Ces deux surfaces jouent des rôles différents: ε_1 habille une surface latérale existante, tandis que ε_2 obture le trou central du diabolo. Le bord du diabolo n'étant pas à arête vive, il existe entre ε_1 et ε_2 une petite bande de la surface V_E qui est respectée par ε_1 et ε_2.

Si nous traçons un vecteur proportionnel à dA, dB, dC, acroissements de A, B, C désiré, les circonstances suivantes pourront se produire:

Si le vecteur perce ε_2, il faudra utiliser successivement des points tels que P, Q.

Si le vecteur perce ε_1, il faudra utiliser successivement des points tels que M et N. (A noter qu'il existe une possibilité de transfert par une seule opération, mais qu'il est moins avantageux.)

Si le vecteur perce ce qui subsiste de V_E, il faudra utiliser un point tel que R.

On voit finalement que les points de V_E utilisés pour le transfert sont confinés dans une région très limitée de cette surface.

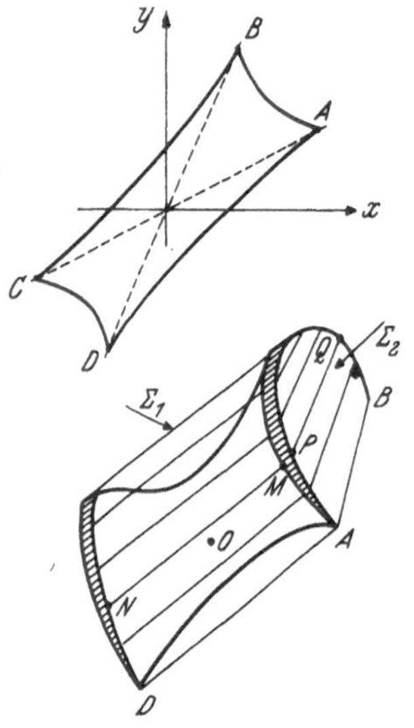

Fig. 6

3. Un cas particulièrement simple est celui où on ne s'intéresse qu'à la forme des orbites et non à leur orientation. Le paramètre C n'intervient donc pas. Or, il se trouve que C ne figure pas dans les deux premières équations (22). La manoeuvrabilité est donc canonique dans l'espace (A, B, φ). Le domaine V_E se réduit à sa trace sur le plan $X\,O\,Y$, qui est entièrement concave. Les seuls points à utiliser pour le transfert sont les points A, B, C, D qui correspondent aux apogées et périgées de l'orbite avec $\lambda = \pi/2$. On retrouve un résultat bien connu et la discussion de la manoeuvre à effectuer pour un transfert quelconque peut s'achever sans difficulté.

4. Les cas d'un transfert quelconque d'orbite elliptique à orbite elliptique fait évidemment appel, sur chaque surface V_E, à la zone limitée que nous venons de définir. Mais au cours d'une dépense finie de vitesse latente, la surface V_E se déforme. Les déplacements du point figuratif sont imposés par la constance de λ et l'évolution de μ avec A et B, Γ restant constant. Quand cette évolution amène le point sur le bord d'une surface ε_1 ou ε_2, il saute brusquement à l'autre extrémité de la bitangente.

B. *Trajectoires hyperboliques*

1. Il y a peu de chose à changer aux équations (22) pour l'adapter au cas d'une orbite hyperbolique.

Posons maintenant:

$$\frac{B^2}{2} = \frac{V^2}{2} - \frac{K}{2}$$

B a une interprétation physique simple. C'est la vitesse à l'infini.

Les quantités α et μ conservent la même définition avec la nouvelle signification de B.

Les formules (22) sont à remplacer par les suivantes:

$$\begin{aligned} X &= \dot{A}\,\frac{B}{A} = \frac{1}{\mu}\sin\lambda \\ Y &= \dot{B} = \varepsilon\sqrt{1 + 2\,\alpha\mu - \mu^2}\cos\lambda + \mu\sin\lambda \\ Z &= \dot{C}\,B\sqrt{1+\alpha^2} = (\alpha-\mu)\cos\lambda + \varepsilon\sqrt{1+2\,\alpha\mu-\mu^2}\left(1+\frac{\alpha}{\mu}\right)\sin\lambda \end{aligned} \tag{23}$$

La surface V_H ainsi définie, a la propriété importante de présenter des *nappes infinies*. En effet, Γ peut maintenant devenir infini, donc μ s'annuler.

X et Z tendent alors vers l'infini, le rapport Z/X tendant vers la valeur $\pm \propto$. (La valeur $- \propto$ correspond à r infini amont.)

Ce fait correspond évidemment à la possibilité, pour le mobile à l'infini amont ou aval, de faire varier la direction de l'asymptote de sortie ou d'entrée sans dépense appréciable de vitesse latente. Une abatée d'angle très faible, effectuée très à l'avance, modifie en effet simultanément A et C sans consommation de masse, la direction de l'asymptote d'entrée restant évidemment inchangée.

2. Le seul problème que nous examinerons concerne le transfert entre orbites hyperboliques.

Considérons un mobile venant de l'infini et se dirigeant vers l'astre attirant. Pour simplifier, nous considèrons celui-ci comme *ponctuel*, les correctifs à apporter pour tenir compte de son rayon étant examinés par la suite.

Le mobile peut profiter de son passage à proximité de l'astre:

Soit pour acquérir gratuitement un gain de vitesse quelconque à condition d'accepter une déviation de 180°.

Soit pour effectuer un changement de direction quelconque sans augmentation de vitesse.

Dans le premier cas, il doit passer infinement près de l'astre et profiter de sa survitesse infinie pour opérer une impulsion infiniment petite que lui communiquera un supplément fini d'énergie.

Dans le second, il suffira qu'il règle sa vitesse aréolaire à une valeur convenable.

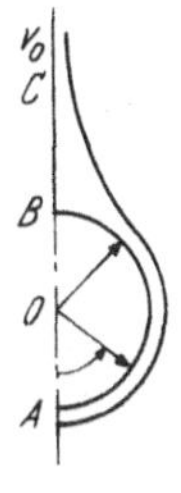

Fig. 7

Si le mobile désire profiter du passage à proximité de l'astre pour acquérir au meilleur compte un supplément de vitesse déterminé et une déviation déterminée quelle doit être sa manoeuvre?

Admettons que l'opération optimale fasse appel à une seule opération propulsive et cherchons à la définir.

3. Il n'est pas licite dans le cas présent de rendre globalement convexe le volume V_H en raison des restrictions qui ont été indiquées à propos des trajectoires non fermées. Il est bien évident par exemple, que l'utilisation d'un point à l'infini de V_H dans la direction $1, 0, \propto$, exclut tout point ultérieur puisqu'elle correspond à $t = + \infty$.

Soit H_- le point à l'infini dans la direction $1, 0, - \propto$ utilisable pour $t = - \infty$.

H_+ le point à l'infini dans la direction $1, 0, \propto$ utilisable pour $t = + \infty$.

K_1 le premier point utilisé après H_-

K_2 le dernier point utilisé avant H_+.

Pour que la combinaison $H_- K_1$ ne soit pas surclassable, il faut évidemment que K_1 fasse partie du contour apparent de V_H suivant la ligne de visée H_- de même pour $H_+ K_2$.

Finalement, *s'il n'y a qu'une seule impulsion motrice, elle commencera en un point de V_H dont le plan tangent contient H_- et finira en un point dont le plan tangent contient H_+.*

La condition précédente s'exprime par la nullité du déterminant.

$$\begin{vmatrix} \dfrac{\partial x}{\partial \lambda} & \dfrac{\partial y}{\partial \lambda} & \dfrac{\partial z}{\partial \lambda} \\ \dfrac{\partial x}{\partial \mu} & \dfrac{\partial y}{\partial \mu} & \dfrac{\partial z}{\partial \mu} \\ 1 & 0 & \eta \alpha \end{vmatrix} = 0 \qquad \text{avec } \eta = \pm 1$$

Le calcul aboutit à la condition suivante

$$-\operatorname{tg}^2\lambda\,\varepsilon\sqrt{1+2\alpha\mu-\mu^2}+2\mu\operatorname{tg}\lambda+\varepsilon\sqrt{1+2\alpha\mu-\mu^2}+\eta=0 \qquad (24)$$

$\varepsilon = 1$ signifie que le point s'éloigne du centre (vitesse radiale > 0). Il faut prendre $\eta = -1$ pour le point correspondant au début de la propulsion $\eta = +1$ pour la fin.

Pendant la propulsion, λ et r restent constants, α et μ passent de $\alpha_1 \mu_1$ à $\alpha_2 \mu_2$, et on a φ étant la vitesse latente dépensée.

$$\varphi\cos\lambda = \delta(V\cos\psi) = \varepsilon_2 B_2\sqrt{1+2\alpha_2\mu_2-\mu_2^2} - \varepsilon_1 B_1\sqrt{1+2\alpha_1\mu_1-\mu_1^2}$$

Mais par définition de α et μ

$$\alpha_1\mu_1 B_1^2 = \alpha_2\mu_2 B_2^2 = \frac{K}{r}$$

d'où

$$\varphi\cos\lambda\sqrt{\frac{r}{K}} = \varepsilon_2\sqrt{\frac{1-\mu_2^2}{\alpha_2\mu_2}+2} - \varepsilon_1\sqrt{\frac{1-\mu_1^2}{\alpha_1\mu_1}+2} \qquad (24)$$

de même de

$$\varphi\sin\lambda = \delta(V\sin\psi)$$

on tire

$$\varphi\sin\lambda\sqrt{\frac{r}{K}} = \sqrt{\frac{\mu_2}{\alpha_2}} - \sqrt{\frac{\mu_1}{\alpha_1}} \qquad (25)$$

Traçons un diagramme dans le plan des vitesses réduites

$$u\sqrt{\frac{r}{K}} = U \qquad \text{et} \qquad v\sqrt{\frac{r}{K}} = V$$

Cela vient à poser

$$U = \varepsilon\sqrt{\frac{1-\mu^2}{\alpha\mu}+2} \qquad \text{d'où} \qquad \varphi\sqrt{\frac{r}{K}}\cos\lambda = U_2 - U_1$$

$$V = \sqrt{\frac{\mu}{\alpha}} \qquad\qquad \varphi\sqrt{\frac{r}{K}}\sin\lambda = V_2 - V_1$$

On a aussi:

$$\frac{1}{\alpha\mu} = U^2 + V^2 - 2$$

$$\frac{\mu}{\alpha} = V^2$$

Dans le plan U, V l'équation (24) représente une hyperbole

$$[U(1-\operatorname{tg}^2\lambda) + 2V\operatorname{tg}\lambda]^2 = U^2 + V^2 - 2$$

Son axe fait avec $\overrightarrow{OV}$ l'angle $2\lambda - \pi$.

Deux points P_1 et P_2 de cette hyperbole se correspondent par la pente de $P_1 P_2$ qui doit être égalé à λ. Le triangle des vitesses $O P_1 P_2$ est figuré à l'échelle, pour un centre d'attraction situé sur la partie négative de $O U$. Comme on a $B^2 \;\; r/K = 1/\alpha\mu = U^2 + V^2 - 2$, la valeur initiale de B fixe le rayon $O P$, et le rayon $O P_2$ fixe sa valeur finale.

Quant à la déviation Δ, elle est donnée par la formule:

$$\Delta = \operatorname{Arctg}\frac{V_2^2-1}{U_2 V_2} - \operatorname{Arctg}\frac{V_1^2-1}{U_1 V_1} + \operatorname{Arctg}\alpha_1 + \operatorname{Arctg}\alpha_2$$

$$\left(\text{avec} \quad \frac{1}{\alpha^2} = \frac{U^2+V^2-2}{V^2}\right)$$

Traitons, à titre d'exemple, le cas où on arrive de l'infini sans vitesse ($B_1 = 0$) et où on se fixe une déviation nulle. On trouve que le triangle des vitesses a la configuration ci-dessous:

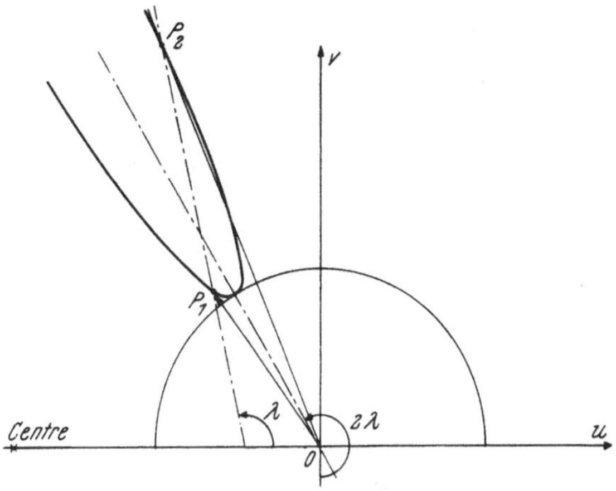

Fig. 8

La manoeuvre consiste donc à diriger le mobile droit sur l'astre attirant et à appliquer à un moment donné une poussée à 45° de la vitesse suffisante pour doubler la vitesse radiale. Si le point a été bien choisi, la vitesse à l'infini se retrouve parallèle à la direction de départ. On voit facilement que

$$B_2 \sqrt{\frac{r}{K}} = 2\sqrt{r}$$

tandis que

$$\varphi \sqrt{\frac{r}{K}} = 2$$

Le gain procuré par cette méthode pour accroître la vitesse est donc de $\sqrt{2}$.

Aperçu d'une généralisation

Parmi les trajectoires ainsi trouvées, certaines seront irréalisables. Ce sont toutes celles qui amèneraient à passer à une distance de l'astre inférieure à son rayon. Seront donc en fait les plus intéressants, les astres qui tolèrent un maximum de survitesse, c'est-à-dire ceux pour lesquels la quantité m/r est la plus grande.

Ceci posé, une approche possible des trajectoires économiques dans le champ d'un système d'astres quelconques consisterait à considérer que l'action de chacun est limitée à une région de l'espace assez petite pour qu'on puisse y

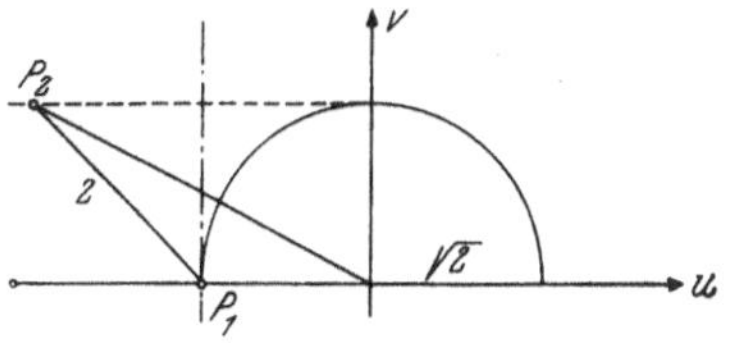

Fig. 9

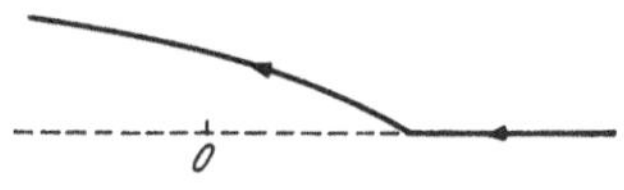

Fig. 10

négliger la variation de l'action des autres astres. Les trajectoires à comparer, dans le système solaire par exemple, seraient des arcs d'ellipses solaires venant rebondir sur les diverses planètes, la manoeuvre au voisinage de la planète étant optimisée par la théorie précédente. Ce point de vue met en évidence l'intérêt d'atteindre le plus tôt possible une planète à "puits de potentiel" très profond, véritable tremplin pour réaliser, à bon compte, de nouveaux accroissements de vitesse.

Discussion

M. Macé demande si l'optimisation étudiée par M. Contensou vise finalement à minimiser pour une mission donnée la masse initiale, l'énergie dépensée, ou la puissance du propulseur.

M. Ducrocq évoque le problème de la trajectoire optimale partant du sol et aboutissant à la mise en orbite d'un satellite. Il demande si la méthode exposée par M. Contensou fournit les conditions les plus avantageuses concernant l'angle de lancement, et la loi de variation et d'inclinaison de la poussée.

M. Contensou précise que les méthodes générales dont l'exposé constitue la première partie de sa communication sont applicables à n'importe quel problème d'optimisation. En revanche l'application aux trajectoires spatiales donnée dans la seconde partie se place dans un cadre beaucoup plus restreint. La variable que l'on cherche à économiser pour une mission donnée est bien, en gros, la masse. Plus précisément il faut se réferer à l'hypothèse fondamentale adoptée et qui est la suivante: on considère un engin dont tous les éléments sont définis à l'exception d'un seul: à savoir la loi suivant laquelle la masse sera éjectée en fonction du temps (la vitesse d'éjection étant par contre une fonction bien définie de la masse actuelle) et c'est cette loi $m = f(t)$ qu'il s'agit d'optimiser. Il est bien évident qu'il s'agit d'un problème schématisé dont la réalité s'écartera plus ou moins. Mais de telles généralisations sont indispensables si on veut atteindre des résultats d'une certaine généralité. Il appartient ensuite à l'ingénieur chargé d'un projet, dont de tels résultats guident utilement les tâtonnements, de pousser les choses plus loin à l'occasion du cas particulier qui l'intéresse.

La schématisation adoptée paraît fournir des résultats directement applicables au cas de la propulsion chimique. Une seconde approximation sera nécessaire pour définir au mieux l'arc de trajectoire propulsée, à durée non nulle et poussée limitée qui encadrera la percussion théorique. Cette schématisation ne convient absolument pas aux véhicules à propulsion électrique pour lesquels la puissance installée est l'élément prédominant et qui nécessitent une théorie différente.

En ce qui concerne les trajectoires de mise en orbite évoquées par M. Ducrocq, M. Contensou convient qu'il n'avait pas spécialement en vue ce problème. La proximité de la terre donne une importance primordiale à la nécessité pour les trajectoires d'être extérieures à la planète. De toutes façons si on ne tient pas compte de l'atmosphère, la solution du problème est bien connue: il convient de donner au départ une vitesse horizontale un peu supérieure à la vitesse circulaire de manière à se trouver, aux antipodes, à l'altitude de satellisation, et de donner à ce moment le supplément de vitesse qui fait passer à l'orbite circulaire.

Si on tient compte de l'atmosphère il est difficile de dire quelque chose de général, étant donnée la variété des formes et des lois de traînée.

Astrodynamics and Planetary Research

By

J. M. J. Kooy[1]

(With 19 Figures)

Abstract — Zusammenfassung — Résumé

Astrodynamics and Planetary Research. In the article the possibility of research of the nearest planets by means of unmanned space vehicles is discussed. In order to obtain an artificial satellite revolving around a planet, a retro-rocket to decelerate near the planet will be required. The problem to direct this retro-rocket has been considered as well as the problem of orbit observation after the action of the retro-rocket. By using a group of four vehicles moving as a cluster to the object planet and being decelerated one after the other, the possibility can be created to obtain all data required for the determination of the orbital motion after deceleration. At the same time, the four satellites can move in different orbit planes around the planet for scanning the spatial distribution of the planetary magnetic field and corresponding regions of radiation. By such a system of artificial satellites also the direct research of the spatial distribution of the high planetary atmosphere will become possible.

Raumfahrt und Planetenforschung. Es wird die Möglichkeit untersucht, wie die erdnächsten Planeten mittels Raumfahrzeugen erforscht werden können. Um einen künstlichen Mond um den Planeten in Umlauf zu bringen, ist eine Bremsrakete erforderlich, die die Bewegung in Planetennähe verzögert. Das Richtproblem der Bremsrakete wie auch die Frage, auf welche Weise die Bewegung nach dem Bremsen durch Beobachtung bestimmt werden kann, werden näher untersucht. Bei Verwendung von vier Raumfahrzeugen, die sich als Gruppe auf den Planeten zu bewegen und nacheinander abgebremst werden, besteht die Möglichkeit, die Bewegung jedes Fahrzeuges nach dem Abbremsen direkt zu bestimmen. Weiters können die Bahnebenen der Umlaufsbahnen der vier künstlichen Monde verschieden gewählt werden, wodurch die direkte Erforschung der räumlichen Verteilung des planetaren magnetischen Feldes und der zugehörigen Strahlungsgürtel sowie der räumlichen Verteilung der äußeren Atmosphäre des Planeten möglich wird.

L'astronautique et l'investigation des planètes. Dans cet article la possibilité de l'investigation des planètes au moyen des véhicules spatiaux sans équipage est discutée. Pour obtenir un satellite artificiel autour de la planète, il faut une fusée de frein pour ralentir près de la planète. Le problème de diriger la fusée de frein dans la direction nécessaire est considéré, comme le problème d'observer le mouvement du véhicule pendant et après l'action de la fusée de frein. Par l'usage d'un groupe de quatre véhicules qui mouvent ensemble vers la planète et qui sont ralentis l'un après l'autre, on peut créer la possibilité d'obtenir tous les renseignements nécessaires du mouvement après le ralentissement et en même temps d'obtenir quatre satellites, mouvant dans des plans divers, nécessaire pour l'investigation de la distribution spatiale du champ magnétique et des régions de radiation correspondant autour de la planète. Avec un tel système des satellites artificiels aussi l'investigation directe de la haute atmosphère de la planète deviendra possible.

[1] N.V.R., Lector of the Royal Military Academy, St. Ignatiusstraat 99 a, Breda, The Netherlands.

I. Introduction

In order to investigate the planets of our solar system by means of space vehicles it will be necessary, in a later state of the development of spaceflight, to establish artificial satellites around these planets. In a further future, these artificial satellites will be manned, but in a nearer future these research vehicles will be unmanned. The first planets which will be objects of research for such unmanned satellites will be Venus and Mars. Nevertheless, also these "near planets" are still very remote for the space flight standards of to-day and the problem must be solved how to telecommand a space vehicle at such an enormous distance.

It will be practically impossible to give the vehicle at the last burn-out-point, as last stage of the multi-step rocket, such initial conditions (hence space coordinates and corresponding speed components) that the vehicle in pure gravitational flight will be "catched" by the object planet as artificial satellite. Practically, it will be always necessary to decelerate it by artificial means, i.e. by a retro-rocket, when the vehicle has come into the neighbourhood of the target planet. Throughout the voyage from the last burn-out-point near the earth up to the environs of the target planet, the vehicle will freely tumble around its center of gravity. This rotational motion, corresponding with initial conditions depending on accidental small disturbances which cannot be foreseen in detail, will be unknown, so that the attitude of the vehicle at the end of the transit voyage will be also unknown. In order to telecommunicate the attitude of the vehicle with respect to the celestial sky a gyro stabilized system of reference (gyro table), mounted in the vehicle, will be required. By disturbances, mainly due to friction, such a gyro table will decline from its initial position about 1/10 degree per hour. Hence for a transit voyage of at least several weeks (in case of the nearest planet Venus) the gyro table cannot be used as non rotating material system of reference.

However, although the attitude of the vehicle at the end of the transit voyage will be unknown, it will be theoretically possible to bring the vehicle into a known attitude, by directing the roll axis parallel with the solar rays and further to use a conspicuous star (for example Sirius) for the complete determination of the position of the vehicle frame as to the celestial sky. Then, a gyro table which was firstly rigidly fixed as to the frame of the vehicle can be used as system of reference for further rotational motions of the vehicle which must be tele-commanded from the terrestrial command station. It will be obvious that, as to the tele-communication of the new attitudes of the vehicle and as to the telecommand of this attitude, difficult technical problems arise.

Further, it will be necessary to observe the orbital motion of the vehicle around the object planet, as obtained after deceleration of the retro-rocket. But in connection with the large distance and the relative smallness of the vehicle, optical detection will be excluded, so that we have to look for another solution. We can equip the vehicle with a radio transmitter of constant wave length and measure the wave length of the signal received at the terrestrial station. Then, by Doppler effect, we can measure the radial speed of the vehicle with respect to the terrestrial station. Considering the motion of the earth and of the object planet around the sun as known, the radial speed measured at a certain instant of time can be considered as a function of the 6 elliptic orbit elements which determine the Kepler motion of the artificial satellite around the object planet, even taking also into account the time of travel of the electromagnetic waves. Hence, by measuring the radial speed at 6 instants of time, we would theoretically obtain 6 relations in the 6 unknown elliptic orbit elements. However, these relations become utterly complex and the numerical determination of the 6 orbit elements from these relations will be scarcely possible. We can meet this difficulty by reversing the problem. If we are assuming a certain KEPLER motion around the planet, corresponding with certain initial conditions (hence with a certain radius vector $\bar{r}_0$ and speed $\bar{v}_0$ as to the planet-centric non-rotating system) we can compute which radial speed will be observed at different instants of time, thereby taking into account the time of signal travel. Doing this for different assumed KEPLER motions around the planet, we can try to interpolate by comparing the measured radial speeds with the computed values. However, in some cases we do not know which KEPLER motion will be most favourable for exploration. In case of Venus, the first object planet coming into consideration, we know nothing about the planetary rotation. If we like to scan the magnetic field which Venus certainly will have, it will be very suitable that the orbit plane of the satellite will approximately coincide with a meridian plane of this field. But because the axis of rotation of the planet is fully unknown, the axis of the magnetic field is fully unknown, too. Hence, it seems very desirable to make arrangements by which we are able to vary the KEPLER motion by telecommand.

Now, in order to meet this difficulty we can launch different vehicles to the object planet which, after deceleration by a retro-rocket become artificial satellites around the object planet. If then their orbit planes are quite different the scanning of the magnetic field and corresponding radiation belts can always be carried out in a satisfactory way. But using different vehicles, we can also meet the difficulty which arises when we wish — as we must do — to detect the KEPLER motions which are obtained after deceleration by the retro-rockets. For, as long as a vehicle is moving only acted upon by gravitational forces, we can compute the future positions with a high degree of precision. For these gravitational forces are only due to the celestial bodies coming into play and the motions of these celestial bodies may be considered as known. Hence, as long as a vehicle has not been decelerated by retro-rocket action, the positions as functions of time are known by celestial mechanics computations. Then, if we launch 4 vehicles with the same object planet as target, and if we equip each vehicle with a radio transmitter of constant wave length and a corresponding radio receiver, it will be possible to control the motion of one of the vehicles, during and after deceleration by retro-rocket thrust, if the three other vehicles continue to move undisturbed by retro-rocket thrust. The three other vehicles serve then only to measure the radial speed of the first vehicle, with respect to the other vehicle in question, by DOPPLER effect, and to telecommunicate these measurements in some code to the terrestrial station. If these measurements begin immediately at the beginning of the retro-rocket braking of the first vehicle, the motion of this first vehicle, during and after the retro-rocket deceleration, can be determined from the telecommunicated measurements of the three other vehicles. Subsequently, the second vehicle can be decelerated by retro-rocket thrust under similar control of the first, third and fourth vehicle, etc. Hence, by decelerating the vehicles consecutively, it will be possible to be directly informed about the motion of all four vehicles, after consecutive deceleration of each vehicle by rocket thrust.

II. Determination of Motion of a Braked Vehicle by the Other Three Not Braked Vehicles

Shortly before the retro-rocket braking, the position of the first vehicle (which will be decelerated) is known. Now, applying the DOPPLER effect we assume that the three other vehicles are measuring the radial speed (as to any of these three vehicles) as functions of time. Thereby, no directing of the antenna's is required. The measurements are transmitted to the terrestrial station in some code. Let us assume that the supervising vehicles are in the direct neighbourhood of the braked vehicle which must be supervised, so that we have only to take into account the time of light travel (= signal travel) from the supervising vehicle to the earth. Then the reception time t_i corresponding with the time t of measurement by the supervizing vehicle i with coordinates $x_i\, y_i\, z_i$ follows from:

$$t + \frac{1}{c}\sqrt{[x_i(t) - x_{\oplus}(t_i)]^2 + [y_i(t) - y_{\oplus}(t_i)]^2 + [z_i(t) - z_{\oplus}(t_i)]^2} = t_i \qquad i = 1, 2, 3, \qquad c = \text{speed of light} \tag{1}$$

in which $x_{\oplus}$, $y_{\oplus}$ and $z_{\oplus}$ denote the coordinates of the earth with respect to the heliocentric system of coordinates, whereas also x_i, y_i, z_i refer to this heliocentric system. The time t corresponding with time t_i can then be solved numerically from (1).

Let us further assume — by example — that Venus be the object planet. In Fig. 1, $x\ y\ z$ be the heliocentric system of reference, $x'\ y'\ z'$ the Venusian system and $\xi\ \eta\ \zeta$ the vehicle i centric system, whereas A be the vehicle which is supervised by the vehicles $i = 1, 2, 3$. $\bar{r}_{\female}$ is the radiusvector of Venus as to the heliocentric system of reference. The three systems of reference $x\ y\ z$, $x'\ y'\ z'$, and $\xi\ \eta\ \zeta$ are parallel with the corresponding axes and do not rotate with respect to the celestial sky.

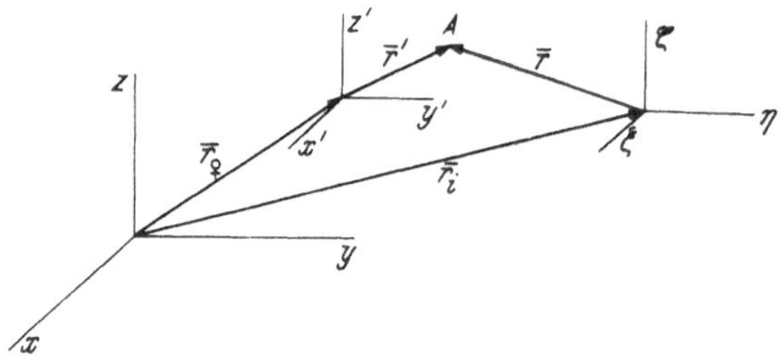

Fig. 1

We then obtain as to system $\xi\ \eta\ \zeta$:

$(\bar{r}\,\bar{v})/r = v_{\mathrm{rad}_i}$, in which $\bar{r}\,\bar{v}$ denotes the scalar product of $\bar{r}$ and the speed $\bar{v}$ of A as to system $\xi\ \eta\ \zeta$, whereas v_{rad_i} denotes the radial speed of A as to the supervising vehicle i.

Now $\bar{r}_{\female} + \bar{r}' = \bar{r}_i + \bar{r}$, so that:

$$\bar{r}_{\female} + \bar{r}' - \bar{r}_i = \bar{r} \qquad \text{and} \qquad \dot{\bar{r}}_{\female} + \dot{\bar{r}}' - \dot{\bar{r}}_i = \bar{v}.$$

Hence we obtain:

$$\left.\begin{aligned} &\frac{(\bar{r}_{\female} + \bar{r}' - \bar{r}_i)\,(\dot{\bar{r}}_{\female} + \dot{\bar{r}}' - \dot{\bar{r}}_i)}{|\bar{r}_{\oplus} + \bar{r}' - \bar{r}_i|} = v_{\mathrm{rad}_i}(t) \\ &\qquad i = 1, 2, 3 \end{aligned}\right\} \tag{2}$$

Writing (2) in non vector notation, we obtain:

$$\frac{[x_{\female}(t) + x' - x_i(t)]\,[\dot{x}_{\female}(t) + \dot{x}' - \dot{x}_i(t)] + [y_{\female}(t) + y' - y_i(t)]\,[\dot{y}_{\female}(t) + \dot{y}' - \dot{y}_i(t)] + [z_{\female}(t) + z' - z_i(t)]\,[\dot{z}_{\female}(t) + \dot{z}' - \dot{z}_i(t)]}{\sqrt{[x_{\female}(t) + x' - x_i(t)]^2 + [y_{\female}(t) + y' - y_i(t)]^2 + [z_{\female}(t) + z' - z_i(t)]^2}} = v_{\mathrm{rad}_i}(t),$$

in which x', y', and z' denote the coordinates of A with respect to the planet centric system $x'\,y'\,z'$. Arranging with respect to $\dot{x}'$, $\dot{y}'$ and $\dot{z}'$ we can write:

$$\left.\begin{gathered}
[x_{♀}(t) + x' - x_i(t)]\,\dot{x}' + [y_{♀}(t) + y' - y_i(t)]\,\dot{y}' + [z_{♀}(t) + z' - z_i(t)]\,\dot{z}' = \\
= v_{\mathrm{rad}_i}(t)\sqrt{[x_{♀}(t) + x' - x_i(t)]^2 + [y_{♀}(t) + y' - y_i(t)]^2 + [z_{♀}(t) + z' - z_i(t)]^2} + \\
+ [x_{♀}(t) + x' - x_i(t)]\,[\dot{x}_i(t) - \dot{x}_{♀}(t)] + [y_{♀}(t) + y' - y_i(t)]\,[\dot{y}_i(t) - \dot{y}_{♀}(t)] + \\
+ [z_{♀}(t) + z' - z_i(t)]\,[\dot{z}_i(t) - \dot{z}_{♀}(t)] \\
i = 1, 2, 3
\end{gathered}\right\} \quad (3)$$

Solving the eqs. (3) with respect to $\dot{x}'$, $\dot{y}'$, and $\dot{z}'$, we obtain:

$$\left.\begin{gathered}
\dot{x}' = \frac{1}{D}\begin{vmatrix} A_1 & y_{♀}(t) + y' - y_1(t) & z_{♀}(t) + z' - z_1(t) \\ A_2 & y_{♀}(t) + y' - y_2(t) & z_{♀}(t) + z' - z_3(t) \\ A_3 & y_{♀}(t) + y' - y_3(t) & z_{♀}(t) + z' - z_2(t) \end{vmatrix} \\
\dot{y}' = \frac{1}{D}\begin{vmatrix} x_{♀}(t) + x' - x_1(t) & A_1 & z_{♀}(t) + z' - z_1(t) \\ x_{♀}(t) + x' - x_2(t) & A_2 & z_{♀}(t) + z' - z_3(t) \\ x_{♀}(t) + x' - x_3(t) & A_3 & z_{♀}(t) + z' - z_2(t) \end{vmatrix} \\
\dot{z}' = \frac{1}{D}\begin{vmatrix} x_{♀}(t) + x' - x_1(t) & y_{♀}(t) + y' - y_1(t) & A_1 \\ x_{♀}(t) + x' - x_2(t) & y_{♀}(t) + y' - y_3(t) & A_2 \\ x_{♀}(t) + x' - x_3(t) & y_{♀}(t) + y' - y_2(t) & A_3 \end{vmatrix} \\
\text{in which:} \\
D = \begin{vmatrix} x_{♀}(t) + x' - x_1(t) & y_{♀}(t) + y' - y_1(t) & z_{♀}(t) + z' - z_1(t) \\ x_{♀}(t) + x' - x_2(t) & y_{♀}(t) + y' - y_2(t) & z_{♀}(t) + z' - z_2(t) \\ x_{♀}(t) + x' - x_3(t) & y_{♀}(t) + y' - y_3(t) & z_{♀}(t) + z' - z_3(t) \end{vmatrix} \\
\text{and} \\
A_i = v_{\mathrm{rad}_i}(t)\sqrt{[x_{♀}(t) + x' - x_i(t)]^2 + [y_{♀}(t) + y' - y_i(t)]^2 + [z_{♀}(t) + z' - z_i(t)]^2} + \\
+ [x_{♀}(t) + x' - x_i(t)]\,[\dot{x}_i(t) - \dot{x}_{♀}(t)] + [y_{♀}(t) + y' - y_i(t)]\,[\dot{y}_i(t) - y_{♀}(t)] + \\
+ [z_{♀}(t) + z' - z_i(t)]\,[\dot{z}_i(t) - \dot{z}_{♀}(t)], \\
i = 1, 2, 3
\end{gathered}\right\} \quad (4)$$

The motion of Venus and the motion of the supervising space vehicles may be considered as known, so that in (4) the functions $x_{♀}(t)$, $y_{♀}(t)$, $z_{♀}(t)$, $x_i(t)$, $y_i(t)$, $z_i(t)$, $i = 1, 2, 3$ are all known functions of time. Hence, we may write the system (4) in symbolic form:

$$\dot{x}' = f_1(x', y', z', t), \qquad \dot{y}' = f_2(x', y', z', t), \qquad \dot{z}' = f_3(x', y', z', t) \tag{4'}$$

Now let the retro-rocket braking begin at time $t = 0$ and let the coordinates of the decelerating vehicle with respect to the planet centric system at that time instant be x_0', y_0', z_0'. These initial coordinates may be considered as known. Then in connection with (4′) we can apply as zero approximation:

$$\begin{gathered}
x'(t)_0 = x_0' + f_1(x_0', y_0', z_0', t = 0)\,t, \qquad y'(t)_0 = y_0' + f_2(x_0', y_0', z_0', t = 0)\,t \\
z'(t)_0 = z_0' + f_3(x_0', y_0', z_0', t = 0)\,t
\end{gathered} \tag{5}$$

Substituting (5) in the right members of (4), these right members become functions in t only, so that we obtain:

$$\dot{x}'(t)_1 = F_1(t)_0, \qquad \dot{y}'(t)_1 = F_2(t)_0, \qquad \dot{z}'(t)_1 = F_3(t)_0 \tag{6}$$

Now for a certain integration interval $t = 0 \to t$, let us represent the right members of (6) as polynomial expressions of LAGRANGE by writing:

$$F_i(t)_0 = \sum_{k=1}^{n} \frac{\prod\limits_{i \neq k} (t - t_i)}{\prod\limits_{i \neq k} (t_k - t_i)} F_i(t_k)_0 = \sum_{k=0}^{n-1} A_k^{(i)} t^k,$$

$$i = 1, 2, 3$$

We then obtain after integrating of (6):

$$\left.\begin{aligned} x'(t)_1 &= \sum_{k=0}^{n-1} \frac{A_k^{(1)}}{k+1} t^{k+1} + x_0' \\ y'(t)_1 &= \sum_{k=0}^{n-1} \frac{A_k^{(2)}}{k+1} t^{k+1} + y_0' \\ z'(t)_1 &= \sum_{k=0}^{n-1} \frac{A_k^{(3)}}{k+1} t^{k+1} + z_0' \end{aligned}\right\} \tag{7}$$

Then we can again substitute the solution in first approximation (7) in the left members of (4), etc. For a time interval $t = 0 \to t$ which is not too large, this procedure will converge and the initial values of x', y' and z' for a following time interval can then be computed with any degree of accuracy.

Of course the system (4), indicated by (4') in symbolic form, can also be integrated by steps according to the method of RUNGE and KUTTA, starting from the initial state $t = 0 \to x_0', y_0', z_0'$. Indicating a time step by h and the corresponding increments of x', y' and z' by k, l and m respectively, we obtain the scheme:

$$\left.\begin{aligned}
& k = \frac{1}{6} k_1 + \frac{1}{3} k_2 + \frac{1}{3} k_3 + \frac{1}{6} k_4, \qquad l = \frac{1}{6} l_1 + \frac{1}{3} l_2 + \frac{1}{3} l_3 + \frac{1}{6} l_4, \\
& m = \frac{1}{6} m_1 + \frac{1}{3} m_2 + \frac{1}{3} m_3 + \frac{1}{6} m_4 \\
& k_1 = f_1(x', y', z', t) \qquad l_1 = f_2(x', y', z', t) \\
& k_2 = f_1\left(x' + \frac{k_1}{2}, y' + \frac{l_1}{2}, z' + \frac{m_1}{2}, t + \frac{h}{2}\right) h \qquad l_2 = f_2\left(x' + \frac{k_1}{2}, y' + \frac{l_1}{2}, z' + \frac{m_1}{2}, t + \frac{h}{2}\right) h \\
& k_3 = f_1\left(x' + \frac{k_2}{2}, y' + \frac{l_2}{2}, z' + \frac{m_2}{2}, t + \frac{h}{2}\right) h \qquad l_3 = f_2\left(x' + \frac{k_2}{2}, y' + \frac{l_2}{2}, z' + \frac{m_2}{2}, t + \frac{h}{2}\right) h \\
& k_4 = f_1(x' + k_3, y' + l_3, z' + m_3, t + h)\, h \qquad l_4 = f_2\left(x' + k_3, y' + \frac{l_3}{2}, z' + m_3, t + h\right) h \\
& m_1 = f_3(x', y', z', t) \\
& m_2 = f_3\left(x' + \frac{k_1}{2}, y' + \frac{l_1}{2}, z' + \frac{m_1}{2}, t + \frac{h}{2}\right) h \\
& m_3 = f_3\left(x' + \frac{k_2}{2}, y' + \frac{l_2}{2}, z' + \frac{m_2}{2}, t + \frac{h}{2}\right) h
\end{aligned}\right\} \quad (8$$

The scheme must be treated in the sequence $k_1\, l_1\, m_1\, k_2\, l_2\, m_2\, k_3\, l_3\, m_3\, k_4\, l_4\, m_4$. Repeating the computation with steps $2\,h$ (hence with half the number of steps) the difference in result of both computations is of an order of magnitude equal to 1/15 of the inaccuracy of the first computation.

III. The Bringing of Each Vehicle in the Neighbourhood of the Object Planet into a Known Attitude as to the Celestial Sky

Let us suppose that each of the 4 vehicles is free to tumble about its center of gravity throughout the free trajectory from the last burn-out point near the earth up to the environs of the object planet. First, we can bring each vehicle into a known attitude as to the celestial sky by directing the roll axis of the vehicle in the local direction of the solar rays and by determining further the attitude by using a conspicuous star — for example Sirius — as reference star.

In the following consideration we shall assume that the roll axis, yaw axis, and pitch axis of the vehicle coincide with the principal inertia axes of the vehicle which pass through the center of gravity of the vehicle. In each vehicle a telescope may be mounted the axis of which is still in one plane with the roll axis of the vehicle, see Fig. 2, whereas the angle η between the telescopic axis and the roll axis can be adjusted.

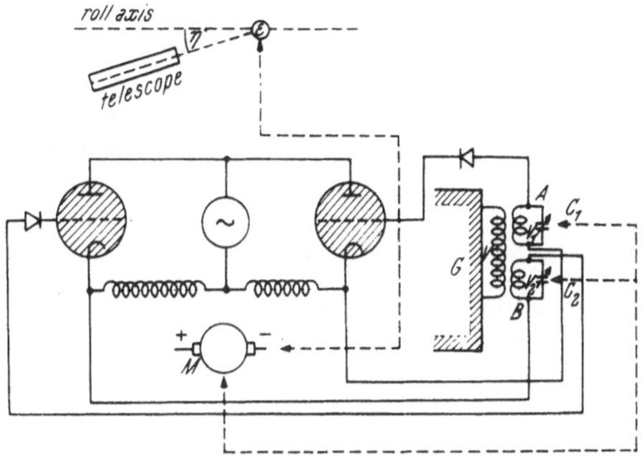

Fig. 2. G = receiving set in vehicle. Windings for clockwise and anticlockwise run energized via thyratrons

The servomotor M, adjusting the angle η, is commanded from the terrestrial command station by a variable modulation frequency ν in such a way that the angle η becomes a function of ν. The dotted lines in Fig. 2 indicate mechanical connections. When the servomotor M is at rest again $\nu_1 > \nu > \nu_2$, ν_1 and ν_2 being the own frequencies of the resonance circuits A and B which depend on the rotary condensers C_1 and C_2 which are mechanically actuated by the servomotor M.

If we use Sirius as reference star η will be the angle included by the lines of sight of the solar center and Sirius as seen from the vehicle. The numerical value of this angle can be determined as follows. If x, y and z be the heliocentric coordinates of the vehicle and the z axis of the heliocentric system is chosen parallel with the earth axis and the x-axis passes through point Ariës, and further α and β be the declination and right ascension of Sirius, the direction cosines of the telescopic axis directed to Sirius become:

$$\lambda_1 = \cos\alpha \cos\beta, \qquad \lambda_2 = \cos\alpha \sin\beta, \qquad \lambda_3 = \sin\alpha.$$

Further, if the roll axis is parallel with the solar rays, the direction of the roll axis coincides with the direction of

$\bar{r} = \bar{\imath}\, x + \bar{\jmath}\, y + \bar{k}\, z$ (= radius vector of the vehicle as to the heliocentric system).

Then:

$$\eta = \arc\cos \frac{x \cos\alpha \cos\beta + y \cos\alpha \sin\beta + z \sin\alpha}{\sqrt{x^2 + y^2 + z^2}} \tag{9}$$

Now, the directing of the roll axis of the vehicle parallel with the solar rays can be effected by an electronic optic system, in combination with a system of three "flywheels", which are mounted in the vehicle with the axes along the principal axes of inertia. The "flywheels" can be driven by electric motors, or can be directly the armatures of these motors. When the three flywheels are at rest, the vehicle may not carry out any rotation about the center of gravity. Hence, the moment of momentum of the vehicle as a whole, with respect to the center of gravity, must be zero. In order to reach this before, always auxiliary jet tubes are required, in order to give external moments around the roll, yaw, and pitch axis for taking away any residual moment of momentum when the flywheels are at rest. These auxiliary jet tubes can again be commanded by servomotors which are steered by variable modulation frequencies from the terrestrial command station. The angular speeds about the three axes of inertia can be measured by angular speed meters and telecommunicated in some code to the terrestrial station. The reduction of the residual moment of momentum to zero can be carried out by telecommand from the terrestrial station when the vehicle is still in the environs of the earth.

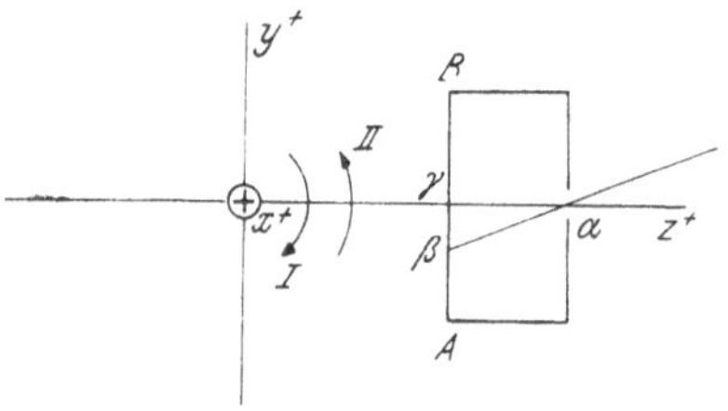

Fig. 3. X^+ directed from observer

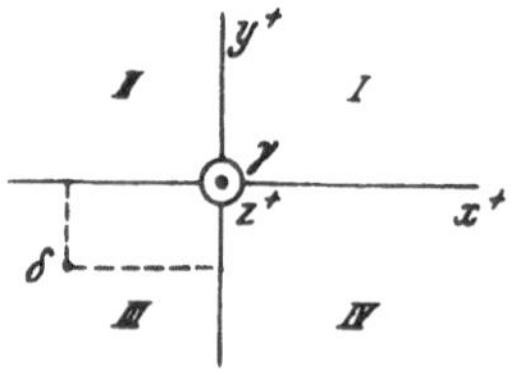

Fig. 4. Z^+ directed to observer

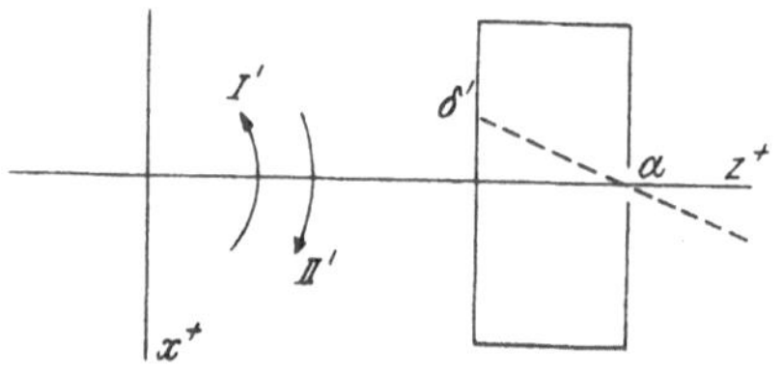

Fig. 5

Now let us indicate the principle system of inertia of the vehicle by $X\,Y\,Z$, the X and Y axis being the yaw and pitch axis, and the Z axis being the roll axis. The electronic optic system mentioned above can in principle be realized as a camera obscura of which the wall $A\,B$ (see Fig. 3) opposite to the orifice α is covered by a screen of photocells. Let us indicate the flywheel-servomotors, which are co-axial with X_+, Y_+, and Z_+, by A_X, A_Y, and A_Z. Then, in order to direct the Z-axis in space, only rotations about the X-axis and Y-axis are required, so that for this purpose only the servomotors A_X and A_Y must be actuated. Now, let us first assume that the vehicle has such an orientation that the solar center lies in the $Y\,Z$-plane. A solar ray passing through the orifice α actuates the photocell β. Thereby the servomotor A_X must come into operation turning the flywheel according to the curved arrow I. Then the space vehicle will turn with respect to the celestial sky according to the curved arrow II. When the ray of the light ultimately strikes the photo cell γ in the center of the screen, the servomotor A_X comes to a stop. The photo cells intermediate between β and γ are consecutively actuated, all giving A_X impulses by which the vehicle turns according to arrow *II*. The photo cells in the plane of Fig. 3 on the Y_+ side of γ will give impulses to A_X by which the space vehicle will turn in reverse direction.

Now let us consider the more general case that, by a solar ray passing the orifice α, the photocell δ be actuated (see Fig. 4). In this figure the figure plane coincides with the photocell screen and X_+ and Y_+ are projected on this plane.

As seen from Fig. 4, the X coordinate and Y coordinate of δ are both negative. Then from Fig. 3 it follows that the flywheel of A_X must turn (as to system XYZ) according to the curved arrow *I* and from Fig. 5 that the flywheel of A_Y must

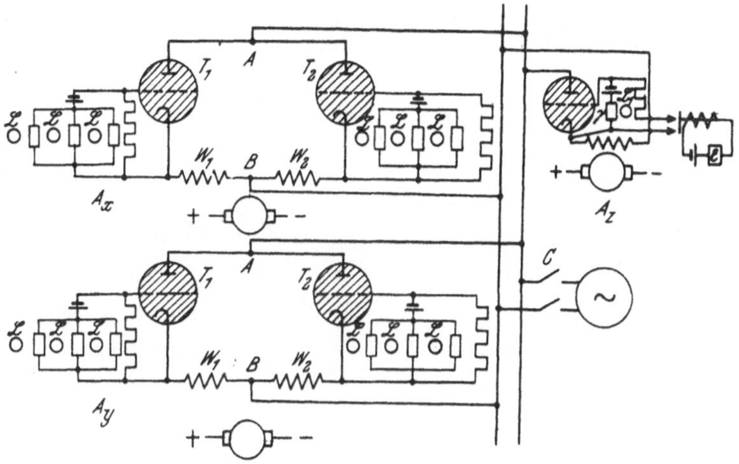

Fig. 6

turn (as to system XYZ) according to the curved arrow I' causing a corresponding rotation of the space vehicle according to curved arrow II'. (In Fig. 5 $\alpha\delta'$ is the projection of the solar ray, actuating the photo cell δ, on the plane XZ.) The actuation of A_X and A_Y can be thought as intermittant, the servomotors turning only as long as a corresponding photo cell is actuated.

Summarizing, the electronical connections of the photo cells in the XY plane of Fig. 4 with the servomotors A_X and A_Y must be such that these servomotors rotate, when some photo cell X, Y is actuated, according to the scheme:

X_+ sense of rotation of A_Y according to curved arrow II' } see Fig. 5
X_- sense of rotation of A_Y according to curved arrow I' }

Y_+ sense of rotation of A_X according to curved arrow II } see Fig. 3
Y_- sense of rotation of A_X according to curved arrow I }

The servomotors A_X and A_Y will stop when the solar ray passing orifice α strikes the photo cell γ. From the foregoing we conclude: for an actuated photo cell lying in one of the quadrants *I*, *II*, *III*, *IV* of Fig. 4, the two servomotors A_X and A_Y must be applied and, as to the sense of rotation, differently in any of these quadrants. Further, for an actuated photo cel on the X-axis or Y-axis, only one servomotor A_X or A_Y must be applied, in different sense of rotation for X_+ or X_-, or Y_+ and Y_-.

Hence, as to the electronical connections of the photo cells with both servo motors A_X and A_Y, we can distinguish 4 groups, corresponding with the quadrants

I, II, III, and *IV* of Fig. 4, and again 4 groups for the photo cells lying on the (projected) half axes X_+, X_-, Y_+ and Y_-.

To meet these requirements, let us equip the servomotors A_X and A_Y with stator windings W_1 and W_2 for clockwise and anticlockwise rotation, with corresponding thyratrons T_1 and T_2. Between A and B an alternating voltage may act whereas the armature is acted upon by a continuous voltage. Thyratron T_1 (as to A_X and A_Y) is connected with 3 photocells as indicated in Fig. 6, as well as thyratron T_2. Further each of these photo cells has an illuminating electric amp L. As soon as this lamp L shines the photocell will be activated. The lamps L then correspond with photocells of the photocell screen in the camera obscura.

Then, in connection with the foregoing considerations we have, as to the photocells in Fig. 6, the following combinations coming into play:

1	3	5		2	4	6	A_X
☐	☐	☐		☐	☐	☐	
1	4	7		2	3	8	A_Y
☐	☐	☐		☐	☐	☐	

clockwise run   anticlockwise run

$$\text{Group } \begin{matrix} 1 \\ 1 \end{matrix} \left\} \begin{matrix} A_X \text{ clockwise} \\ A_Y \text{ clockwise} \end{matrix} \right. \qquad \text{Group } \begin{matrix} 2 \\ 2 \end{matrix} \left\} \begin{matrix} A_X \text{ anticlockwise} \\ A_Y \text{ anticlockwise} \end{matrix} \right.$$

$$\text{Group } \begin{matrix} 3 \\ 3 \end{matrix} \left\} \begin{matrix} A_X \text{ clockwise} \\ A_Y \text{ anticlockwise} \end{matrix} \right. \qquad \text{Group } \begin{matrix} 4 \\ 4 \end{matrix} \left\} \begin{matrix} A_X \text{ anticlockwise} \\ A_Y \text{ clockwise} \end{matrix} \right.$$

Group 5 A_X only, clockwise   Group 6 A_X only, anticlockwise

Group 7 A_Y only, clockwise   Group 8 A_Y only, anticlockwise

Further we can distinguish a "group" 9, for which no servomotor A_X or A_Y is applied. Now in case of the four "quadrant" groups, in which the two servomotors A_X and A_Y are simultaneously applied, we have in Fig. 6 for any of these

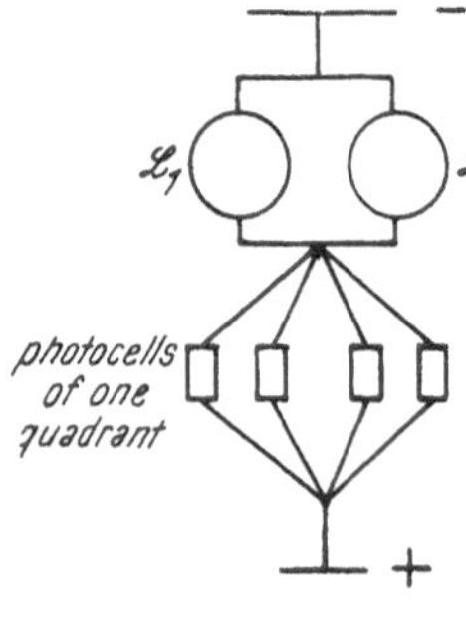

Fig. 7   Fig. 8   Fig. 9

groups two corresponding lamps L. These lamps L_1 and L_2 can be switched in parallel and with all the photocells of the corresponding quadrant in parallel, in series (see Fig. 7). Then if one of these quadrant photocells is activated by the solar ray coming through the orifice of the camera obscura, L_1 and L_2 will burn and the corresponding photocells in Fig. 6 will be activated.

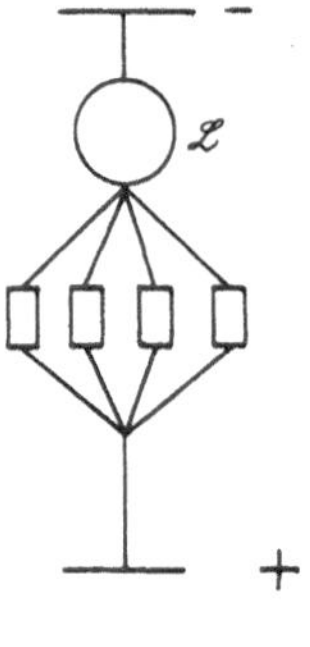

In similar way the lamp L in Fig. 8 corresponding with the group 5, 6, 7 or 8, can be switched in series with the photocells of the photocell screen in parallel, situated on the corresponding half axis in Fig. 4.

Further, when the photocell γ is actuated, the servomotors A_X and A_Y will not be activated, simply because this photocell is *not* connected with any lamp L of A_X or A_Y, but is switched in series (Fig. 9) with the lamp L_1 of A_Z (Fig. 6), by which the roll motor (with flywheel axis along the roll axis) is engaged. The space vehicle will then rotate around the roll axis, until the light of Sirius (the reference star), passing through the telescope of Fig. 4, will activate a photocell ε (see Fig. 6). Then an electromagnetic switch will open the circuit of L_1 by which the roll motor stops. In order to initiate the whole process of giving the vehicle an initial known attitude as to the celestial sky by means of the solar rays and a reference star, the switch C only must be closed. [The general opening and closing of all kinds of switches can be achieved by a robot finger which can move in cylinder polar coordinate fashion and which can be telecommanded from the terrestrial station by 3 variable modulation frequencies.]

IV. To Determine the Euler Angles, Determining the Attitude of the Vehicle as to the Heliocentric System of Reference, with Roll Axes Parallel to the Solar Rays and Sirius (the Reference Star) in the Telescopic Field Center

Let us again indicate the vehicle rigid system (system of principal axes of inertia passing through the center of gravity) by $X\ Y\ Z$ and let again the Z-axis coincide with the roll-axis. $\underline{x}\ \underline{y}\ \underline{z}$ be a rectangular system parallel with the heliocentric system $x\ y\ z$ (unit vector $\bar{\imath}, \bar{\jmath}, \bar{k}$) with the z-axis parallel to the earth axis and the x-axis passing through point Ariës. Let us further assume that the telescopic axis is situated in the plane $X\,Z$. Passing from $\underline{x}\ \underline{y}\ \underline{z}$ to $X\ Y\ Z$ we have the following consecutive rotations:

$$\underbrace{\underline{x}\ \underline{y}\ \underline{z} \rightarrow x'\,y'\,z}_{\varphi \text{ about } z}\quad \underbrace{}_{\theta \text{ about } y'}\quad x''\,y'\,Z \quad \underbrace{}_{\psi \text{ about } Z}\quad X\,Y\,Z, \qquad \text{in which}$$

φ, θ and ψ be the EULER angles in general and φ_0, θ_0 and ψ_0 the angles of EULER defining the initial attitude.

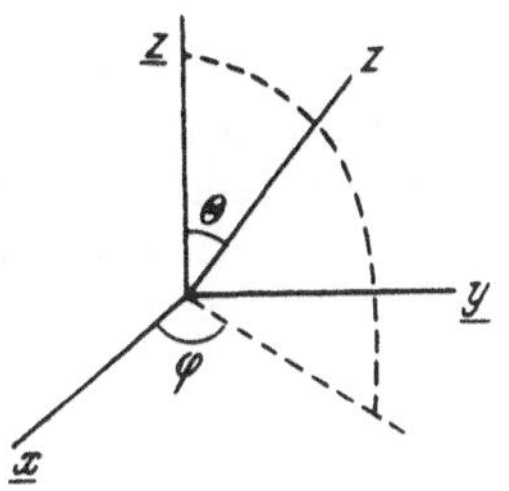

Fig. 10

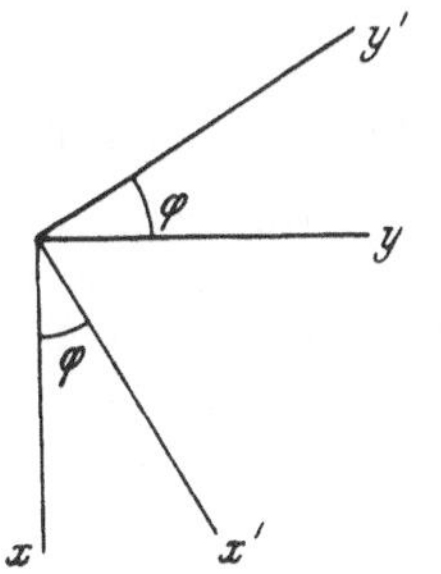

Fig. 11

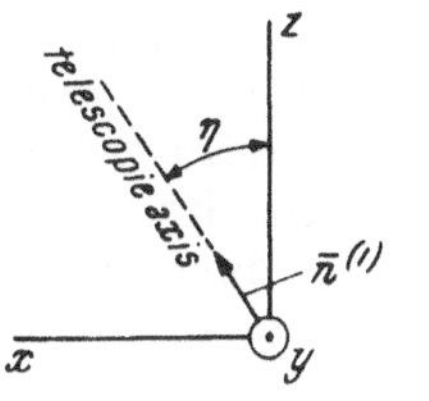

Fig. 12. $Y+\perp$ to paper, directed to observer

Then, if x, y and z are the heliocentric coordinates of the vehicle, we read directly from Fig. 10:

$$\theta_0 = \arccos \frac{z}{\sqrt{x^2 + y^2 + z^2}} \tag{10}$$

and

$$\varphi_0 = \text{arc tg} \frac{y}{x} \tag{11}$$

Further, $\psi = \angle(y', Y)$. If we indicate the unit vector in the direction of y'_+ by $\bar{k}'$, we read from Fig. 11:

$$\bar{k}' = -\bar{\imath} \sin \varphi + \bar{\jmath} \cos \varphi \tag{12}$$

Further, as we have supposed that the telescope is mounted in the vehicle in such a way that the telescopic axis is still (at any η) situated in the plane XZ (Fig. 12) and the direction of Z_+ coincides with the direction of $\bar{r} = \bar{\imath}x + \bar{\jmath}y + \bar{k}z$, (13), the direction of Y_+ coincides with the direction of the vector $\bar{r} \times \bar{n}^{(1)}$, in which $\bar{n}^{(1)}$ denotes a unit vector in the direction of Sirius (the reference star). Hence, if α and β are the declination and right ascension of Sirius,

$$\bar{n}^{(1)} = \bar{\imath} \cos\alpha \cos\beta + \bar{\jmath} \cos\alpha \sin\beta + \bar{k} \sin\alpha. \tag{14}$$

Hence

$$\psi = \angle\, y', Y = \text{arc cos} \frac{\bar{k}' (\bar{r} \times \bar{n}^{(1)})}{|\bar{r} \times \bar{n}^{(1)}|} \tag{15}$$

Now, from (13) and (14) we obtain:

$$\bar{r} \times \bar{n}^{(1)} = \begin{vmatrix} \bar{\imath} & \bar{\jmath} & \bar{k} \\ x & y & z \\ \cos\alpha\cos\beta & \cos\alpha\sin\beta & \sin\alpha \end{vmatrix}.$$

Hence from (15) we obtain:

$$\psi_0 = \text{arc cos} \frac{-\sin\varphi_0 (y \sin\alpha - z \cos\alpha \sin\beta) + \cos\varphi_0 (z \cos\alpha \cos\beta - x \sin\alpha)}{\sqrt{(y \sin\alpha - z\cos\alpha\sin\beta)^2 + (z\cos\alpha\cos\beta - x\sin\alpha)^2 + (x\cos\alpha\sin\beta - y\cos\alpha\cos\beta)^2}} \tag{16}$$

in which, from (11),

$$\sin\varphi_0 = \frac{y}{\sqrt{x^2 + y^2}} \quad \text{and} \quad \cos\varphi_0 = \frac{x}{\sqrt{x^2+y^2}}.$$

Hence the "initial attitude" of the vehicle is defined by θ_0, φ_0, and ψ_0.

V. Determination of the Euler Angles Determining the Required Attitude for Firing the Retro-Rocket

We can prescribe that the retro rocket thrust will act in the negative speed direction of the motion of the vehicle as to the heliocentric system. Assuming that the retro-rocket is mounted in the vehicle in such a way that the nozzle axis coincides with the Z-axis (= roll-axis), the direction of Z_+ must then coincide with the direction of $\bar{v} = \bar{\imath}\dot{x} + \bar{\jmath}\dot{y} + \bar{k}\dot{z}$. Then, if we indicate the Euler angles of the required attitude by θ_1, φ_1, and ψ_1, we obtain (see Fig. 13)

$$\theta_1 = \angle\,(Z_+, z_+) = \angle\,(\bar{k}, \bar{v}).$$

Hence

$$\theta_1 = \text{arc cos} \frac{\dot{z}}{\sqrt{\dot{x}^2 + \dot{y}^2 + \dot{z}^2}} \tag{17}$$

Further we read from Fig. 13:

$$\varphi_1 = \text{arc tg} \frac{\dot{y}}{\dot{x}} \tag{18}$$

If we would prescribe that the retro thrust must act in the negative direction of the *relative* motion of the vehicle as to the object planet (hence of the motion as to the planet centric system $x'\ y'\ z'$ in Fig. 1) we would obtain, with Venus as target planet, instead of (17) and (18):

$$\theta_1 = \text{arc cos}\, \frac{\dot{z} - \dot{z}_♀}{\sqrt{(\dot{x} - \dot{x}_♀)^2 + (\dot{y} - \dot{y}_♀)^2 + (\dot{z} - \dot{z}_♀)^2}} \qquad (17')$$

and

$$\varphi_1 = \text{arc tg}\, \frac{\dot{y} - \dot{y}_♀}{\dot{x} - \dot{x}_♀} \qquad (18')$$

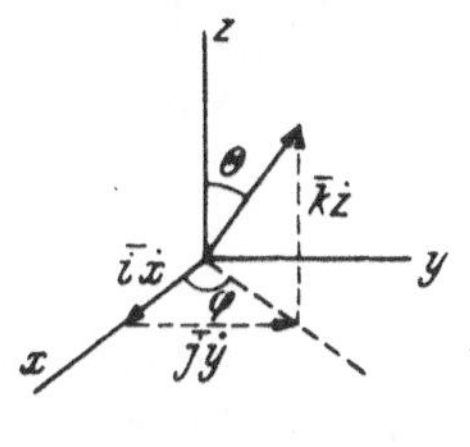

Fig. 13. (Erratum: for θ and φ read θ_1 and φ_1)

For the rest, the angle ψ has no influence on the direction of the rocket thrust, so that we are quite free as to the choice of ψ_1.

VI. The Telecommand of the Required Attitude for Firing the Retro-Rocket

The transference of attitude: $\theta_0 \to \theta_1$, $\varphi_0 \to \varphi_1$, $\psi_0 \to \psi_1$ must be carried out by telecommand at planetary distance, in a time lapse as short as possible. Therefore it is necessary to give a sequence (train) of commands throughout some short time lapse, which is received throughout a corresponding time lapse which as a whole is r/c seconds later, if r is the distance from the terrestrial command station to the vehicle and c is the speed of light. At the end of the reception time lapse, the EULER angles must have the required values for firing the retro-rocket. We have now to discuss the question how this can be arranged.

In general we can write for the moment of momentum of the vehicle as to the non-rotating system $\underline{x}\ \underline{y}\ \underline{z}$ with the origin at the center of gravity of the vehicle:

$$\bar{B} = \sum m\,(\bar{r} \times \bar{v}) \qquad (19)$$

In (19), m denotes the mass of an arbitrary particle of the vehicle and $\bar{r}$ and $\bar{v}$ the radius vector and speed of this particle as to the system $\underline{x}\ \underline{y}\ \underline{z}$, and the summation must be taken over all parts of the vehicle.

Now let us write:

$$\bar{v} = \bar{v}_s + \bar{v}',$$

in which $\bar{v}_s$ = tow speed of particle m due to the rotary motion of the vehicle as solid body and $\bar{v}'$ = relative speed of particle m as to a vehicle rigid system.

Then:

$$\bar{B} = \sum m\,(\bar{r} \times \bar{v}_s) + \sum m\,(\bar{r} \times \bar{v}'),$$

in which

$$\sum m\,(\bar{r} \times \bar{v}_s) = \bar{B}'$$

is the moment of momentum of the whole vehicle as rigid body as to the non rotating system $\underline{x}\ \underline{y}\ \underline{z}$, whereas $\Sigma m\,(\bar{r} \times \bar{v}') = \bar{B}''$ is the moment of momentum as to the vehicle rigid system $X\,Y\,Z$, of moving engine parts within the vehicle. Let us assume that the only moving parts coming into play in this respect are the three "flywheels" with axes along the axes of inertia X, Y and Z of the vehicle.

If the vehicle is not acted upon by external moments, we obtain:

$$\frac{d\bar{B}}{dt} + (\bar{u} \times \bar{B}) = 0,$$

in which $\bar{u}$ denotes the angular speed vector of the vehicle and $d\bar{B}/dt$ the time derivative of $\bar{B}$ as to the vehicle rigid system $X\,Y\,Z$.

Now we can write $\bar{B} = \bar{B}' + \bar{B}''$, so that:

$$\frac{d\bar{B}'}{dt} + (\bar{u} \times \bar{B}') = -\frac{d\bar{B}''}{dt} - (\bar{u} \times \bar{B}'') \tag{19}$$

If $\bar{\imath}_1$, $\bar{\imath}_2$, and $\bar{\imath}_3$ be the unit vectors in X_+, Y_+ and Z_+ direction, and J_X, J_Y and J_Z the moments of inertia of the vehicle as to X, Y and Z axis (principal axes of inertia) we obtain:

$$\bar{B}' = \bar{\imath}_1 u_X J_X + \bar{\imath}_2 u_Y J_Y + \bar{\imath}_3 u_Z J_Z \tag{20}$$

in which u_X, u_Y and u_Z are the components of $\bar{u}$ in X_+, Y_+ and Z_+ direction. If ω_X, ω_Y and ω_Z are the angular speeds (as to vehicle rigid system $X\,Y\,Z$) and θ_X, θ_Y and θ_Z the corresponding moments of inertia of the three flywheels, we obtain further:

$$\bar{B}'' = \bar{\imath}_1 \omega_X \theta_X + \bar{\imath}_2 \omega_Y \theta_Y + \bar{\imath}_3 \omega_Z \theta_Z \tag{21}$$

Then from (20) and (21):

$$\bar{u} \times \bar{B}' = \begin{vmatrix} \bar{\imath}_1 & \bar{\imath}_2 & \bar{\imath}_3 \\ u_X & u_Y & u_Z \\ u_X J_X & u_Y J_Y & u_Z J_Z \end{vmatrix}$$

$$\bar{u} \times \bar{B}'' = \begin{vmatrix} \bar{\imath}_1 & \bar{\imath}_2 & \bar{\imath}_3 \\ u_X & u_Y & u_Z \\ \omega_X \theta_X & \omega_Y \theta_Y & \omega_Z \theta_Z \end{vmatrix}$$

Hence the component equations of (19) become:

$$\left.\begin{aligned} J_X \frac{du_Y}{dt} + u_Y u_Z (J_Z - J_Y) &= -\theta_X \frac{d\omega_X}{dt} - u_Y \omega_Z \theta_Z + u_Z \omega_Y \theta_Y \\ J_Y \frac{du_Y}{dt} + u_X u_Z (J_X - J_Z) &= -\theta_Y \frac{d\omega_Y}{dt} - u_Z \omega_X \theta_X + u_X \omega_Z \theta_Z \\ J_Z \frac{du_Z}{dt} + u_X u_Y (J_Y - J_X) &= -\theta_Z \frac{d\omega_Z}{dt} - u_X \omega_Y \theta_Y + u_Y \omega_X \theta_X \end{aligned}\right\} \tag{22}$$

Further, we read from the Figs. 15 and 16:

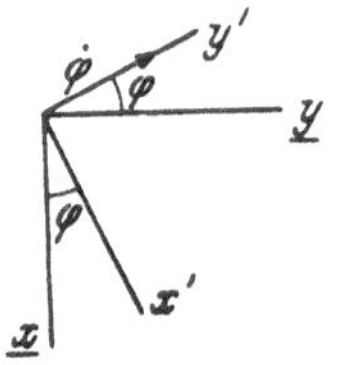

Fig. 14. $\underline{z}^+$ perpendicular to plane of paper, directed to observer $\dot{\varphi} \| \underline{z}^+$ (erratum: for $\dot{\varphi}$ read $\dot{\theta}$)

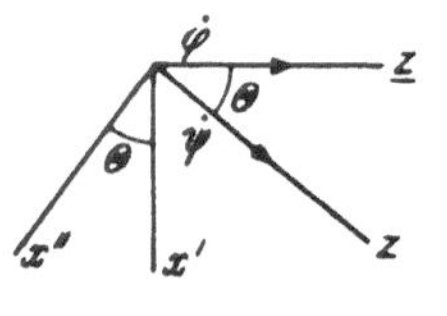

Fig. 15. y'^+ perpendicular to plane of paper, directed from observer $\dot{\theta} \| y'^+$

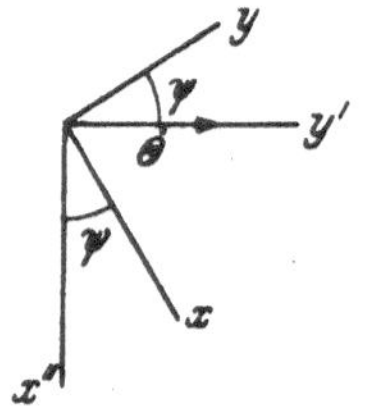

Fig. 16. Z^+ perpendicular to plane of paper, directed to observer $\dot{\psi} \| Z^+$

$$\left.\begin{aligned} u_X &= -\dot{\varphi}\sin\theta\cos\psi + \dot{\theta}\sin\psi \\ u_Y &= \quad\dot{\varphi}\sin\theta\sin\psi + \dot{\theta}\cos\psi \\ u_Z &= \dot{\psi} + \dot{\varphi}\cos\theta \end{aligned}\right\} \tag{23}$$

Now we can prescribe θ, φ and ψ as functions of time throughout the time interval $t = 0 \to t_1$ in which the transference of attitude:

$$\theta_0 \to \theta_1, \qquad \varphi_0 \to \varphi_1, \qquad \psi_0 \to \psi_1$$

must occur. This prescription must then be arranged in such a way that at $t = 0$, $\dot{\theta}_0 = 0$, $\dot{\varphi} = 0$ and $\dot{\psi} = 0$, whereas at time t_1 these time derivatives must again become equal to zero. Further, if we put $\psi_0 = \psi_1$, the function ψ can be reduced to a constant $\psi = \psi_0$, so that, through the interval $t = 0 \to t_1$, $\dot{\psi} = 0$ and $\sin\psi = \sin\psi_0$ and $\cos\psi = \cos\psi_0$ are constants.

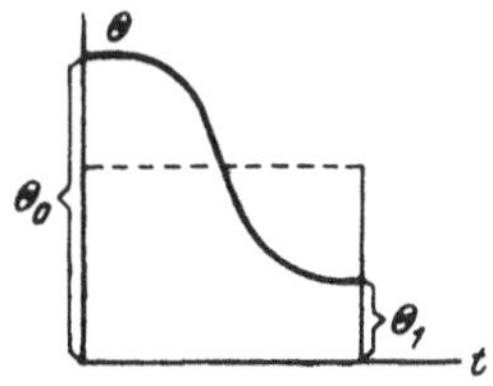

Fig. 17

[Possible prescribed functions for θ, φ and ψ. Let us assume by way of example that $\theta_0 > \theta_1$. It will be necessary that $\dot{\theta}_{t=0} = 0$ and $\dot{\theta}_{t=t_1} = 0$. For we can choose as possible prescribed function for θ:

$$\theta = \theta_1 + \frac{1}{2}(\theta_0 - \theta_1) + \frac{1}{2}(\theta_0 - \theta_1)\cos\frac{\pi}{t_1}t \tag{A}$$

Then

$$\theta_{t=0} = \theta_1 + \frac{1}{2}(\theta_0 - \theta_1) + \frac{1}{2}(\theta_0 - \theta_1) = \theta_1 + \theta_0 - \theta_1 = \theta_0,$$

and

$$\theta_{t_1} = \theta_1 + \frac{1}{2}(\theta_0 - \theta_1) - \frac{1}{2}(\theta_0 - \theta_1) = \theta_1.$$

Differentiating (A) we obtain:

$$\dot{\theta} = -\frac{\pi}{2t_1}(\theta_0 - \theta_1)\sin\frac{\pi}{t_1}t.$$

Hence

$$\dot{\theta}_{t=0} = 0 \qquad \text{and} \qquad \dot{\theta}_{t_1} = 0.$$

Further:

$$\ddot{\theta} = -\frac{\pi^2}{2t_1^2}(\theta_0 - \theta_1)\cos\frac{\pi}{t_1}t,$$

so that

$$\ddot{\theta}_{t=0} = -\frac{\pi^2}{2t_1^2}(\theta_0 - \theta_1),$$

and

$$\ddot{\theta}_{t_1} = \frac{\pi^2}{2t_1^2}(\theta_0 - \theta_1).$$

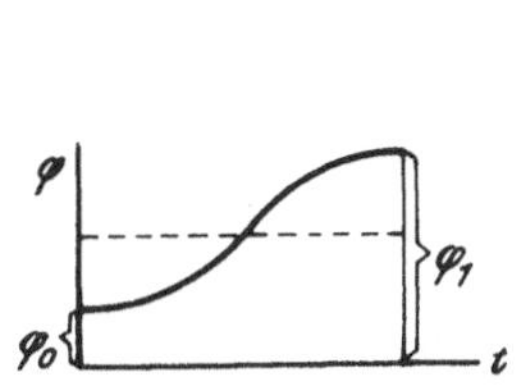

Fig. 18

Let us assume by way of example that $\varphi_0 < \varphi_1$. Then we can choose as possible prescribed function of φ:

$$\varphi = \varphi_0 + \frac{1}{2}(\varphi_1 - \varphi_0) - \frac{1}{2}(\varphi_1 - \varphi_0)\cos\frac{\pi}{t_1}t \tag{B}$$

Then

$$\varphi_{t=0} = \varphi_0 \quad \text{and} \quad \varphi_{t_1} = \varphi_0 + \frac{1}{2}(\varphi_1 - \varphi_0) + \frac{1}{2}(\varphi_1 - \varphi_0) = \varphi_0 + \varphi_1 - \varphi_0 = \varphi_1$$

Further,

$$\dot{\varphi} = \frac{\pi}{2\,t_1}(\varphi_1 - \varphi_0)\sin\frac{\pi}{t_1}t.$$

Hence

$$\dot{\varphi}_{t=0} = 0 \qquad \text{and} \qquad \dot{\varphi}_{t_1} = 0.$$

$$\ddot{\varphi} = \frac{\pi^2}{2\,t_1^2}(\varphi_1 - \varphi_0)\cos\frac{\pi}{t_1}t,$$

so that

$$\ddot{\varphi}_{t=0} = \frac{\pi^2}{2\,t_1^2}(\varphi_1 - \varphi_0)$$

and

$$\ddot{\varphi}_{t_1} = -\frac{\pi}{2\,t_1^2}(\varphi_1 - \varphi_0).$$

For the rest ψ may be chosen as constant ψ_0, so that then $\psi \equiv \psi_0$. (C)]

Then by the prescription of the EULER angles as functions of time, we obtain also u_X, u_Y and u_Z as functions of time from (23). Introducing these expressions in (22), we obtain a set of differential equations of the form:

$$\frac{d\omega_X}{dt} = f_1(\omega_Y, \omega_Z, t), \qquad \frac{d\omega_Y}{dt} = f_2(\omega_X, \omega_Z, t), \qquad \frac{d\omega_Z}{dt} = f_3(\omega_X, \omega_Y, t) \qquad (24)$$

[If we choose as prescribed functions (A), (B), and (C), we obtain after substitution in (23):

$$u_X = +\frac{\pi}{2\,t_1}(\varphi_1 - \varphi_0)\sin\frac{\pi}{t_1}t\cdot\sin\left[\theta_1 + \frac{1}{2}(\theta_0 - \theta_1) + \frac{1}{2}(\theta_0 - \theta_1)\cos\frac{\pi}{t_1}t\right]\cos\psi_0 +$$

$$-\frac{\pi}{2\,t_1}(\theta_0 - \theta_1)\sin\frac{\pi}{t_1}t\sin\psi_0$$

$$u_Y = +\frac{\pi}{2\,t_1}(\varphi_1 - \varphi_0)\sin\frac{\pi}{t_1}t\cdot\sin\left[\theta_1 + \frac{1}{2}(\theta_0 - \theta_1) + \frac{1}{2}(\theta_0 - \theta_1)\cos\frac{\pi}{t_1}t\right]\sin\psi_0 +$$

$$-\frac{\pi}{2\,t_1}(\theta_0 - \theta_1)\sin\frac{\pi}{t_1}t\cos\psi_0$$

$$u_Z = +\frac{\pi}{2\,t_1}(\varphi_1 - \varphi_0)\sin\frac{\pi}{t_1}t\cdot\cos\left[\theta_1 + \frac{1}{2}(\theta_0 - \theta_1) + \frac{1}{2}(\theta_0 - \theta_1)\cos\frac{\pi}{t_1}t\right]$$

Further:

$$\dot{u}_X = \frac{\pi^2}{2\,t_1^2}(\varphi_1 - \varphi_0)\cos\frac{\pi}{t_1}t\cdot\sin\left[\theta_1 + \frac{1}{2}(\theta_0 - \theta_1) + \frac{1}{2}(\theta_0 - \theta_1)\cos\frac{\pi}{t_1}t\right]\cos\psi_0 +$$

$$-\frac{\pi}{2\,t_1}(\varphi_1 - \varphi_0)\sin\frac{\pi}{t_1}t\cdot\cos\left[\theta_1 + \frac{1}{2}(\theta_0 - \theta_1) + \frac{1}{2}(\theta_0 - \theta_1)\cos\frac{\pi}{t_1}t\right]\cdot$$

$$\cdot\frac{\pi}{2\,t_1}(\theta_0 - \theta_1)\sin\frac{\pi}{t_1}t\cdot\cos\psi_0 - \frac{\pi^2}{2\,t_1^2}(\theta_0 - \theta_1)\cos\frac{\pi}{t_1}t\cdot\sin\psi_0$$

$$\dot{u}_Y = \frac{\pi^2}{2\,t_1^2}(\varphi_1 - \varphi_0)\cos\frac{\pi}{t_1}t \cdot \sin\left[\theta_1 + \frac{1}{2}(\theta_0 - \theta_1) + \frac{1}{2}(\theta_0 - \theta_1)\cos\frac{\pi}{t_1}t\right]\sin\psi_0 +$$

$$-\frac{\pi}{2\,t_1}(\varphi_1 - \varphi_0)\sin\frac{\pi}{t_1}t \cdot \cos\left[\theta_1 + \frac{1}{2}(\theta_0 - \theta_1) + \frac{1}{2}(\theta_0 - \theta_1)\cos\frac{\pi}{t_1}t\right] \cdot$$

$$\cdot \frac{\pi}{2\,t_1}(\theta_0 - \theta_1)\sin\frac{\pi}{t_1}t \cdot \sin\psi_0 - \frac{\pi^2}{2\,t_1^2}(\theta_0 - \theta_1)\cos\frac{\pi}{t_1}t \cdot \cos\psi_0$$

$$\dot{u}_Z = \frac{\pi^2}{2\,t_1^2}(\varphi_1 - \varphi_0) \cdot \cos\frac{\pi}{t_1}t \cdot \cos\left[\theta_1 + \frac{1}{2}(\theta_0 - \theta_1) + \frac{1}{2}(\theta_0 - \theta_1)\cos\frac{\pi}{t_1}t\right] +$$

$$+\frac{\pi}{2\,t_1}(\varphi_1 - \varphi_0)\sin\frac{\pi}{t_1}t \cdot \sin\left[\theta_1 + \frac{1}{2}(\theta_0 - \theta_1) + \frac{1}{2}(\theta_0 - \theta_1)\cos\frac{\pi}{t_1}t\right] \cdot$$

$$\cdot \frac{\pi}{2t_1}(\theta_0 - \theta_1)\sin\frac{\pi}{t_1}t.$$

These expressions in t for u_X, u_Y, u_Z, $\dot{u}_X$, $\dot{u}_Y$ and $\dot{u}_Z$ must be substituted in the eqs. (22), yielding equations of the form (24).]

Starting from the initial state:

$$t = 0 \rightarrow \omega_X = 0, \qquad \omega_Y = 0, \qquad \omega_Z = 0,$$

the system (24) can be integrated by steps according to the method of RUNGE and KUTTA. If k, l, and m are the increments of ω_X, ω_Y, and ω_Z, corresponding with a step $\Delta t = h$, we obtain the scheme:

$$k_1 = f_1(\omega_Y, \omega_Z, t)\,h \qquad l_1 = f_2(\omega_X, \omega_Z, t)\,h$$

$$k_2 = f_1\left(\omega_Y + \frac{l_1}{2}, \omega_Z + \frac{m_1}{2}, t + \frac{h}{2}\right)h \qquad l_2 = f_2\left(\omega_X + \frac{k_1}{2}, \omega_Z + \frac{m_1}{2}, t + \frac{h}{2}\right)h$$

$$k_3 = f_1\left(\omega_Y + \frac{l_2}{2}, \omega_Z + \frac{m_2}{2}, t + \frac{h}{2}\right)h \qquad l_3 = f_2\left(\omega_X + \frac{k_2}{2}, \omega_Z + \frac{m_2}{2}, t + \frac{h}{2}\right)h$$

$$k_4 = f_1(\omega_Y + l_3, \omega_Z + m_3, t + h)\,h \qquad l_4 = f_2(\omega_X + k_3, \omega_Z + m_3, t + h)\,h$$

$$m_1 = f_3(\omega_X, \omega_Y, t)$$

$$m_2 = f_3\left(\omega_X + \frac{k_1}{2}, \omega_Y + \frac{l_1}{2}, t + \frac{h}{2}\right)h$$

$$m_3 = f_3\left(\omega_X + \frac{k_2}{2}, \omega_Y + \frac{l_2}{2}, t + \frac{h}{2}\right)h$$

$$m_4 = f_3(\omega_X + k_3, \omega_Y + l_3, t + h)\,h$$

Of course (23) can also be solved by applying the method of successive approximations, starting from the solution in zero approximation:

$$\omega_X = \left(\frac{d\omega_X}{dt}\right)_0 t, \qquad \omega_Y = \left(\frac{d\omega_Y}{dt}\right)_0 t, \qquad \omega_Z = \left(\frac{d\omega_Z}{dt}\right)_0 t,$$

in which $(d\omega_X/dt)_0$, $(d\omega_Y/dt)_0$ and $(d\omega_Z/dt)_0$ are found by substituting $t = 0$, $\omega_X = 0$, $\omega_Y = 0$, $\omega_Z = 0$ in the eqs. (24).

For a time lapse $t = 0 \rightarrow t'$ which is not too large, this procedure will converge, allowing to find the new initial conditions:

$$t = t' \rightarrow \omega_X', \omega_Y', \omega_Z'$$

for a following time lapse, then using as solution in zero approximation:

$$\omega_X = \omega_X' + \left(\frac{d\omega_X}{dt}\right)_{t'} t, \qquad \omega_Y = \omega_Y' + \left(\frac{d\omega_Y}{dt}\right)_{t'} t, \qquad \omega_Z = \omega_Z' + \left(\frac{d\omega_Z}{dt}\right)_{t'} t,$$

in which

$$\left(\frac{d\omega_X}{dt}\right)_{t'}, \quad \left(\frac{d\omega_Y}{dt}\right)_{t'} \quad \text{and} \quad \left(\frac{d\omega_Z}{dt}\right)_{t'}$$

are found by substituting t', ω_X', ω_Y', ω_Z' in (24).

Having determined ω_X, ω_Y and ω_Z as functions of t, corresponding with the angles of EULER as prescribed functions of t, we have to proceed to the problem how the corresponding telecommand can be technically realized.

The three flywheels which are driven by the motors A_X, A_Y and A_Z mentioned above can be driven also by three corresponding electric motors M_X, M_Y, and M_Z, which are directly coupled with A_X, A_Y, and A_Z respectively, in such a way that the armatures of A_X, M_X, and A_Y, M_Y, and A_Z, M_Z constitute with the corresponding flywheel one rotating rigid body. When A_X, A_Y, or A_Z are energized, M_X, M_Y and M_Z only are driven, and vice-versa. During the process of directing the retro-rocket, M_X (and similarly M_Y and M_Z) has to carry out a number of rotations in the time lapse $t = 0 \to t_1$ in accordance with the predetermined function $\omega_X = \omega_X(t)$. Now the armature of M_X (and similarly M_Y and M_Z) can be acted upon by a continuous voltage (Fig. 19), whereas the number of applied stator windings can be made dependent on the position of a slide contact S. The stationary angular speed of M_X will then be a function of the position of S. Now we can displace the slide contact S by means of a servo-electric motor which is telecommanded from the terrestrial station by a variable modulation frequency ν_X, in similar way as the servomotor varying the angle η in Fig. 2. Then the position of S, and hence also the corresponding stationary angular speed of M_X, will become a function of this modulation frequency. If the whole design is made such that the rate of increase and decrease of ω_X throughout the process of directing remains sufficiently small, ω_X will be a unique function of the received modulation frequency.

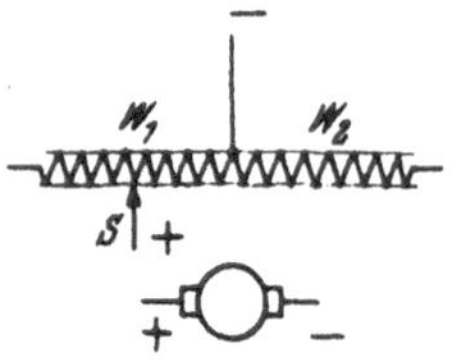

Fig. 19. W_1 = stator windings for clockwise run, W_2 = stator windings for anticlockwise run

Hence the required $\nu_X = \nu_X(t)$ will be known and the modulation frequency must be varied accordingly throughout a time lapse $t' = 0 \to t_1$. Then, if r be the distance of the vehicle from the terrestrial station and c the speed of light, the corresponding sequence of required rotations will be carried out r/c seconds later. Thereby, if required, also a correction for the DOPPLER effect can be made.

VII. Supervising Observation of the Rotary Motion of the Space Vehicle

During the procedure of directing the retro-rocket, no supervising observation is required. Such an observation *during* the process would be of little practical importance because the retardation due to the finite speed of the signals excludes a direct recoil between observation and telecommand. Nevertheless, it will be always suitable when after accomplishment the result can be detected by observation. For such a detection the angular speed meters mentioned above can be used by directly measuring the angular speeds u_X, u_Y, and u_Z. These angular speeds can then be translated in modulation frequencies emitted by the vehicle, commanding a servomotor at the terrestrial station, which moves a pointer along a dial, indicating directly the angular speed. Then, after reception of this information, the angular speeds u_X, u_Y, and u_Z become known as functions of time throughout the process of directing, so that the left members of (23) are given as functions of time. Starting from the initial state $\theta_0\,\varphi_0\,\psi_0$, these equations can again be integrated by steps according to the method of RUNGE and KUTTA. The eqs. (23) become:

$$\begin{aligned} u_X(t) &= -\dot\varphi \sin\theta \cos\psi + \dot\theta \sin\psi \\ u_Y(t) &= +\dot\varphi \sin\theta \sin\psi + \dot\theta \cos\psi \\ u_Z(t) &= \dot\psi + \dot\varphi \cos\theta \end{aligned}$$

The system determinant then becomes:

$$D = \begin{vmatrix} 0 & -\sin\theta\cos\psi & \sin\psi \\ 0 & +\sin\theta\sin\psi & \cos\psi \\ 1 & \cos\theta & 0 \end{vmatrix} = $$

$$= -\sin\theta\cos^2\psi - \sin\theta\sin^2\psi = -\sin\theta.$$

Hence:

$$\left.\begin{aligned} \dot\psi &= \frac{-1}{\sin\theta}\begin{vmatrix} u_X(t) & -\sin\theta\cos\psi & \sin\psi \\ u_Y(t) & +\sin\theta\sin\psi & \cos\psi \\ u_Z(t) & \cos\theta & 0 \end{vmatrix} \\ \dot\varphi &= \frac{-1}{\sin\theta}\begin{vmatrix} 0 & u_X(t) & \sin\psi \\ 0 & u_Y(t) & \cos\psi \\ 1 & u_Z(t) & 0 \end{vmatrix} = \frac{u_Y(t)\sin\psi - u_X(t)\cos\psi}{\sin\theta} \\ \dot\theta &= \frac{-1}{\sin\theta}\begin{vmatrix} 0 & -\sin\theta\cos\psi & u_X(t) \\ 0 & +\sin\theta\sin\psi & u_Y(t) \\ 1 & \cos\theta & u_Z(t) \end{vmatrix} = u_X(t)\sin\psi + u_Y(t)\cos\psi \end{aligned}\right\} \quad (25)$$

Let us write (25)

$$\dot\psi = \xi(\theta, \psi, t), \qquad \dot\varphi = \eta(\theta, \psi, t), \qquad \dot\theta = \zeta(\psi, t) \qquad (25')$$

Let the increments of θ, φ and ψ, corresponding with $\Delta t = h$, be k, l and m. Then solving the system according to the method of RUNGE and KUTTA, we obtain the scheme:

$$k = \frac{1}{6}k_1 + \frac{1}{3}k_2 + \frac{1}{3}k_3 + \frac{1}{6}k_4, \qquad l = \frac{1}{6}l_1 + \frac{1}{3}l_2 + \frac{1}{3}l_3 + \frac{1}{6}l_4,$$

$$m = \frac{1}{6}m_1 + \frac{1}{3}m_2 + \frac{1}{3}m_3 + \frac{1}{6}m_4$$

$$k_1 = \zeta(\psi, t)\,h \qquad\qquad l_1 = \eta(\theta, \psi, t)$$

$$k_2 = \zeta\left(\psi + \frac{m_1}{2}, t + \frac{h}{2}\right)h \qquad l_2 = \eta\left(\theta + \frac{k_1}{2}, \psi + \frac{m_1}{2}, t + \frac{h}{2}\right)h$$

$$k_3 = \zeta\left(\psi + \frac{m_2}{2}, t + \frac{h}{2}\right)h \qquad l_3 = \eta\left(\theta + \frac{k_2}{2}, \psi + \frac{m_2}{2}, t + \frac{h}{2}\right)h$$

$$k_4 = \zeta(\psi + m_3, t + h)\,h \qquad l_4 = \eta(\theta + k_3, \psi + m_3, t + h)\,h$$

$$m_1 = \xi(\theta, \psi, t)\,h$$

$$m_2 = \xi\left(\theta + \frac{k_1}{2}, \psi + \frac{m_1}{2}, t + \frac{h}{2}\right)h$$

$$m_3 = \xi\left(\theta + \frac{k_2}{2}, \psi + \frac{m_2}{2}, t + \frac{h}{2}\right)h$$

$$m_4 = \xi(\theta + k_3, \psi + m_3, t + h)\,h$$

VIII. Grazing the Planetary Atmosphere by Means of a Globular Satellite

In order to investigate the atmosphere of the object planet it is possible, by sufficient deceleration of one vehicle by retro-thrust, to *graze* the planetary atmosphere. During this operation, after braking, a globular satellite can be released, in order to obtain a simple aerodynamic drag. In order to detect the motion of this globular satellite, after release, it must again carry a transmitter of constant wave-length, which can be the same as used before. Then by studying the disturbance due to the aerodynamic drag — which in case of a globular satellite will be always in negative direction of the speed of the vehicle relative to the atmosphere, we can not only learn something concerning the density distribution with the altitude above the planetary surface, but also concerning the planetary rotation, for the atmosphere will rotate with the planet as a whole. (This will be especially of interest in case of Venus, of which the planetary rotation is still wholy unknown.)

If we denote the speed of the vehicle as to the planetary centric not rotating system of reference by $\bar{v}$, the local tow speed of the atmosphere due to planetary rotation by $\bar{v}_s$, and the local atmosphere density by $\varrho(x, y, z)$, we may write for the aerodynamic drag in vector notation:

$$\bar{W} = -\frac{\bar{v} - \bar{v}_s}{|\bar{v} - \bar{v}_s|}\,\varrho(x, y, z)\,k\,|\bar{v} - \bar{v}_s|^2 = -(\bar{v} - \bar{v}_s)\,\varrho(x, y, z)\,k\,|\bar{v} - \bar{v}_s| \tag{26}$$

in which k is a constant. If further $\bar{u}$ be the angular speed of the planet, $\bar{r}$ the radius vector of the globular satellite and $\bar{i}$, $\bar{j}$ and $\bar{k}$ unit vectors in x_+, y_+ and z_+ direction of the planetary centric non rotating system of reference, we may write:

$$\bar{v}_s = \bar{u} \times \bar{r} = \begin{vmatrix} \bar{i} & \bar{j} & \bar{k} \\ u_x & u_y & u_z \\ x & y & z \end{vmatrix} = \bar{i}\,(u_y z - u_z y) + \bar{j}\,(u_z x - u_x z) + \bar{k}\,(u_x y - u_y x) \tag{27}$$

in which u_x, u_y and u_z are the components of $\bar{u}$ in $\bar{i}$, $\bar{j}$ and $\bar{k}$ direction. If the direction cosines of $\bar{u}$ as to the planetary centric system $x\,y\,z$ be p, q and s, we may write:

$$u_x = p\,u \qquad u_y = q\,u, \qquad u_z = s\,u \tag{28}$$

Then if g_0 be the acceleration due to gravity at the planetary surface and R be the planetary radius, we obtain from (26), (27) and (28) as differential equations of motion of our spherical satellite grazing the planetary atmosphere:

$$\left.\begin{aligned}
\ddot{x} &= -g_0 R^2 \frac{x}{(x^2 + y^2 + z^2)^{3/2}} - (\dot{x} - q\,u\,z + s\,u\,y)\,\varrho\,(x, y, z) \times \\
&\times k \sqrt{(\dot{x} - q\,u\,z + s\,u\,y)^2 + (\dot{y} - s\,u\,x + p\,u\,z)^2 + (\dot{z} - p\,u\,y + q\,u\,x)^2} \\
\ddot{y} &= -g_0 R^2 \frac{y}{(x^2 + y^2 + z^2)^{3/2}} - (\dot{y} - s\,u\,x + p\,u\,z)\,\varrho\,(x, y, z) \times \\
&\times k \sqrt{(\dot{x} - q\,u\,z + s\,u\,y)^2 + (\dot{y} - s\,u\,x + p\,u\,z)^2 + (\dot{z} - p\,u\,y + q\,u\,x)^2} \\
\ddot{z} &= -g_0 R^2 \frac{z}{(x^2 + y^2 + z^2)^{3/2}} - (\dot{z} - p\,u\,y + q\,u\,x)\,\varrho\,(x, y, z) \times \\
&\times k \sqrt{(\dot{x} - q\,u\,z + s\,u\,y)^2 + (\dot{y} - s\,u\,x + p\,u\,z)^2 + (\dot{z} - p\,u\,y + q\,u\,x)^2}
\end{aligned}\right\} \tag{29}$$

If we assume that magnetic axis and rotation axis sufficiently coincide, we may consider p, q and s as known, following from the structure of the magnetic field which can be scanned before.

Then, assuming an atmospheric density distribution $\varrho(x, y, z)$ and angular speed of rotation u, and starting from initial conditions:

$$t = 0 \rightarrow x_0, y_0, z_0, \dot{x}_0, \dot{y}_0, \dot{z}_0,$$

the system (29) can be integrated by steps.

Applying the method of Runge and Kutta, let us write (29):

$$\left.\begin{aligned} \dot{v}_x &= F_x(x, y, z, v_x, v_y, v_z) \qquad & \dot{x} &= v_x \\ \dot{v}_y &= F_y(x, y, z, v_x, v_y, v_z) \qquad & \dot{y} &= v_y \\ \dot{v}_z &= F_z(x, y, z, v_x, v_y, v_z) \qquad & \dot{z} &= v_z \end{aligned}\right\} \tag{30}$$

If the increments of x, y, z, v_x, v_y, v_z, corresponding with the step $\Delta t = h$ be:

$$k, l, m, \alpha, \beta, \gamma \text{ respectively,}$$

we obtain the scheme:

$$k = \frac{1}{6} k_1 + \frac{1}{3} k_2 + \frac{1}{3} k_3 + \frac{1}{6} k_4,$$

and similarly l, m, α, β and γ.

$$k_1 = v_x h, \qquad k_2 = \left(v_x + \frac{k_1}{2}\right) h, \qquad k_3 = \left(v_x + \frac{k_2}{2}\right) h, \qquad k_4 = (v_x + k_3)\, h$$

and similarly $l_1, l_2, l_3, l_4, m_1, m_2, m_3, m_4$.

$$\alpha_1 = F_x(x, y, z, v_x, v_y, v_z)\, h$$

$$\alpha_2 = F_x\left(x + \frac{k_1}{2}, y + \frac{l_1}{2}, z + \frac{m_1}{2}, v_x + \frac{\alpha_1}{2}, v_y + \frac{\beta_1}{2}, v_z + \frac{\gamma_1}{2}\right) h$$

$$\alpha_3 = F_x\left(x + \frac{k_2}{2}, y + \frac{l_2}{2}, z + \frac{m_2}{2}, v_x + \frac{\alpha_2}{2}, v_y + \frac{\beta_2}{2}, v_z + \frac{\gamma_2}{2}\right) h$$

$$\alpha_4 = F_x(x + k_3, y + l_3, z + m_3, v_x + \alpha_3, v_y + \beta_3, v_z + \gamma_3)\, h$$

and similarly $\beta_1, \beta_2, \beta_3, \gamma_1, \gamma_2, \gamma_3$.

The scheme must be passed through in the sequence:

$$k_1\, l_1\, m_1\, \alpha_1\, \beta_1\, \gamma_1\, k_2\, l_2\, m_2\, \alpha_2\, \beta_2\, \gamma_2\, k_3\, l_3\, m_3\, \alpha_3\, \beta_3\, \gamma_3\, k_4\, l_4\, m_4\, \alpha_4\, \beta_4\, \gamma_4$$

Repeating the computation with steps $\Delta t = 2\,h$, 1/15th of the difference in result of both computations indicates the order of magnitude of the remaining inaccuracy, corresponding with the first computation. The *observation* of the motion of the globular satellite, grazing the atmosphere, and again leaving the atmosphere, can be carried out by the 3 other vehicles, in the way as indicated above. By comparing the observed and computed motion the assumptions made about u and $\varrho(x, y, z)$ can be tested.

IX. The Application of Radio Relay Vehicles

If the planetary distance becomes too large for direct radio contact which will be certainly the case when we like in a more remote future to scan the larger planets by means of unmanned space vehicles, we can meet this difficulty by inserting radio relay space vehicles between the earth and the object planet. It is possible in principle to arrange a number of such radio relay vehicles in the same elongated elliptical orbit around the sun, in similar way as some swarms of meteorites. The relay station can then be spaced equally throughout the orbit. If the vehicles are equipped with solar batteries, they can be charged when the stations pass the perigee of the orbit. The major axis of the elongated orbit must then have the direction of the location of the object planet at the planned time of scanning.

In this way, after having built up a system of such relay station swarms, telecommunication and telecommand of unmanned space vehicles will become possible throughout the solar system. However, the building up of such an interplanetary relay system will take considerable time.

Discussion

Mr. Stern: It is very interesting what you have said, but I would like to know how the three other vehicles are controlling the motion of the first vehicle, during and after deceleration by rocket thrust. You say the three other vehicles are measuring the radial speed to be observed with respect to each of these vehicles. What do you call the radial speed in this discussion?

Mr. Kooy: The radial speed is the speed into the direction of the line of sight, as seen from the controlling vehicle.

Mr. Stern: How is this radial speed then measured?

Mr. Kooy: By means of Doppler effect. The signal of constant wave length, emitted by the vehicle to be observed, is received by the controlling vehicle. The frequency ν' of the received signal is measured and telemetered in some code to the terrestrial station. Then if ν be the known emission frequency of the signal, we obtain the relation:

$$\frac{\nu'}{\nu} = 1 - \frac{v_r}{c}$$

in which v_r is the radial speed of recession and c the speed of light.

Mr. Hilton: If I remember correctly, the linear speed of the earth and of Venus in their orbits is about 20 miles/sec., so that these four satellites would pass through the sphere of Venusian gravitational influence in less than 24 hours. The programme of braking, and observation and triangulation would be very hurried, particularly as all information would have a delay time of the order of 5–10 minutes.

Mr. Kooy: Indeed. Therefore, in order to hurry up the operations, a set of consecutive commands as a whole is emitted by the terrestrial control station, which is travelling through space as one concentrated wave group and which is received in due course by the vehicle to be controlled. Then during this reception all steering manipulations are carried out in practically the same confined time lapse.

Mr. Hilton: Yes, but wouldn't you think that some form of navigation would have to be monitored from the space vehicle itself, instead of being monitored from the earth? This would involve a high degree of sophistication in each vehicle.

Mr. Kooy: Of course, we can equip the vehicle itself with a self maintaining system, wholy operating according to a precalculated enclosed time program, or coming automatically into action by influence of the target planet, when the vehicle has come sufficiently in the neighbourhood. But I believe that it will be of great advantage to monitore from the earth. Then we have still the possibility to readjust the guidance program if this would be required.

Mr. Fraeijs de Veubeke: Will it not be suitable to apply methods of measurement, so that the measure results can directly be expressed in terms of the astronomical unit?

Mr. Kooy: I believe that in the underlying problems, there is no special reason to prefer such methods.

Mr. Kovalevsky: I wanted to make a remark in what Mr. Fraeijs de Veubeke said: I think it is very misleading of talking about the ill defined ratio between the astronomical unit and the meter. As soon as you are in interplanetary space it seems more plausible that you are working in the same system as the astronomers use, based on the astronomical unit. But then angular measurements must be used, as already applied in the attitude control, giving much better results than any thing you can get by radar and Doppler measurements.

Mr. Kooy: I regret that I cannot share the opinion of the last speaker as to the applicability of radar and Doppler measurements. If we express a distance in astronomical units, our lack of knowledge remains that the astronomical unit is only known up to the accuracy of 100,000 km. And just in space navigational problems, this lack of knowledge will be a great handicap, for in such problems a difference of some thousands of km in location will often be of enormous importance. This lack of knowledge, however, can just be decreased by applying such a more recent method as radar. In this connection it may be of interest to mention that by recent Russian radar measurements of the distance of Venus, during the youngest period of closed approach to the earth, the astronomical unit was found to be 149,457,000 km. This result demonstrates the great capability of radar as a measuring tool. For the rest I believe that in futural space flight practice the km will be a more suitable unit of length than the astronomical unit, also in case of interplanetary flight, when approaching the target planet.

On Guidance and Landing Accuracy Requirements in Re-Entry Trajectories[1]

By

Luigi Broglio[2]

(With 11 Figures)

Abstract — Zusammenfassung — Résumé

On Guidance and Landing Accuracy Requirements in Re-Entry Trajectories. In Section II problems referring to guidance requirements are solved following the general approach of [1]. Errors in velocity and angles, represented by means of errors in limiting conic parameters, produce variations in maximum deceleration, total heat transferred, angular ranges, etc. Charts to evaluate such variations are described.

In Section III perturbations with respect to the simplifying assumptions are considered, and a general small-perturbations theory is developed. Any perturbation gives rise to changes in the limiting conic parameters; by considering the effect of a unit perturbation, the effect of a distributed set of them is obtained by simple integration.

Über Genauigkeitsforderungen von Rückkehrbahnen. In Kapitel II werden Lenkungsprobleme behandelt. Die Untersuchungen folgen einer früheren Arbeit [1]. Es werden Geschwindigkeits- und Richtungsfehler, welche durch Fehler in den konischen Parametern bewirkt werden, sowie deren Auswirkungen auf die Verzögerungsspitze, auf den totalen Wärmeübergang, auf Winkeländerungen usw. untersucht. Karten für die Bestimmung derartiger Abweichungen werden beschrieben.

In Kapitel III werden Störungen betrachtet und es wird eine allgemeine Theorie kleiner Störungen entwickelt. Jede Störung bewirkt Änderungen der begrenzenden konischen Parameter. Der Einfluß einer ganzen Gruppe von Störungen kann durch einfache Integration erhalten werden, wenn die Auswirkungen von Einheitsstörungen bekannt sind.

Précisions requises pour le guidage et l'atterrissage dans les trajectoires de rentrée. Suite à un travail précédent [1], des problèmes de guidage sont traités. Les erreurs de vitesse et d'angle, ayant leur répercussion sur les paramètres de la conique limite, produisent des modifications de la décélération maximum, de la quantité de chaleur transférée, de la distance angulaire couverte, etc. Des abaques permettant leur calcul sont présentées.

Dans une troisième section une théorie générale de petites perturbations est développée en rapport avec les hypothèses simplificatrices. Toute perturbation modifie les paramètres de la conique limite. Une intégration sur des perturbations unitaires permet d'évaluer l'effet de perturbations réparties.

---

[1] Paper presented at the International Symposium on "Space Flight and Re-Entry Trajectories" organized by the International Academy of Astronautics of the IAF, Louveciennes, 19—21 June 1961.

[2] Rome University, Rome, Italy.

Glossary of Symbols

a_n	normal acceleration (in Part II coefficient of Eq. (26))
a_t	tangent acceleration
b_n	coefficients of expansion (26)
c_D	drag coefficient
g	gravity acceleration
h	gravitational constant
k	$(c_D A)/m$
m	body mass
n	ratio of deceleration to gravity
q	heat rate
r	radius from earth center
t	time
w	areal velocity
x	nondimensional density
z	altitude
A	main cross-area
C_P	nondimensional value of generic P-property
D	drag
E	total energy
L	lift
P	point of trajectory
Q	total heat transferred
V	velocity
α	planetary atmosphere constant
β	α/R
ε	nondimensional total energy
φ	function defining lift modulation (Eq. (5))
η	nondimensional k (Eq. (5))
κ	limiting conic parameter
λ	parameter of lift modulation (Eq. (5))
θ	flight path angle
ψ	rotation of flight plane
ϱ	air density
ξ	nondimensional altitude (Eq. (4))
χ	nondimensional conic parameter
Δa_n, Δa_t, Δa_b	components of perturbing acceleration on normal, tangent, binormal, respectively

Subscripts

*	denotes values at DP
∞	values at infinity
1	reference values
C	values of limiting conic
f	final values

I. Introduction

Safe landing of a spacecraft requires, as well known, that some prescribed bounding values of heat transferred, maximum deceleration, range traveled, be not overcome. This can be achieved by prescribing an adequate program of lift and drag modulation: but — once this has been done — a major problem is a guidance problem, i.e., to give the proper velocity vector to the spacecraft at a given point.

Following a preceding work of the Author [1], instead of the velocity vector, and without any need to define a "re-entry altitude", the parameters of the limiting, or approach conic can be selected. Thus, the problem arises to evaluate the effects on engineering quantities (such as heat, deceleration, range, etc.) of errors on such limiting conic parameters; or, conversely, to evaluate maximum tolerable errors in limiting conic parameters so as to prevent undue overcoming of engineering quantities boundaries. This can be done if a complete set of results, connecting engineering quantities to conic parameters is available; if a graphical presentation is prepared, very simple constructions allow a ready answer to the most of guidance requirements problems.

Some errors in the evaluation of such guidance problems can arise from the simplified scheme upon which the theoretical analysis is performed. Secondary causes can sometimes produce noticeable effects, which are to be calculated. If the "cause" is a first-order one, a linear theory of small-perturbations can be developed. Disturbances arising from nonsphericity of the earth, variation of gravity, mass changes during flight (ablation) can be adequately represented by disturbing forces acting along the various points of the trajectory; and the basic singularity is then the effect of a unit pulse acting at a generic point of the trajectory. By using the approach of limiting conic parameters, the said disturbing pulse can be regarded as that producing a change in the limiting conic parameters of the portion of trajectory following it. And, so, each point of the disturbed trajectory is considered as belonging to a (fictitious) trajectory whose limiting conic is changed with respect to the undisturbed one; the change is given by the sum of those produced by the disturbances preceding the point under concern.

II. Guidance Requirements for Re-Entry

1. Introductory Remarks and Recalls on Similar Solutions

It has been shown in [1] that re-entry trajectories of spacecrafts are described by the two simultaneous differential equations:

$$r^2 \frac{d}{dr}\left(\frac{1}{r^2 \cos^2\theta}\right) + \frac{h}{w^2} - \frac{L}{D}\frac{k\varrho}{\cos^3\theta} = 0 \tag{1}$$

$$\frac{d \log \dfrac{h}{w^2}}{dr} + \frac{k\varrho}{\sin\theta}\left(1 - \frac{L}{D}\tan\theta\right) = 0$$

Here the independent variable r is the distance from the earth's center, whereas the unknowns are θ (flight path angle) and w (areal velocity $= \{\text{velocity } V\} \times \times r\cos\theta$); h is the gravitational constant ($\{\text{gravity acceleration } g\} \cdot 2\,r^2$), ϱ is the density. The parameter k is equal to $(C_D A)/m$, (where C_D is the drag coefficient, A is the frontal area, m is the body mass), and can be modulated along the trajectory, as well as the lift/drag ratio (L/D). Modulation can also include variation of C_D vs. (L/D) [2].

Eqs. (1) were solved in [1] for the density law $\varrho\, r^\alpha = \text{const.}$, ($\alpha = 900$, for the earth). It should be pointed out that such law is practically coincident with the commonly accepted one $\varrho\, e^{\beta z} = \text{const.}$ [with $z =$ altitude, $\beta = (\alpha/\text{earth radius})$] and lends itself much better for similarity purposes.

In order to obtain similarity laws it is interesting to take as reference values the quantities at the point where the total deceleration reaches its maximum value, or deceleration peak (DP). If there is more than one deceleration peak, reference can be made to any one of them. Quantities at DP (denoted by a star) are connected by the equation:

$$k_* \varrho_* r_* = \alpha_* \sin\theta_* \tag{2}$$

where: k_* is the value of k at DP; and:

$$\alpha_* = \frac{\alpha - \left\{\dfrac{d}{d(r/r_*)}\left[\dfrac{c_D A}{m}\sqrt{1 + \left(\dfrac{L}{D}\right)^2}\right]\right\}_*}{1 - \dfrac{\sin\theta_*}{n_*}\sqrt{1 + \left(\dfrac{L}{D}\right)^2_*}} \tag{3}$$

Thus, if the non-dimensional abscissa is introduced:

$$\xi = 1 - \frac{r_*}{r} \tag{4}$$

and it is set:

$$k = k_* \eta(\xi); \qquad \frac{L}{D} = \lambda\varphi(\xi) \tag{5}$$

Eqs. (1) can be reduced, in nondimensional form, to the only equation:

$$(1-\xi)^2 \frac{d}{d\xi} \log \left\{ \frac{\eta\,\lambda\varphi(\alpha_* \sin\theta_*)\,(1-\xi)^\alpha}{\cos^3\theta} - \frac{d}{d\xi}\left(\frac{(1-\xi)^2}{\cos^2\theta}\right) \right\} +$$
$$+ \frac{(\alpha_* \sin\theta_*)\cdot\eta(1-\xi)^\alpha}{\sin\theta}(1-\lambda\varphi\tan\theta) = 0 \qquad (6)$$

The boundary conditions to Eqs. (6) can be established in two different ways. According to the first way, values of maximum deceleration ($n_*\, g_*$) and angle θ_* can be prescribed at DP ($\xi = 0$). According to the alternative way, it is seen that the limiting form of Eqs. (1) — as ϱ approaches zero — yield the two properties of the keplerian conic: areal velocity $w = \text{const.}$; total energy $E = V^2 - h/r = \text{const.}$ And so, the boundary conditions to Eqs. (6) may also prescribe the nondimensional limiting value of the conic in-variants:

$$\text{as } \xi \to 1 \begin{cases} \lim \dfrac{w^2}{h\,r_*} = \kappa = \text{const.} \\[2ex] \lim \dfrac{V^2 - (h/r)}{h/r_*} = \varepsilon = \text{const.} \end{cases} \qquad (7)$$

Since, however, the two alternative ways of assigning boundary conditions to Eq. (6) must provide — obviously — the same solution — there must be obviously a correspondence of the type:

$$\begin{array}{ll} \varepsilon = \varepsilon(n_*, \theta_*) & n_* = n_*(\varepsilon, \kappa) \\ \kappa = \kappa(n_*, \theta) & \theta_* = \theta_*(\varepsilon, \kappa) \end{array} \qquad (8)$$

The problem is so established in a similar form (since the ballistic parameter is eliminated), with boundary conditions at the deceleration peak, and a continuous linkage to the outer space. For all quantities of interest for engineering purposes a similarity law will hold, and [1] provides the law of similarity for each of them.

2. *Limiting Conic Parameters*

The limiting conic parameters chosen in [1] are, as said above, the total energy:

$$E = V^2 - \frac{h}{r} \qquad (9)$$

and the areal velocity (or angular momentum):

$$w = V\,r\cos\theta \qquad (10)$$

They can also be made nondimensional, and so the two independent parameters ε, κ can be defined:

$$\varepsilon = \frac{E_\infty}{h/r_*}$$
$$\kappa = \frac{w_\infty{}^2}{h\,r_*} \qquad (11)$$

The above choice, although correct in principle, needs a reconsideration in view of guidance problems.

In fact, an adequate choice should lead to a pair of parameters which be sufficiently uncoupled, that is: one depending almost entirely on the velocity at a prescribed height, and the other almost entirely on the angle at the same height.

In this way re-entry conditions are well described by two such parameters. Now it is easily seen that ε is depending only on V; this is not the case with κ with respect to the angle θ.

The equation of nondimensional limiting conic may be written [1]:

$$1 + \tan^2\theta = \frac{1 - \xi + \varepsilon}{\kappa(1 - \xi)^2}$$

The value χ of $\tan^2\theta$ at DP ($\xi = 0$) is consequently given by:

$$\chi = \frac{1 + \varepsilon}{\kappa} - 1 \tag{12}$$

It should be pointed out that χ is not the angle of the actual trajectory at DP (its value has been denoted by θ_*) nor is it necessarily a positive value: in fact, if the limiting conic perigee is higher than r_*, the angle at r_* is imaginary, and χ is negative.

In any case its value is sufficiently characteristic of the behavior of the limiting conic from the point of view of the angle θ, whereas ε is most adequately representing the velocity V. The parameters ε and χ will be chosen throughout this work for guidance analysis purposes.

3. Guidance Errors and Corresponding Variations of Limiting Conic

Errors in guidance arise from a wrong estimation of angle θ, velocity V, at a given radius r.

Obviously, such errors correspond to undue variations in limiting conic parameters with respect to design values. Thus:

$$\begin{aligned} \Delta E_\infty &= \Delta(V^2) \\ \Delta w_\infty &= r\Delta(V\cos\theta) \end{aligned} \tag{13}$$

whence also the errors in nondimensional parameters $\Delta\varepsilon$ and $\Delta\chi$ can be calculated:

$$\begin{aligned} \Delta\varepsilon &= \Delta\varepsilon(\Delta V, \Delta\theta) \\ \Delta\chi &= \Delta\chi(\Delta V, \Delta\theta) \end{aligned} \tag{14}$$

It is of particular interest to consider the case that quantities such as ΔV, $\Delta\theta$ may be regarded as relatively small changes about given values V and θ. It is thus possible to obtain $\Delta\varepsilon$ and $\Delta\chi$ simply as:

$$\begin{aligned} \Delta\varepsilon &= \frac{\partial\varepsilon}{\partial V}\Delta V + \frac{\partial\varepsilon}{\partial\theta}\Delta\theta \\ \Delta\chi &= \frac{\partial\chi}{\partial V}\Delta V + \frac{\partial\chi}{\partial\theta}\Delta\theta \end{aligned} \tag{15}$$

The above written partial derivatives, according to Eqs. (9), (10), (11), have the expression:

$$\begin{aligned} \frac{\partial\varepsilon}{\partial V} &= \frac{r_*}{h} 2\,V \\ \frac{\partial\varepsilon}{\partial\theta} &= 0 \\ \frac{\partial\chi}{\partial V} &= \left(\frac{r_*}{r}\right)^2 (1 + \tan^2\theta)\, 2\left(\frac{h}{r} - \frac{h}{r_*}\right)\frac{1}{V^3} \\ \frac{\partial\chi}{\partial\theta} &= \left(\frac{r_*}{r}\right)^2 \frac{2\tan\theta}{\cos^2\theta}\left[1 - \frac{(h/r) - (h/r_*)}{V^2}\right] \end{aligned} \tag{16}$$

The foregoing expressions can be transformed so as to depend on ε, χ and from ξ (which means that the dependency of guidance errors on limiting conic parameters is influenced — although slightly, as it will be seen now — by the altitude).

In view of the application to similar solutions, reference is here made to nondimensional quantities, such as C_V [1], given by:

$$C_V = \frac{2\,r_*}{h}\,V \tag{17}$$

and so the following relationships are obtained:

$$\begin{aligned}
\frac{\partial \varepsilon}{\partial C_V} &= 2\sqrt{\varepsilon + 1 - \xi} \\
\frac{\partial \varepsilon}{\partial \theta} &= 0 \\
\frac{\partial \chi}{\partial C_V} &= -\frac{2^{1/2}\,\xi(\chi+1)}{(\varepsilon+1)\,(\varepsilon+1-\xi)^{1/2}} \\
\frac{\partial \chi}{\partial \theta} &= 2(\chi+1)\sqrt{\frac{2(\chi+1)\,(\varepsilon+1-\xi)}{(\varepsilon+1)\,(1-\xi)^2} - 1}
\end{aligned} \tag{18}$$

It is therefore easily shown that:

(i) since $\partial\varepsilon/\partial\theta = 0$, ε is insensitive to angle variations;

(ii) since ξ has generally small values, χ is essentially depending on θ;

(iii) the above indicated partial derivatives are depending — although slightly — on the value of ξ, i.e., on the altitude at which the errors are considered.

The above-said conclusions hold also for finite errors, although the dependency is not so easily shown.

4. *Effects on Trajectory Quantities*

In [1], Part I, it is shown — as said — that entry conditions are represented by the nondimensional parameters of the limiting conic (in this paper the quantities ε and χ have been selected as typical quantities). It is also shown that all the quantities related to the trajectory, such as total heat Q, heat rate q; velocity V, time t, angular range φ, etc., can be represented in a "similar" form, i.e., as the product of a unit dimensional quantity (depending on the characteristics of a particular body being considered) by a nondimensional quantity (which is related only to ε and χ, i.e., independent on the body itself). Thus, f.i.:

$$Q = Q_1(\text{body}) \cdot C_Q(\varepsilon, \chi) \tag{19}$$

A complete summary of quantities such as Q_1, q_1, V_1, etc., and of the coefficients C_Q, C_q, C_V, etc., is given in Table I, [1].

It is also shown in [1], as said above, that an alternative approach to the problem is to prescribe ratio n_* of maximum deceleration (at DP, of course) to local gravity and angle θ_* at the same point; such values can be used as initial conditions (for $\xi = 0$) and so the relationship between entry and DP conditions is simply given by the equations:

$$\begin{aligned}
n_* &= n_*(\varepsilon, \chi) \\
\theta_* &= \theta_*(\varepsilon, \chi)
\end{aligned} \tag{20}$$

There exist no simpler means to establish Eqs. (20) than a step-by-step integration of Eq. (6). Approximate analytical formulas will be obtained later.

The question now arises in which practical way, the effects of the variations $\Delta\varepsilon$, $\Delta\chi$ of the foregoing chapter on n_*, θ_*, C_Q, C_q, C_V, etc., can be evaluated.

For this purpose the following procedure can be used. On a ε, χ cartesian diagram the curves $n_* =$ const., $\theta_* =$ const., $C_Q =$ const., etc., are plotted; so at every point (describing a particular entry condition), the values of n_*, θ_*, C_Q, C_q, etc., can be read (Fig. 1). (Hereafter the generic property is denoted by P, and its nondimensional symbol by C_p.)

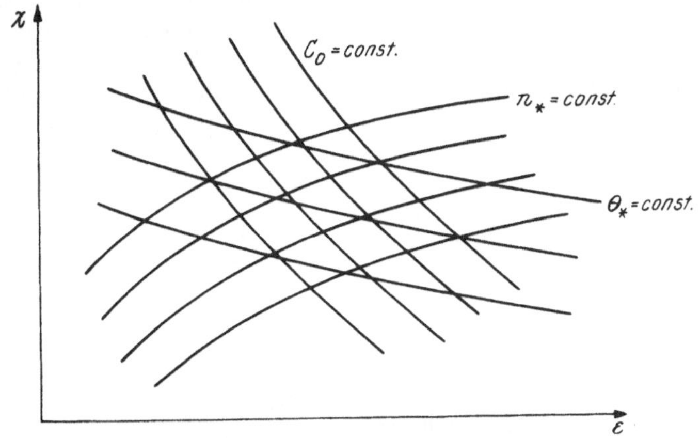

Fig. 1. Charts for similar solutions

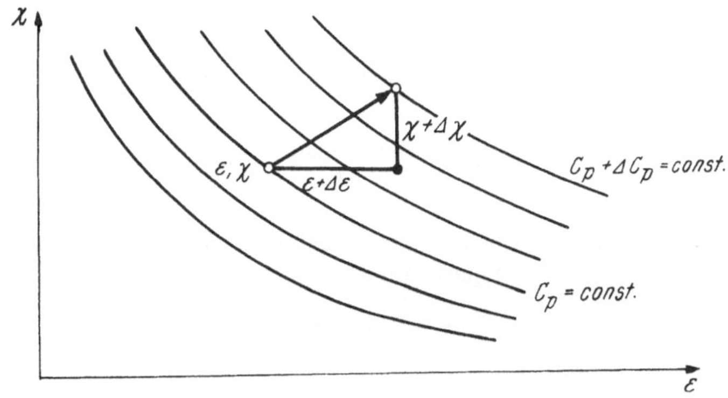

Fig. 2. Effects of changes of limiting conic parameters

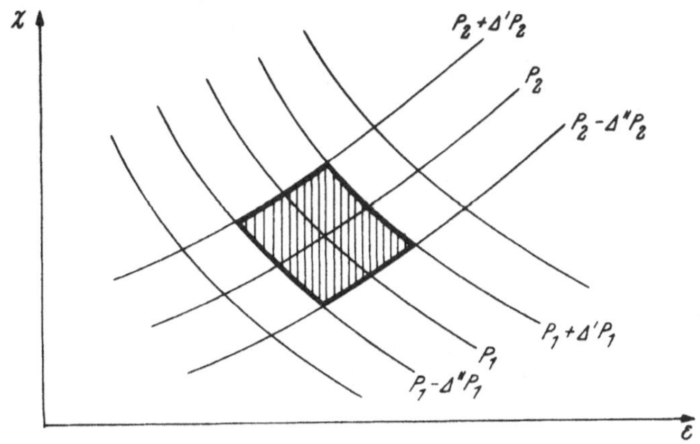

Fig. 3. Evaluation of maximum allowable errors

When charts similar to that of Fig. 1 are ready and available, any guidance problem can be solved:

(i) Evaluation of effects such as the ΔC_P corresponding to prescribed variations $\Delta\varepsilon$, $\Delta\chi$ is simply made by considering the point of coordinates $\varepsilon + \Delta\varepsilon$, $\chi + \Delta\chi$ and reading on the new points the changed quantity $C_p + \Delta C_p$ (Fig. 2);

(ii) If one considers two quantities P_1, P_2 (such as, f.i., C_Q and n_*), the question arises to find guidance requirements so as to allow maximum changes, about the design conditions, of value $+\Delta' P_1$; $-\Delta'' P_1$; $+\Delta' P_2$; $-\Delta'' P_2$ (Fig. 3). In this case the shaded area describes the range of permitted variation of ε, χ about the starting point;

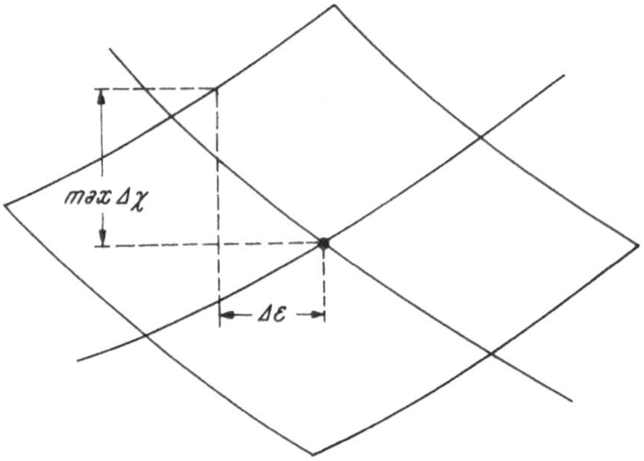

Fig. 4. max $\Delta\chi$ for a prescribed $\Delta\varepsilon$

(iii) It is also possible to obtain relationships connecting maximum allowable $\Delta\chi$ to a prescribed $\Delta\varepsilon$ (and vice versa); this can be accomplished by associating to each $\Delta\chi$ (Fig. 4) the corresponding value of $\Delta\varepsilon$. If $\Delta\varepsilon$ and $\Delta\chi$ are transformed into ΔC_V and $\Delta\theta$ as described in the previous chapter, a complete guidance requirement analysis can be performed.

5. *Conic-DP Charts*

There appears to be a particular interest in quantities at DP (θ_* and n_*). The importance of n_* is stressed by its engineering meaning itself: whereas θ_* is important mainly in regard to the fact that it determines scale length (Eq. (2)).

Approximate relationship can be derived in this field. Their validity is enhanced by the fact that — in contrast to local approximate relationships — they are of average, or global, kind, and so are reasonably little influenced by the approximations themselves.

The general equations (6) are firstly re-written for sake of clearness:

$$\frac{d \log (h/w^2)}{dr} + \frac{k\varrho}{\sin\theta}\left(1 - \frac{L}{D}\tan\theta\right) = 0$$
$$r^2 \frac{d}{dr}\left(\frac{1}{r^2\cos^2\theta}\right) + \frac{h}{w^2} - \frac{L}{D}\frac{k\varrho}{\cos^3\theta} = 0 \qquad (21)$$

(for symbols, see Glossary).

It is well known that, as ϱ approaches zero, w^2 and E approach constant values, say $w_\infty{}^2$ and E_∞; and the flight path angle θ_C follows the law:

$$\tan^2\theta_C = \frac{E_\infty}{w_\infty{}^2} r^2 + \frac{h}{w_\infty{}^2} r - 1 \qquad (22)$$

Introducing the nondimensional density:

$$x = \left(\frac{r_*}{r}\right)^a \tag{23}$$

and letting:

$$\tan^2\theta = \tan^2\theta_C + \phi(x) \tag{24}$$

$\phi(x)$ must vanish at infinity ($x = 0$), and must satisfy the differential equation:

$$\alpha\, x \frac{d\phi}{dx} + 2\,\phi = r\left(\frac{h}{w^2} - \frac{h}{w_\infty{}^2}\right) - \frac{L}{D}\,\varphi\,\frac{\alpha_*\sin\theta_*}{\cos^3\theta}\,x^{1-(1/\alpha)} \tag{25}$$

With a proper choice of the coefficients it is always possible to set:

$$r\left(\frac{h}{w^2} - \frac{h}{w_\infty{}^2}\right) = \sum_0^\infty{}_n\, a_n\, x^n \qquad (a_0 = 0)$$

$$-\frac{L}{D}\,\varphi\,\frac{a_*\sin\theta_*}{\cos^3\theta} \simeq -\frac{L}{D}\,\varphi\,\frac{a_*\sin\theta_*}{\cos^3\theta_*} = \sum_0^\infty{}_n\, b_n\, x^n \tag{26}$$

and so, it is obtained from Eq. (25):

$$\phi = \sum_0^\infty{}_n \left(\frac{a_n}{2+\alpha\, n}\,x^n + \frac{b_n}{\alpha(n+1)+1}\,x^{n+1-(1-\alpha)}\right) \simeq \frac{1}{\alpha}\sum_0^\alpha{}_n \left(\frac{a_n\, x^n}{n} + \right.$$

$$\left. + \frac{b_n}{n+1}\,x^{n+1-(1/\alpha)}\right) \tag{27}$$

The problem is thus reduced to find the coefficients a_n and b_n. The first set of them (the a_n's) is obtained approximately by stopping the expansion at $n = 2$, and by imposing at DP the following conditions:

$$\left.\begin{aligned} w^2 &= w_*{}^2 \\ \frac{dw^2}{dr} &= \left(\frac{dw^2}{dr}\right)_* \\ \left(\frac{d^2w^2}{dr^2}\right) &= \left(\frac{d^2w^2}{dr^2}\right)_* \end{aligned}\right\} \quad \text{as obtained from the equation of motion and by the conditions at DP}$$

Simple algebra (Appendix) then provides the relationships:

$$\chi - \frac{11}{6\alpha}\frac{\chi+1}{\varepsilon+1} = F_1 \tag{28}$$

$$-\log\frac{\chi+1}{\varepsilon+1} - \frac{1}{3}\frac{\sin\theta_*}{\sqrt{\dfrac{\chi+1}{\varepsilon+1}\dfrac{\varepsilon+1-\xi_e}{(1-\xi_e)^2} - 1}} = F_2 \tag{29}$$

where:

$$F_1 = F_1(\theta_*, n_*)$$
$$F_2 = F_2(\theta_*, n_*) \tag{30}$$

are functions of quantities at DP, given in Appendix. ξ_e is the point at which density is sensibly zero. The system (28), (29) provides approximate conic-DP relationships.

6. *Typical Results*

Fig. 5 shows the conic-DP charts for an unmodulated re-entry with lift. The cases $L/D = 0$; $L/D = 0.1$; $L/D = 0.2$; $L/D = 0.5$ were considered.

It is seen that curves $n_* =$ const. can cut each other: this simply means that the same entry condition is corresponding to more than one deceleration peak. In the same plane it is also very easy to determine the kind of limiting conic (according to the sign of ε). On the same plot curves such as skipping limit, re-exit on a circular orbit, homeless re-exit into space may be shown also. (Here only the $L/D = 0$ case is presented).

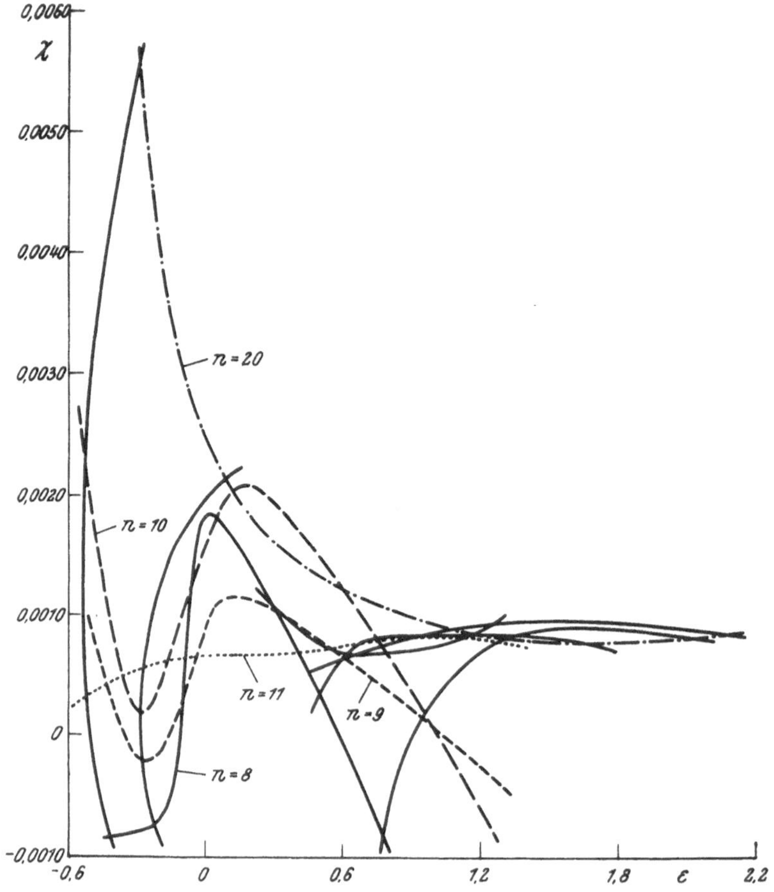

Fig. 5. Conic-DP relationships $(L/D) = 0$

Similar results are obtained for other values of L/D: it is seen however that increase of L/D gives rise to a softer trend in the above-said curves. Furthermore no cases of more than one DP have been found.

III. Small-Perturbations Theory in Re-Entry Trajectories

1. *General Considerations*

The analytical model which was hitherto considered is a simplified one, i.e., a nonrotating radially symmetric planet and a radially symmetric atmosphere where the density is varying according to a prescribed law.

A more realistic picture of the phenomena must, however, be considered sometimes. The effects of the simplifying assumptions can be evaluated by considering the effects of such assumptions as disturbances superimposed to the base-trajectory. Likewise, maneuvers during re-entry, in order to improve the accuracy of a safe landing can also be studied as perturbations superimposed to the base-trajectory.

The following chapters are devoted to evaluate the effects of a generic perturbation. It is assumed that such perturbation be a first-order one, so that a linearized theory can be applied.

The theory developed in [1] is based on Eqs. (1). They can be resolved along the tangent t to the trajectory and along its normal n in terms of accelerations. They can be written in symbolic form:

$$[a_t] = 0$$
$$[a_n] = 0 \tag{31}$$

whereas on the bi-normal the equation provides identically $0 = 0$.

A disturbing acceleration of components Δa_t, Δa_n, Δa_b is now considered, small enough to allow to discard changes in terms such as $[a_t]$ $[a_n]$. The equation of motion is consequently written as:

$$\begin{aligned} [a_t] + \Delta a_t &= 0 \\ [a_n] + \Delta a_n &= 0 \\ \Delta a_b &= 0 \end{aligned} \tag{32}$$

If, as said above, first-order effects are considered, the evaluation of the effects due to Δa_t, Δa_n, Δa_b, can be made by considering a "unit" solution, and then superimposing the effects of the various perturbations.

2. *Unit Perturbation*

A generic point P (defined by its distance r from the planet center) of a re-entry trajectory is considered (Fig. 6); the perturbation acting there is re-solved

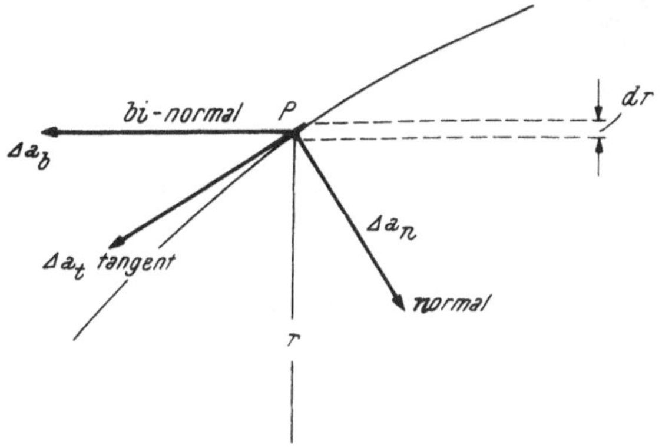

Fig. 6. Unit perturbation

along the three lines (tangent, normal, bi-normal). The effect of Δa_t is an abrupt change in the total velocity given by:

$$dV = \Delta a_t \cdot dt \tag{33}$$

(where dt is the time elapsed to travel the distance dr). The effect of Δa_n is an abrupt change in flight path:

$$d\theta = \frac{\Delta a_t}{V} dt \tag{34}$$

The effect of Δa_b is a rotation of the plane of the trajectory:

$$d\psi = \frac{\Delta a_b \, dt}{V} \tag{35}$$

Since $dt = dr/(V \sin \theta)$, the foregoing equations can be written:

$$\begin{aligned} dV &= \frac{\Delta a_t}{V \sin \theta} dr \\ d\theta &= \frac{\Delta a_n}{V^2 \sin \theta} dr \\ d\psi &= \frac{\Delta a_b \, dr}{V^2 \sin \theta} \end{aligned} \tag{36}$$

Since no perturbations are acting after P, the equations of the disturbed trajectory after such point are the same of the undisturbed one, with the conditions that velocity, flight path angle, and plane rotation at P are $V + dV$, $\theta + d\theta$, $d\psi$.

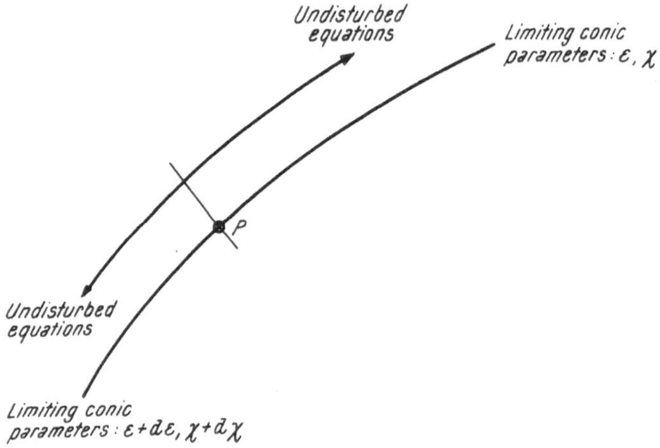

Fig. 7. Change in limiting conic parameters

This is equivalent to say that the portion of trajectory following the point of perturbation is an undisturbed trajectory such that:

(a) the plane is rotated by $d\psi$;

(b) the limiting conic parameters are changed by such quantities $d\varepsilon$, $d\chi$ that the variations dV, $d\theta$ are produced at P.

In other words, the solution corresponding to the perturbation is the one indicated in Fig. 7. Before r, the equations of the trajectory are the undisturbed ones and (ε, χ) are the limiting conic parameters; after r the equations are still the undisturbed ones, and $(\varepsilon + d\varepsilon, \chi + d\chi)$ are the limiting conic parameters. Besides that, the plane of the trajectory is rotated by an angle $d\psi$ about the focal axis of the limiting conic.

Naturally the question arises as how to relate $d\varepsilon$, $d\chi$ to dV, $d\theta$, and vice-versa. This is very simply done by calculating three trajectories: one corresponding to

(ε, χ); one corresponding to $(\varepsilon + \delta\varepsilon, \chi)$; one to $(\varepsilon, \chi + \delta\chi)$. If $\delta\varepsilon$, $\delta\chi$ are sufficiently small, by comparing quantities corresponding to the same ξ, one has, f.i.:

$$\frac{\partial(V/V_*)}{\partial\varepsilon} \cong \frac{(V/V_*)_{\varepsilon+\delta\varepsilon,\chi} - (V/V_*)_{\varepsilon,\chi}}{\delta\varepsilon} \tag{37}$$

and so for all the other derivatives, such as:

$$\frac{\partial(V/V_*)}{\partial\chi}, \quad \frac{\partial\theta}{\partial\varepsilon}, \quad \frac{\partial\theta}{\partial\chi}$$

whence, by simple inversion, also partial derivatives such as:

$$\frac{\partial\varepsilon}{\partial(V/V_*)}; \quad \frac{\partial\varepsilon}{\partial\theta}; \quad \frac{\partial\chi}{\partial(V/V_*)}; \quad \frac{\partial\chi}{\partial\theta}$$

are evaluated.

Thus, coming back to the unit solution, the limiting conic of the points following p has parameters $\varepsilon + d\varepsilon$, $\chi + d\chi$ where:

$$d\varepsilon = \frac{\partial\varepsilon}{\partial(V/V_*)} \frac{\Delta a_t}{V\,V_* \sin\theta}\,dr + \frac{\partial\varepsilon}{\partial\theta} \frac{\Delta a_n}{V^2 \sin\theta}\,dr$$

$$d\chi = \frac{\partial\chi}{\partial(V/V_*)} \frac{\Delta a_t}{V\,V_* \sin\theta}\,dr + \frac{\partial\chi}{\partial\theta} \frac{\Delta a_n}{V^2 \sin\theta}\,dr \tag{38}$$

and a rotation $d\psi$ of its plane:

$$d\psi = \frac{\Delta a_b\,dr}{V^2 \sin\theta} \tag{39}$$

3. Distributed Perturbations

If a continuous distribution Δa_t, Δa_n, Δa_b of perturbations is applied to the trajectory, each element of it can be considered as a portion of a trajectory having as limiting conic $\varepsilon + \Delta\varepsilon$, $\chi + \Delta\chi$ where $\Delta\varepsilon$ and $\Delta\chi$ are the sum of the quantities $d\varepsilon$, $d\chi$ preceding the point under concern. And, so:

$$\Delta\varepsilon = \int_{inf.}^{r} \left(\frac{\partial\varepsilon}{\partial(V/V_*)} \frac{\Delta a_t}{V\,V_* \sin\theta} + \frac{\partial\varepsilon}{\partial\theta} \frac{\Delta a_n}{V^2 \sin\theta} \right) dr$$

$$\Delta\chi = \int_{inf.}^{r} \left(\frac{\partial\chi}{\partial(V/V_*)} \frac{\Delta a_t}{V\,V_* \sin\theta} + \frac{\partial\chi}{\partial\theta} \frac{\Delta a_n}{V^2 \sin\theta} \right) dr \tag{40}$$

$$\Delta\psi = \int_{inf.}^{r} \frac{\Delta a_b\,dr}{V^2 \sin\theta}$$

which is the general answer to the small perturbations theory in re-entry trajectories. This means that each element can be thought as that of limiting conic $\Delta\varepsilon$, $\Delta\chi$. To calculate $\Delta\varepsilon$ and $\Delta\chi$ it is necessary to know:

(i) the four partial derivatives $\partial\varepsilon/\partial(V/V_*)$, etc.;

(ii) the base trajectory;

(iii) and — of course — the disturbances.

It is also seen that the small-perturbations theory can be applied to similar solutions.

Indeed, Eqs. (40) can be written:

$$\Delta\varepsilon = \int_1^{\xi} \left\{ \frac{\partial \varepsilon}{\partial (V/V_*)} \frac{r_* \Delta a_t}{V_*^2} \frac{1}{(V/V_*) \sin\theta} + \frac{\partial \varepsilon}{\partial \theta} \frac{r_* \Delta a_n}{V^2_*} \frac{1}{(V/V_*)^2 \sin\theta} \right\} \frac{d\xi}{(1-\xi)^2}$$

$$\Delta\chi = \int_1^{\xi} \left\{ \frac{\partial \chi}{\partial (V/V_*)} \frac{r_* \Delta a_t}{V_*^2} \frac{1}{(V/V_*) \sin\theta} + \frac{\partial \chi}{\partial \theta} \frac{r_* \Delta a_n}{V_*^2} \frac{1}{(V/V_*)^2 \sin\theta} \right\} \frac{d\xi}{(1-\xi)^2}$$

$$\Delta\psi = \int_1^{\xi} \frac{r_* \Delta a_b}{V_*^2} \frac{1}{(V/V_*)^2 \sin\theta} \frac{d\xi}{1-\xi^2} \tag{41}$$

where only similar quantities are appearing. Obviously, new parameters must be considered, which arise from the actual expressions of quantities such as Δa_t, Δa_n, Δa_b.

Once the quantities $\Delta\varepsilon$, $\Delta\chi$, $\Delta\psi$ have been calculated, the perturbed velocities and angles at every point of the trajectory are given by:

$$\Delta\theta = \frac{\partial \theta}{\partial \varepsilon} \Delta\varepsilon + \frac{\partial \theta}{\partial \chi} \Delta\chi$$

$$\Delta(V/V_*) = \frac{\partial (V/V_*)}{\partial \varepsilon} \Delta\varepsilon + \frac{\partial (V/V_*)}{\partial \chi} \Delta\chi \tag{42}$$

4. *Application to Non-Conductive Ablation*

An application of the above theory can be made to the case of bodies re-entry with a limited amount of ablation. The difference from the non-ablating case is now in the fact that the mass m is no longer constant, but is varying along the trajectory according to the law:

$$\frac{dm}{dt} = -\frac{q}{K} \tag{43}$$

where K denotes the ablation coefficient [4].

If K is a constant:

$$m(\xi) = m_0 - \frac{Q(\xi)}{K} \tag{44}$$

where $Q(\xi)$ is the total heat transferred to the body from the beginning up to the point being considered.

The aforesaid mass variation can be considered, for every point of the trajectory, as a perturbation of components:

$$\Delta a_n = \Delta a_b = 0; \qquad \Delta a_t = -\frac{1}{2} c_D \varrho A V^2 \left(\frac{1}{m} - \frac{1}{m_0} \right) = \\ = \frac{\frac{1}{2} c_D \varrho A V^2 (m - m_0)}{m m_0} = \\ = \frac{1}{2} k \varrho V^2 \frac{m - m_0}{m_0} \tag{45}$$

and so:

$$\Delta a_t = -\frac{1}{2} \frac{k \varrho V^2 Q}{K m_0} \tag{46}$$

Relative ablation corresponding to the total heat Q_f transferred to the body up to landing is given by (with $\bar{m}$ = final mass):

$$\frac{\bar{m} - m_0}{m_0} = -\frac{Q_f}{K\, m_0} \tag{47}$$

and so:

$$\Delta a_t = \frac{1}{2} k\, \varrho\, V^2 \frac{\bar{m} - m_0}{m_0} \frac{Q}{Q_f} = \frac{1}{2} k\, \varrho_*\, V_*^2 \frac{\varrho}{\varrho_*} \left(\frac{V}{V_*}\right)^2 \frac{\bar{m} - m_0}{m_0} \frac{Q}{Q_f} \tag{48}$$

Therefore Eqs. (41) of the foregoing Article provide:

$$\frac{\Delta \varepsilon}{\dfrac{\bar{m} - m_0}{m_0}} = \int_1^{\xi} \frac{\alpha_* \sin\theta_*}{2} \left(\frac{\varrho}{\varrho_*}\right) \left(\frac{V}{V_*}\right)^2 \frac{\partial \varepsilon}{\partial (V/V_*)} \frac{d\xi}{(1-\xi)^2 \sin\theta}$$

$$\frac{\Delta \chi}{\dfrac{\bar{m} - m_0}{m_0}} = \int_1^{\xi} \frac{\alpha_* \sin\theta_*}{2} \left(\frac{\varrho}{\varrho_*}\right) \left(\frac{V}{V_*}\right)^2 \frac{\partial \chi}{\partial (V/V_*)} \frac{d\xi}{(1-\xi)^2 \sin\theta}$$

$$\Delta \psi = 0 \tag{49}$$

which express the variations of limiting conic parameters corresponding to a total ablation ($\bar{m} = 0$). For $\bar{m} \neq 0$ it is enough to multiply the corresponding quantities by the ablation rate.

5. *Numerical Example*

A numerical case was calculated with the data of Fig. 8. In the same plot the velocity variation $v\,s \cdot \xi$ is represented for a trajectory having a circle as

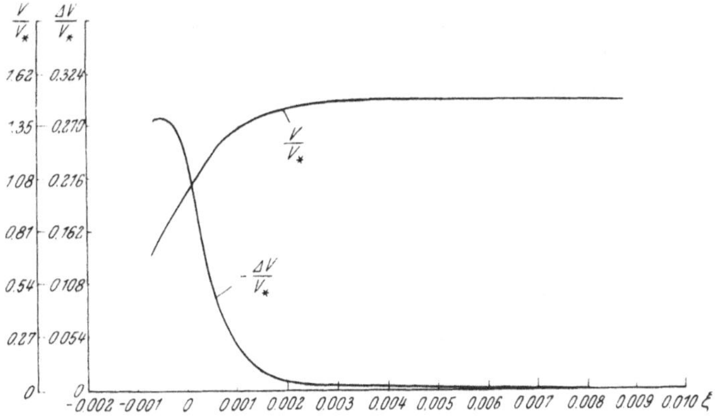

Fig. 8. Variation of velocity for total ablation
$\varepsilon = -0.493\,33$; $\chi = 0.002\,572\,4$; $L/D = 0$
(circular re-entry with $n_* = 10$)

limiting conic. The values of ΔV refer to the case of complete ablation: in the case of partial ablation it is enough — as said above — to multiply the values by the percentage of ablated weight. Thus, f.i., at DP ($V/V_* = 1$), the ordinate of the plot is 0.24, and, for an ablation of 30%, one would have $\Delta V/V_* \simeq -0.08$.

Similar considerations apply to the case of escape velocity (Fig. 9). In this case the change of velocity for an ablation of 30% is ~ -0.07. Both cases refer to a maximum deceleration of 10 g.

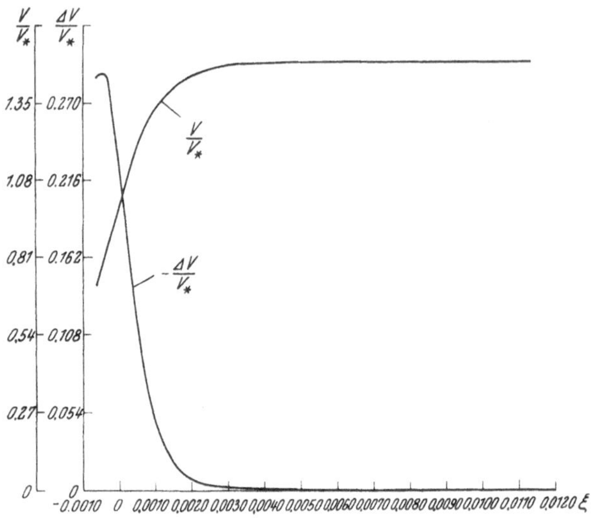

Fig. 9. Variation of velocity for total ablation
$\varepsilon = 0$; $\chi = -0.000\,294\,13$; $L/D = 0$
(parabolic re-entry with $n_* = 10$)

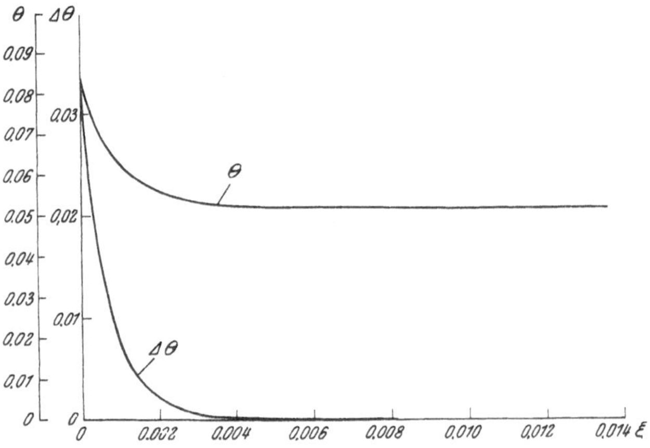

Fig. 10. Variation of flight path angle for total ablation
$\varepsilon = -0.493\,33$; $\chi = -0.002\,572\,4$; $L/D = 0$
(circular re-entry with $n_* = 10$)

Fig. 10 and Fig. 11 show respectively the flight path angle variation for a complete ablation.

In this case the effect is to increase the angle. For the orbital speed, an ablation of 30% gives a change of the angle of about 1/8 of the unperturbed value: the corresponding value for the escape velocity is about 8%.

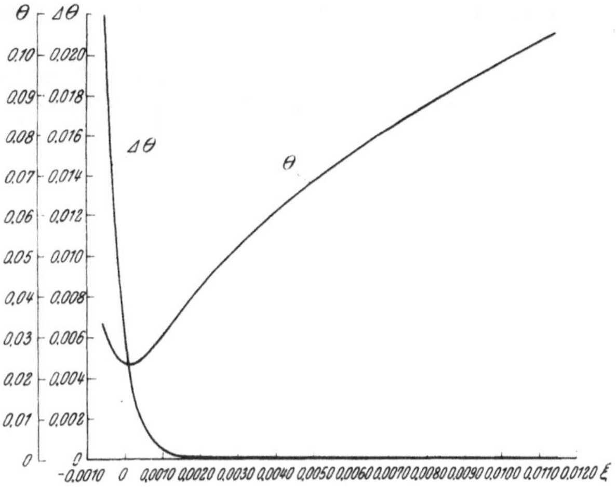

Fig. 11. Variation of flight path angle for total ablation
$\varepsilon = 0$; $\chi = -0.000\,294\,13$; $L/D = 0$
(parabolic re-entry with $n_* = 10$)

Appendix

Determination of Approximate Conic-DP Relationships

1. Let:

$$z = \frac{h\, r_*}{w^2} \tag{A 1}$$

Remembering (23), the first of Eqs. (26) is written:

$$x^{-(1/\alpha)(z - z_\infty)} = a_1\, x + a_2\, x^2 + a_3\, x^3 \tag{A 2}$$

Denoting by a dot derivatives with respect to x, the conditions on a_1, a_2, a_3 for identifying Eqs. (A 2) and its first and second derivative at DP, are:

$$z_* = a_1 + a_2 + a_3 + z_\infty$$

$$\dot{z}_* - \frac{1}{\alpha} z_* = a_1 + 2\, a_2 + 3\, a_3$$

$$\ddot{z}_* - \frac{2}{\alpha}\dot{z}_* + \frac{1}{\alpha}\left(\frac{1}{a} + 1\right) z_* = 2\, a_2 + 6\, a_3$$

whence:

$$a_1 = \frac{\ddot{z}_*}{2} - \left(2 + \frac{1}{\alpha}\right)\dot{z}_* + \left\{\left(1 + \frac{1}{\alpha}\right)\left(2 + \frac{1}{\alpha}\right) + 1\right\}(z_* - z_\infty)$$

$$a_2 = -\ddot{z}_* + \left(3 + \frac{2}{\alpha}\right)\dot{z}_* - \left\{\left(1 + \frac{1}{\alpha}\right)\left(3 + \frac{1}{\alpha}\right)\right\}(z_* - z_\infty)$$

$$a_3 = \frac{\ddot{z}_*}{2} - \left(1 + \frac{1}{\alpha}\right)\dot{z}_* + \left\{\left(1 + \frac{1}{\alpha}\right)\left(2 + \frac{1}{\alpha}\right)\right\}\frac{(z_* - z_\infty)}{2}$$

If quantities of the order of $1/\alpha$ are neglected as compared to unity, one has finally:

$$\begin{aligned} a_1 &= \frac{1}{2}\ddot{z}_* - 2\dot{z}_* + 3(z_* - z_\infty) \\ a_2 &= -\ddot{z}_* + 3\dot{z}_* - 3(z_* - z_\infty) \\ a_3 &= \frac{\ddot{z}_*}{2} - \dot{z}_* + (z_* - z_\infty) \end{aligned} \tag{A 3}$$

2. Evaluation of $\dot{z}_*$, $\ddot{z}_*$ is now performed.

From the second of Eqs. (1), remembering Eqs. (23), (2), (5), of Section II:

$$\frac{dz}{dx} = \frac{\alpha_* \sin_* \theta}{\alpha} \frac{z\eta\, x^{-1/\alpha}}{\sin\theta} [1 - \lambda\varphi \tan\theta]$$

whence:

$$\dot{z}_* = \frac{\alpha_*}{\alpha} z_* [1 - \lambda\varphi_* \tan\theta_*] \tag{A 4}$$

The second derivative is now performed. It is obtained:

$$\ddot{z} = \frac{\alpha_* \sin\theta_*}{\alpha}\left[z\eta\, x^{-1/\alpha}\left\{-\left(\frac{\cos\theta}{\sin^2\theta} + \lambda\varphi\frac{\sin\theta}{\cos^2\theta}\right)\dot{\theta} - \frac{\lambda\dot{\varphi}}{\cos\theta}\right\} + \right.$$
$$\left. + \frac{(1-\lambda\varphi\tan\theta)}{\sin\theta}\left(\dot{z}\eta\, x^{-1/\alpha} + z\dot{\eta}\, x^{-1/\alpha} - \frac{1}{\alpha} z\eta\, x^{-(1/\alpha)-1}\right)\right]$$

The value of $\dot{\theta}$ at DP is obtained from the first of Eqs. (1); and it is easily obtained:

$$\dot{\theta}_* = -\frac{\cot\theta_*}{\alpha} + \frac{z_*\cos^3\theta_*}{2\alpha\sin\theta_*} - \frac{\lambda\varphi_*\alpha_*}{2\alpha}$$

and so:

$$\ddot{z}_* = z_*\frac{\alpha_*\sin\theta_*}{\alpha}\left[\left(\frac{\cot\theta_*}{\alpha} - \frac{z_*\cos^3\theta_*}{2\alpha\sin\theta_*} + \frac{\lambda\varphi_*\alpha_*}{2\alpha}\right)\left(\frac{\cos\theta_*}{\sin^2\theta_*} + \right.\right.$$
$$\left. + \lambda\varphi_*\frac{\sin\theta_*}{\cos^2\theta_*}\right) - \frac{\lambda_*\dot{\varphi}_*}{\cos\theta_*} + \left(\frac{1-\lambda\varphi_*\tan\theta_*}{\sin\theta_*}\right)\left(\dot{\eta}_* + \right.$$
$$\left.\left. + \frac{\alpha_*}{\alpha}(1-\lambda\varphi_*\tan\theta_*)\right)\right] \tag{A 5}$$

3. Eqs. (24) and (27) yield at DP:

$$\chi = \tan^2\theta_* - \frac{1}{\alpha}\left(\sum_0^\infty{}_n \frac{a_n}{n} + \sum_0^\infty{}_n \frac{b}{n+1}\right)$$

and, for $n = 3$ in the sum of a_n's remembering (A 3):

$$\chi = \tan^2\theta_* - \frac{1}{\alpha}\left(\frac{1}{6}\ddot{z}_* - \frac{5}{6}\dot{z}_* + \frac{11}{6}z_*\right) + \frac{11}{6\alpha}z_\infty \tag{A 6}$$

Since:

$$z_* = \frac{h r_*}{w_*} = \frac{\alpha_*\sin\theta_*}{n_*\cos^2\theta_*}; \qquad z_\infty = \frac{1}{\kappa} = \frac{\chi+1}{\varepsilon+1}$$

Eq. (A 6) can be written:

$$\chi - \frac{11}{6\alpha}\frac{\chi+1}{\varepsilon+1} = F_1(\theta_*, n_*) \tag{A 7}$$

where:

$$F_1(\theta_*, n_*) = \tan^2\theta_* - \frac{1}{6\alpha}(\ddot{z}_* - 5\dot{z}_* + 11 z_*) \tag{A 8}$$

is a function depending only on quantities at DP.

In a similar way, from the second of Eqs. (1), by integrating from DP up to infinity (and neglecting $L/D \tan\varphi$, and variations of k):

$$\log\frac{w_\infty^2}{w_*^2} = \frac{2}{3} - \frac{1}{6}\frac{1 - \dfrac{2n_*}{\alpha_*\sin\theta_*\sqrt{1+(L/D)_*^2}}}{\alpha_*\tan\theta_*} + \frac{1}{3}\frac{\sin\theta_*}{\sin\theta_e} - $$

$$- \frac{\cos^2\theta_*}{\alpha_*}\sum_1^\infty{}_n \frac{b_n}{n-1} \tag{A 9}$$

where θ_e is the lowest height at which $\theta = \theta_c$. By letting $\sin\theta_c \simeq \tan\theta_e$:

$$\sin\theta_e \simeq \tan\theta_e = \sqrt{\frac{\chi+1}{\varepsilon+1}\frac{\varepsilon+1-\xi_e}{(1-\xi_e)^2} - 1} \tag{A 10}$$

whence:

$$-\log\frac{\chi+1}{\varepsilon+1} - \frac{1}{3}\frac{\sin\theta_*}{\sqrt{\dfrac{\chi+1}{\varepsilon+1}\dfrac{\varepsilon+1-\xi_e}{(1-\xi_e)^2} - 1}} = F_2(n_*, \theta_*)$$

with:

$$F_2 = \frac{2}{3} - \frac{1}{6}\frac{1 - \dfrac{2n_*}{\alpha_*\sin\theta_*\sqrt{1+(L/D)_*^2}}}{\alpha_*\tan\theta_*} - \frac{\cos^2\theta_*}{\alpha_*}\sum_1^\infty{}_n \frac{b_n}{n-1} - \log z_* \tag{A 11}$$

Eqs. (A 7), (A 11) are the same as Eqs. (28), (29) of Section II.

References

1. L. Broglio, An Exact Similarity Law and a Method of Integration for Re-Entry Trajectories. SIARgraph No. 61, Contract No. AF-61(052)-198, TN 1, September 1961, Rome.
2. A. Ferri and L. Ting, Practical Aspects of Re-Entry Problems. PIBAL Report 705, July 1961.

Monotypesatz und Druck von Berger & Schwarz, Zwettl, NÖ.